Advances in Computing Science

C. Brink
W. Kahl
G. Schmidt (eds.)

Relational Methods in Computer Science

SpringerWienNewYork

Prof. Dr. Chris Brink
Department of Mathematics and Applied Mathematics, University of Cape Town,
Cape Town, South Africa

Dr. Wolfram Kahl
Prof. Dr. Gunther Schmidt
Fakultät für Informatik, Universität der Bundeswehr München,
Neubiberg, Federal Republic of Germany

Typesetting: Camera ready by authors
Printing: Druckerei Novographic, A-1238 Wien
Binding: Fa. Papyrus, A-1100 Wien

Graphic design: Ecke Bonk

Printed on acid-free and chlorine-free bleached paper

With 30 Figures

ISSN 1433-0113

ISBN 3-211-82971-7 Springer-Verlag Wien New York

Advances in Computing Science

Springer Wien New York presents the new book series "Advances in Computing Science". Its title has been chosen to emphasize its scope: "computing science" comprises all aspects of science and mathematics relating to the computing process, and has thus a much broader meaning than implied by the term "computer science". The series is expected to include contributions from a wide range of disciplines including, for example, numerical analysis, discrete mathematics and system theory, natural sciences, engineering, information science, electronics, and naturally, computer science itself.

Contributions in the form of monographs or collections of articles dealing with advances in any aspect of the computing process or its application, are welcome. They must be concise, theoretically sound and written in English; they may have resulted, e.g., from research projects, advanced workshops and conferences. The publications (of the series) should address not only the specialist in a particular field but also a more general scientific audience.

An International Advisory Board which will review papers before publication will guarantee the high standard of volumes published within this series.

If you are interested in submitting work for this series please contact us:

Springer-Verlag
Editorial Department
Sachsenplatz 4-6
A-1200 Wien
Austria
Tel.: x43/1/3302415/532
Facsimile: x43/1/330242665
email: silvia.schilgerius@springer.co.at

The Publisher

We dedicate this book
to the memory of Ernst Schröder (1841–1902),
in celebration of the first centenary of his
Algebra und Logik der Relative
of 1895

Preface

The calculus of relations has been an important component of the development of logic and algebra since the middle of the nineteenth century, when Augustus De Morgan observed that since a horse is an animal we should be able to infer that the head of a horse is the head of an animal. For this, Aristotelian syllogistic does not suffice: We require relational reasoning.

George Boole, in his *Mathematical Analysis of Logic* of 1847, initiated the treatment of logic as part of mathematics, specifically as part of algebra. Quite the opposite conviction was put forward early this century by Bertrand Russell and Alfred North Whitehead in their *Principia Mathematica* (1910 – 1913): that mathematics was essentially grounded in logic. Logic thus developed in two streams.

On the one hand algebraic logic, in which the calculus of relations played a particularly prominent part, was taken up from Boole by Charles Sanders Peirce, who wished to do for the "calculus of relatives" what Boole had done for the calculus of sets. Peirce's work was in turn taken up by Schröder in his *Algebra und Logik der Relative* of 1895 (the third part of a massive work on the algebra of logic). Schröder's work, however, lay dormant for more than 40 years, until revived by Alfred Tarski in his seminal paper "On the calculus of binary relations" of 1941 (actually his presidential address to the Association for Symbolic Logic). Tarski's paper is still often referred to today as the best introduction to the calculus of relations. It gave rise to a whole field of study, that of relation algebras, and more generally Boolean algebras with operators. These formed part of the study of algebraic logic, which in its modern form overlaps with universal algebra, model theory, nonclassical logics, and more recently program semantics and program development.

In the other stream of development, the theory of relations was accorded a key role in *Principia Mathematica*. This important work defined much of the subsequent development of logic in the 20th century, completely eclipsing for some time the development of algebraic logic. In this stream of development, relational calculus and relational methods appear with the development of universal algebra in the 1930's, and again with model theory from the 1950's onwards. In so far as these disciplines in turn overlapped with the development of category theory, relational methods sometimes appear in this context as well.

It is fair to say that the role of the calculus of relations in the interaction between algebra and logic is by now well understood and appreciated, and that relational methods are part of the toolbox of the mathematician and the logician. Over the past twenty years relational methods have however also become of fundamental importance in computer science. For example, much of the theory of nonclassical logics was used (though sometimes re-invented) in the new so-called *program logics*. These arose from the realisation that a program may be thought of as an input-output relation over some state space: an accessibility relation, in the sense of modal logic, saying that from a given initial state certain final states are accessible. Not every relation over a state space can be thought of as a program, but any relation can be thought of as a *specification* of a program. This point of view, that the calculus of relations is fundamental to programs, was clearly enunciated by the Oxford group of Tony Hoare in an influential paper on "Laws of programming" in 1987. Also during the 1980's, much of the equational theory of relation algebras were already being applied to program semantics and program development. For example, the book *Relations and Graphs* by Schmidt and Ströhlein (in German, 1989; in English, 1993) started from the basis of programs as graphs with Boolean matrices. On the other hand, a topic such as relational databases was an essentially new development, owing little to previous mathematical work on or with relations. Thus computer science, as the new application field for relational methods, has both drawn from and contributed to previous logico/mathematical work; this is a sign of healthy development.

However, the role of relational methods within computer science is not yet as well-known or well-understood as it is in logic and mathematics. It is the aim of this book to help address this situation, by giving an overview of relational methods in computer science. We have not tried to give an exhaustive survey – indeed, part of what we claim is that the topic is already too large for an exhaustive survey to fit into one volume. We hope, however, that the topics included will suffice to convey an impression of the power and wide-ranging applicability of relational methods.

Just as we have attempted to bring together under one heading various disparate endeavours in using relational methods in computer science, so we have also attempted to bring together the researchers involved. It became clear to us in the early 1990's that a core group of researchers using or working on relational methods was turning up repeatedly at various specialist conferences. It seemed a good idea, therefore, to organise a meeting specifically on relational methods, so as to give all those who developed or used them in some area of computer science a chance to meet and talk across disciplinary boundaries. The outcome of this idea was the International Seminar on Relational Methods in Computer Science (later known as RelMiCS 1), held as Seminar 9403 at Schloß Dagstuhl in the Saarland in January 1994. We gratefully acknowledge the assistance of the *Internationales Begegnungs- und Forschungszentrum für Informatik* in organising and funding this Seminar. In our preface to the Seminar Report we said:

> Since the mid-1970's it has become clear that the calculus of relations is a fundamental conceptual and methodological tool in computer science just as much as in mathematics. A number of seemingly distinct areas of research have in fact this much in common that their concepts and/or techniques come from the calculus of relations. However, it has also become clear that many opportunities for cross-pollination are being lost simply because there was no organised forum of discussion between

> researchers who, though they use the same concepts and methods, nonetheless perceive themselves as working in different fields.
>
> The aim of this Dagstuhl Seminar was, therefore, to bring together researchers from various subdisciplines of computer science and mathematics, all of whom use relational methods in their work, and to encourage the creation of an active network continuing after the Seminar to exchange ideas and results.

This has become a manifesto of sorts for what is now known as the RelMiCS Group. The idea of producing this book first originated at Dagstuhl, where a number of authors were recruited. Some subsequent organisational work was done from Cape Town and München, but the book only really took shape in Rio de Janeiro: first at a planning session in July 1994, then at the RelMiCS 2 meeting in July 1995. We are pleased to acknowledge our indebtedness to Armando Haeberer for organising these extremely pleasant and productive events. We gratefully acknowledge the financial assistance of CNPq (Conselho Nacional de Desenvolvimento Científico e Tecnológico) for the planning session of July 1994, and of FAPERJ (Fundação de Amparo à Pesquisa do Estado do Rio de Janeiro) and IBM Brazil for RelMiCS 2 in July 1995. It gave us the chance to "workshop" the contents of the book so as to increase the homogeneity of notation and terminology that we felt was desirable.

In consequence, this is a multi-author book. However, though each chapter is written by a different set of authors, it has been our intention that the book should be more than just a collection of papers: it should have unity of style and presentation. In this way we hope that the spirit of bringing together topics and researchers is reflected also in the structure of the book itself.

Chapter 1 sets out the background material needed in the rest of the book, and the notation to be used. The problem of proliferation of different notations and terminology has been an impediment to communication about relational methods between various groups of researchers; it seems justified, therefore, to try, at least in a book like this, for a measure of coherence. After Chapt. 1, which forms the introductory Part I of the book, there are four more parts. Part II deals with current versions of algebras of relations, Part III with applications to logic, Part IV with applications to programming, and Part V with other application areas. For lack of space the Bibliography only contains works actually referred to in the text; a more comprehensive bibliography on the calculus of relations and its applications is available electronically.

Acknowledgements. In the original submissions for this book almost all authors acknowledged an indebtedness to other authors; we cover this by making here a blanket acknowledgement of thanks for mutual cooperation. We personally owe a large debt of thanks to Thomas Ströhlein, who undertook and carried out the arduous task of assembling a bibliography of relational methods based on submissions from several coauthors. The bibliography at the end of this book is a small subset of this comprehensive bibliography. In addition, we gratefully acknowledge work done on the manuscript by Arne Bayer, Thomas Gritzner, and Peter Kempf.

We are grateful to Rudolf Albrecht for establishing contact with the publishers. Of course our largest debt is to the authors who actually wrote the book, and who submitted themselves with good grace to the editorial requirements we imposed on them.

With over 30 contributors to this book we believe it possible that some errors may remain; for these we crave the indulgence of the reader, and would be pleased to receive feedback.

October 1996

Chris Brink
University of Cape Town

Wolfram Kahl, Gunther Schmidt
Universität der Bundeswehr München

Contents

V. Other Application Areas

Contributors

Roland Backhouse, TU Eindhoven

Gabriel Baum, Univ. La Plata

Nadir Belkhiter, Univ. Laval Québec

Rudolf Berghammer, Univ. Kiel

Patrick Blackburn, U. des Saarlandes

Michael Böttner, MPI Nijmegen

Chris Brink, Univ. Cape Town

Jules Desharnais, Univ. Laval Québec

Henk Doornbos, TU Eindhoven

Marcelo Frias, PUC Rio de Janeiro

Antonetta van Gasteren, TU Eindhoven

Armando Haeberer, PUC Rio de Janeiro

Claudia Hattensperger, UniBw München

Wolfgang Heinle, Univ. Bern

Bernard Hodgson, Univ. Laval Québec

Ryszard Janicki, McMaster U. Hamilton

Ali Jaoua, Univ. Tunis

Peter Jipsen, Univ. Cape Town

Wolfram Kahl, UniBw München

Burghard von Karger, Univ. Kiel

Roger Maddux, ISU Ames, Iowa

Ali Mili, Univ. Ottawa

Théodore Moukam, U. Laval, Québec

John Mullins, Univ. Ottawa

Thanh Tung Nguyen, SIGRAPA, Belgium

Ewa Orlowska, Inst. of Tel., Warsaw

Habib Ounalli, Univ. Tunis

David Parnas, McMaster U. Hamilton

Maarten de Rijke, Univ. Warwick

Holger Schlingloff, TU München

Gunther Schmidt, UniBw München

Thomas Ströhlein, TU München

Paulo Veloso, PUC Rio de Janeiro

Yde Venema, Free Univ. Amsterdam

Michael Winter, UniBw München

Jeffery Zucker, McMaster U. Hamilton

A detailed list of addresses and affiliations can be found on page 271.

Chapter 1

Background Material

Peter Jipsen, Chris Brink[1]*, Gunther Schmidt*

This chapter serves the rest of the book: all later chapters presuppose it. It introduces the calculus of binary relations, and relates it to basic concepts and results from lattice theory, universal algebra, category theory and logic. It also fixes the notation and terminology to be used in the rest of the book. Our aim here is to write in a way accessible to readers who desire a gentle introduction to the subject of relational methods. Other readers may prefer to go on to further chapters, only referring back to Chapt. 1 as needed.

1.1 The calculus of sets

Our approach to set theory is, for the most part, informal. We use the capital letters X, Y, Z to denote sets, and lower case x, y, z to denote elements. The symbols "$x \in X$" express the fact that x is an *element* (or *member*) of the *set* X. The *empty set* is the unique set $\emptyset$ that contains no elements. *Singletons* are sets with one element, *unordered pairs* are sets that contain exactly two elements and *unordered n-tuples* $\{x_0, \ldots, x_{n-1}\}$ contain n distinct elements. When convenient, a universal set U may be defined to represent the largest collection under consideration.

Given a set X and a property P, the set of all elements of X that satisfy P is denoted by

$$\{x \in X : P(x)\} \qquad \text{or} \qquad \{x : x \in X \text{ and } P(x)\}.$$

The property P is often described informally, but is understood to be an abbreviation for a precise expression in some formal language.

There are several obvious relations and operations defined on sets:

- *equality:* $X = Y$ if X and Y contain the same elements,
- *inclusion:* $X \subseteq Y$ if every element of X is also an element of Y,
- *union:* $X \cup Y =$ the set of elements in X or Y,
- *intersection:* $X \cap Y =$ the set of elements in both X and Y,
- *difference:* $X - Y =$ the set of elements in X that are not in Y,

[1]Chris Brink gratefully acknowledges the longstanding financial support of the South African Foundation for Research Development.

- *complement:* $\overline{X} = U - X$ provided that a universal set U has been fixed,
- *powerset:* $\mathcal{P}(X) =$ the set of all subsets of X.

Two sets are said to be *disjoint* if their intersection is the empty set.

The expression "calculus of sets" refers to the many interactions between these relations and operations. The main ones are captured by the following observations: For any universal set U and $X, Y, Z \subseteq U$

- $X \subseteq X$ (reflexivity of $\subseteq$),
- $X \subseteq Y$ and $Y \subseteq Z$ imply $X \subseteq Z$ (transitivity of $\subseteq$),
- $X \subseteq Y$ and $Y \subseteq X$ imply $X = Y$ (antisymmetry of $\subseteq$),
- $X \cup Y = Y \cup X$ and $X \cap Y = Y \cap X$ (commutativity of $\cup$, $\cap$),
- $(X \cup Y) \cup Z = X \cup (Y \cup Z)$ and $(X \cap Y) \cap Z = X \cap (Y \cap Z)$ (associativity of $\cup$, $\cap$),
- $X \cap (Y \cup Z) = (X \cap Y) \cup (X \cap Z)$ and $X \cup (Y \cap Z) = (X \cup Y) \cap (X \cup Z)$ (distributivity of $\cup$, $\cap$) and
- $X \cup \overline{X} = U$ and $X \cap \overline{X} = \emptyset$.

Readers familiar with mathematical terminology may note that the first three properties show $\subseteq$ is a partial order on $\mathcal{P}(U)$, and the remaining ones imply that $(\mathcal{P}(U), \cup, \cap, \overline{}, \emptyset, U)$ is a Boolean algebra.

If I is any set, and $\mathcal{X} = \{X_i : i \in I\}$ is a collection of sets indexed by I then the union and intersection of $\mathcal{X}$ are defined by

$$\bigcup \mathcal{X} \triangleq \bigcup_{i \in I} X_i \triangleq \{x : x \in X_i \text{ for some } i \in I\}^2,$$

$$\bigcap \mathcal{X} \triangleq \bigcap_{i \in I} X_i \triangleq \{x : x \in X_i \text{ for all } i \in I\}.$$

From objects x and y we obtain the *ordered pair* or simply *pair* (x, y) with the characteristic property that $(x, y) = (x', y')$ if and only if $x = x'$ and $y = y'$. *Sequences of length* n or n*-tuples* $(x_0, \ldots, x_{n-1})$ are characterized by the analogous property. The *Cartesian product* of sets $X_0, \ldots, X_{n-1}$ is given by

$$X_0 \times \ldots \times X_{n-1} = \prod_{i=0}^{n-1} X_i \triangleq \{(x_0, \ldots, x_{n-1}) : x_i \in X_i \text{ for all } i < n\}.$$

If all the sets X_i are equal to X, we write X^n instead of $\prod_{i=0}^{n-1} X$.

We also fix the following notation for various standard sets:

- $\mathbb{N} = \{0, 1, 2, 3, \ldots\}$, the set of natural numbers.
- $\mathbb{Z} = \{\ldots, -2, -1, 0, 1, 2, \ldots\}$, the set of integers.
- $\mathbb{Q} = \{\frac{m}{n} : m \in \mathbb{Z} \text{ and } 0 \neq n \in \mathbb{N}\}$, the set of rationals.
- $\mathbb{R}$ the set of reals.
- $\mathbb{B} = \{\mathsf{true}, \mathsf{false}\}$, the set of booleans.

---

[2] The symbol $\triangleq$ signifies that the equalities hold by definition.

1.2 The calculus of binary relations

Informally, a binary relation is simply a collection of ordered pairs. More precisely, a *binary relation* R *from a set* X *to a set* Y is a subset of the set of all pairs (x, y) where $x \in X$ and $y \in Y$. In symbols

$$R \subseteq X \times Y.$$

If $X = Y$, we say that R is a *relation over* X. Instead of $(x, y) \in R$, we usually write xRy. By virtue of being sets, binary relations are partially ordered by inclusion. The smallest relation is just the empty set $\emptyset$. Depending on the context, we may also fix a largest relation, called the *universal relation*, which is denoted by V (for example $V = U^2$). Another special relation defined for each set X is the *identity relation*

$$I_X = \{(x, x) : x \in X\}.$$

Since relations are sets, all the set-theoretic operations apply. However, relations are more than just sets, and the structure of their elements allows the definition of many other operations. The most common ones are

- *domain:* $domR = \{x : \text{there exists } y \text{ such that } xRy\}$,
- *range:* $ranR = \{y : \text{there exists } x \text{ such that } xRy\}$,
- *converse:* $R^{\smile} = \{(x, y) : yRx\}$,
- *composition:* $R;S = \{(x, y) : \text{there exists } z \text{ such that } xRz \text{ and } zSy\}$,
- *right residual:* $R\backslash S = \{(x, y) : \text{for all } z,\ zRx \text{ implies } zSy\}$,
- *left residual:* $R/S = \{(x, y) : \text{for all } z,\ ySz \text{ implies } xRz\}$,
- *Peirce product:* $R : Y = \{x : \text{there exists } y \in Y \text{ such that } xRy\}$,
- *image set:* $R(x) = \{y : xRy\}$,
- *exponentiation:* $R^0 = I_U$ where U is a fixed universal set, and $R^{n+1} = R;R^n$.

One of the strengths of the relational calculus is that many properties can be expressed very compactly. This is demonstrated throughout this book, beginning with the list below of common conditions used to classify relations. A relation R is said to be

- *reflexive* if $I_U \subseteq R$ (i.e., xRx for all $x \in U$),
- *transitive* if $R;R \subseteq R$ (i.e., xRy and yRz imply xRz for all $x, y, z \in U$),
- *symmetric* if $R = R^{\smile}$ (i.e., xRy implies yRx for all $x, y \in U$),
- *antisymmetric* if $R \cap R^{\smile} \subseteq I_U$ (i.e., xRy and yRx imply $x = y$),
- a *preorder* if it is reflexive and transitive (also called a *quasi-order*),
- an *equivalence relation* if it is a symmetric preorder,
- a *partial order* if it is an antisymmetric preorder.

Since the intersection of transitive relations is again transitive, any relation R is contained in a smallest transitive relation, called the *transitive closure* of a relation

R. This relation can be defined directly by $R^+ = \bigcap\{T : R \subseteq T \text{ and } T;T \subseteq T\}$, and it is easy to prove that

$$R^+ = \bigcup_{i \in \mathbb{N}} R^i = R \cup R^2 \cup R^3 \cup \cdots$$

The closely related *reflexive transitive closure* of R is defined as $R^* = I_U \cup R^+$. Both operations are of particular interest to computer science since they can model the behaviour of programs with loops.

When several operations appear in the same expression, parentheses are used to indicate the order in which the operations are performed. To avoid proliferation of parentheses the following convention is adopted:

Priority of operations: Unary (superscript) operations ($\overline{}$, $\breve{}$, $^+$, *, ^{n}) are performed first, followed by the binary relation operations ($;$, $:$, $/$, $\backslash$) and finally the binary set operations ($\cup$, $\cap$).

Algebraic properties of relation operations

The most obvious properties of operations defined on relations are listed below.

- Composition is *associative*: $(R;S);T = R;(S;T)$.
- I_U is an *identity* for the composition of relations on U: $I_U;R = R = R;I_U$.
- Composition *distributes* over union: $R;(S \cup T) = R;S \cup R;T$ and $(R \cup S);T = R;T \cup S;T$.
- Conversion *distributes* over union: $(R \cup S)^{\smile} = R^{\smile} \cup S^{\smile}$.
- Conversion is an *involution*: $(R^{\smile})^{\smile} = R$.
- Conversion *antidistributes* over composition: $(R;S)^{\smile} = S^{\smile};R^{\smile}$.
- Composition satisfies what are known as the *Schröder equivalences* (which will later be related to *right-* and *left conjugates*):

$$R;S \cap T = \emptyset \iff R^{\smile};T \cap S = \emptyset \iff T;S^{\smile} \cap R = \emptyset .$$

- The reflexive transitive closure satisfies

$$R^* = I_U \cup R;R^* \qquad \text{and} \qquad R;S \subseteq S \implies R^*;S \subseteq S .$$

Many other relationships hold, but the ones mentioned above have the distinction that they form the basis for an abstract treatment of relations.

Viewing relations as sets of ordered pairs is appealing in its simplicity, but for many applications in computer science it is more useful to "type" a relation $R \subseteq X \times Y$ by explicitly recording its *source* X and *target* Y. So we define a *typed relation from* X *to* Y to be a triple (R, X, Y), where $R \subseteq X \times Y$. The set of all typed relations from X to Y is denoted by $[X \leftrightarrow Y]$, and instead of $T \in [X \leftrightarrow Y]$ we also write $T : X \leftrightarrow Y$. When working only with typed relations, the adjective "typed" is usually omitted. All the operations and properties of (untyped) relations apply here as well, with the understanding that conversion interchanges the source and target, and the binary operations are only defined when the sources and targets of the respective relations are compatible. For the operations $\cup$, $\cap$ the relations must have the same sources and the same targets, and for the other operations, if $R : X \leftrightarrow Y$, $S : Y \leftrightarrow Z$ and $T : X \leftrightarrow Z$ then

$$R^{\smile} : Y \leftrightarrow X, \quad R;S : X \leftrightarrow Z, \quad R\backslash T : Y \leftrightarrow Z \text{ and } T/S : X \leftrightarrow Y.$$

The additional structure of typed relations leads to further useful definitions. A relation $R : X \leftrightarrow Y$ is said to be

- *univalent* if $R^{\smile};R \subseteq I_Y$ (i.e. xRy and xRy' imply $y = y'$ for all x, y, y'),
- *total* if $I_X \subseteq R;R^{\smile}$ (i.e. for all $x \in X$ there exists $y \in Y$ such that xRy),
- a *function* if it is univalent and total,
- *injective* if $R;R^{\smile} \subseteq I_X$ (i.e. xRy and $x'Ry$ imply $x = x'$ for all x, x', y),
- *surjective* if $I_Y \subseteq R^{\smile};R$ (i.e. for all $y \in Y$ there exists $x \in X$ such that xRy),
- a *bijection* if it is an injective and surjective function (i.e. $R;R^{\smile} = I_X$ and $R^{\smile};R = I_Y$).

If $R : X \leftrightarrow Y$ is a function, we write $R : X \to Y$. If R is not total but univalent then it is called a *partial function.*

In either case we will normally use the symbols f, g, h instead of relational symbols, and the notation "$y = f(x)$" instead of "xfy". For two functions $f : X \to Y$ and $g : Y \to Z$, the *composite function* $g \circ f$ is defined by $g \circ f = f;g$, with the result that $(g \circ f)(x) = g(f(x))$.

The notion of "binary relation" can obviously be generalized to higher dimensions: an *n-ary relation based on sets* $X_0, \ldots, X_{n-1}$ is simply a subset of the n-ary Cartesian product $X_0 \times \ldots \times X_{n-1}$. Although less pervasive than binary relations, these generalizations arise naturally in logic, algebra, database theory and many other areas. In the above definition, the elements of an n-ary relation are n-tuples indexed by the set $H = \{0, \ldots, n-1\}$. In applications to computer science it is often convenient to allow arbitrary index sets. For example, if the 3-tuple $t =$ (Smith, Jane, 1234567) represents an entry in a phonebook, a suitable index set would be $H = \{$lastname, firstname, phonenumber$\}$. Such descriptive indices are also called *attributes* in database theory. Associated with each index i is a *domain of values* $\mathcal{D}(i)$, e.g. $\mathcal{D}$(lastname) would be the collection of lastnames valid for a phonebook.

An *H-tuple* t is now simply a function from H to $\bigcup_{i \in H} \mathcal{D}(i)$ such that $t(i) \in \mathcal{D}(i)$, and the set of all H-tuples is the Cartesian product $\prod_{i \in H} \mathcal{D}(i)$. A relation with index set H is a subset of this Cartesian product. Note that if $H = \{0, \ldots, n-1\}$ then an H-tuple is just an n-tuple and a relation with index set H is an n-ary relation.

Given a Cartesian product $\prod_{i \in H} X_i$ there is for each $i \in H$ a *projection function*

$$\pi_i : \prod_{i \in H} X_i \to X_i \quad \text{defined by} \quad \pi_i(t) \triangleq t(i).$$

For a subset J of H, each H-tuple t gives rise to a J-tuple $t[J]$ defined by restricting the function t to values from J. This provides a generalized projection function $\pi_J : \prod_{i \in H} X_i \to \prod_{j \in J} X_j$ given by $\pi_J(t) \triangleq t[J]$.

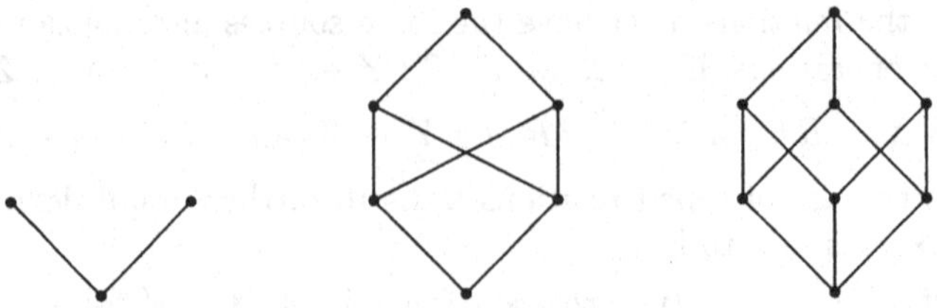

(a) (b) (c)

Fig. 1.1 Some posets and their Hasse diagrams

Using the above notions, any relation over an index set H can be decomposed into a binary relation. A particular decomposition is determined by a subset J of H, and each H-tuple t is mapped to the pair $(t[J], t[\overline{J}])$, where complementation is with respect to H. Conversely, given disjoint sets J, K and a binary relation in which all pairs have J-tuples as first component and K-tuples as second component, each pair (x, y) can be mapped to a unique $J \cup K$-tuple t defined by $t[J] = x$ and $t[K] = y$.

1.3 Partially ordered structures

As mentioned in Sect. 1.2, a binary relation R on a set X is a *partial order* if it is reflexive, transitive and antisymmetric. The pair (X, R) is called a *partially ordered set* or *poset.* Usually "R" is replaced with a more suggestive symbol "$\sqsubseteq$" (read "less-or-equal"), in which case $R^{\smile}$ is written $\sqsupseteq$ (read "greater-or-equal"). If X has only a few elements then it is quite instructive to represent $(X, \sqsubseteq)$ by a *Hasse diagram* where an element x is connected by a straight line to a distinct element y higher up on the page if and only if $x \sqsubseteq y$ and there exists no $z \in X$ distinct from x and y with $x \sqsubseteq z \sqsubseteq y$. Some examples of posets and their Hasse diagrams are given in Fig. 1.1.

In a poset $(X, \sqsubseteq)$ an element x is an *upper bound* of a subset Y of X if $x \sqsupseteq y$ for all $y \in Y$. The *least upper bound* of Y, also called the *join* of Y and denoted by $\bigsqcup Y$, is the (unique) upper bound that is less than every other upper bound of Y. Of course upper bounds and least upper bounds don't necessarily exist for all subsets (see Fig. 1.1 (a), (b)). In particular, since every element is an upper bound of the empty set, $\bigsqcup \emptyset$ exists if and only if $(X, \sqsubseteq)$ has a least element, denoted by $\perp\!\!\!\perp$ (read "bottom"). The largest element of a poset, if it exists, is denoted by $\top\!\!\!\top$ (read "top").

A nonempty subset C of X is a *chain* if it is linearly ordered, i.e. $x \sqsubseteq y$ or $y \sqsubseteq x$ for all $x, y \in C$. This definition allows us to define a *complete partial order* (*cpo* for short) as a poset $(X, \sqsubseteq)$ in which every chain has a least upper bound. In computer science cpos are mainly used as models of denotational semantics for programs. A poset endowed with extra structure is naturally thought of as a *partially ordered structure.* Algebraic versions of the calculus of sets and the calculus of relations are in a natural way represented as such structures.

Lattices

The *greatest lower bound* or *meet* of a subset Y in a poset $(X, \sqsubseteq)$ is denoted by $\bigsqcap Y$, and can be defined as the least upper bound of Y in the *dual poset* $(X, \sqsupseteq)$. A *lattice* is a nonempty poset in which every two-element subset $\{x, y\}$ has a join $x \sqcup y = \bigsqcup\{x, y\}$ and a meet $x \sqcap y = \bigsqcap\{x, y\}$. These two operations are

- *associative:* $(x \sqcup y) \sqcup z = x \sqcup (y \sqcup z)$ and $(x \sqcap y) \sqcap z = x \sqcap (y \sqcap z)$,
- *commutative:* $x \sqcup y = y \sqcup x$ and $x \sqcap y = y \sqcap x$,
- *idempotent:* $x \sqcup x = x$ and $x \sqcap x = x$, and satisfy the
- *absorptive laws:* $(x \sqcup y) \sqcap x = x$ and $(x \sqcap y) \sqcup x = x$.

The equational properties listed above characterize lattices in the following sense: Given a set X with two binary operations $\sqcup$ and $\sqcap$ that satisfy the above equations for all $x, y, z \in X$, define a relation $\sqsubseteq$ on X by

$$x \sqsubseteq y \quad \iff \quad x \sqcup y = y.$$

Then $(X, \sqsubseteq)$ turns out to be a lattice in which the join and meet operations agree with $\sqcup$ and $\sqcap$. Hence a lattice may be viewed as either a poset $(X, \sqsubseteq)$ or an algebraic structure $(X, \sqcup, \sqcap)$. It is easy to give examples of a set of sets closed under union and intersection; this would be an example of a lattice.

In a lattice the least upper bound of any finite nonempty subset can be found by repeated application of the join operation. A lattice $(X, \sqsubseteq)$ is said to be

- *bounded* if it has a smallest element $\perp\!\!\!\perp$ $(= \bigsqcup \emptyset = \bigsqcap X)$ and a largest element $\top\!\!\!\top$ $(= \bigsqcup X = \bigsqcap \emptyset)$,
- *complete* if every subset Y has a join $\bigsqcup Y$. In that case Y also has a meet $\bigsqcap Y = \bigsqcup\{\text{all lower bounds of } Y\}$,
- *complemented* if it is bounded and every $x \in X$ has a *complement* $\overline{x}$ such that $x \sqcup \overline{x} = \top\!\!\!\top$ and $x \sqcap \overline{x} = \perp\!\!\!\perp$,
- *distributive* if $x \sqcup (y \sqcap z) = (x \sqcup y) \sqcap (x \sqcup z)$ and $x \sqcap (y \sqcup z) = (x \sqcap y) \sqcup (x \sqcap z)$ for all $x, y, z \in X$.

We note that in a distributive lattice the complement of an element is unique (whenever it exists).

Boolean algebras

With the above terminology, a *Boolean algebra* is defined as a complemented distributive lattice $\mathcal{A} = (A, \sqcup, \sqcap, \overline{}, \perp\!\!\!\perp, \top\!\!\!\top)$. Since it can be proved that Boolean algebras satisfy *De Morgan's laws* $\overline{x \sqcup y} = \overline{x} \sqcap \overline{y}$ and $\overline{x \sqcap y} = \overline{x} \sqcup \overline{y}$, it is actually enough to list only the operations $\sqcup$, $\overline{}$ and $\perp\!\!\!\perp$. The others are recovered as

$$x \sqcap y = \overline{\overline{x} \sqcup \overline{y}} \qquad \text{and} \qquad \top\!\!\!\top = \overline{\perp\!\!\!\perp}.$$

The theory of Boolean algebras is an abstract algebraic version of the calculus of sets (Sect. 1.1) and propositional logic (Sect. 1.6). Standard examples of Boolean algebras are the collection of all subsets of a universal set U:

$$\mathcal{B} = (\mathcal{P}(U), \cup, \overline{}, \emptyset).$$

Figure 1.1 (c) shows the Hasse diagram of such an algebra when U is a 3-element set. Other examples are obtained by considering subcollections $C \subseteq \mathcal{P}(U)$ with the property that $\emptyset \in C$ and for all $X, Y \in C$ we also have both $X \cup Y \in C$ and $\overline{X} \in C$. Then $\mathcal{C} = (C, \cup, \overline{}, \emptyset)$ is also a Boolean algebra[3], referred to as a *subalgebra* of $\mathcal{B}$. In fact, by a fundamental result of Stone, dating back to 1935, every Boolean algebra can be obtained in this way (up to isomorphism).

To make this precise, we need a few definitions. An *atom* in a Boolean algebra is a minimal non-$\perp\!\!\!\perp$ element. In the algebra $\mathcal{B}$ above, the atoms are the singleton subsets of U. A subset F of a Boolean algebra is

- *meet-closed* if $x, y \in F$ implies $x \sqcap y \in F$,
- an *upset* if $x \in F$ and $x \sqsubseteq y$ imply $y \in F$,
- a *filter* if it is a meet-closed upset,
- an *ultrafilter* if it is a filter, $\perp\!\!\!\perp \notin F$ and for all x in the algebra either $x \in F$ or $\overline{x} \in F$.

The last condition is equivalent to saying that F is a maximal proper filter. For example, given an atom a in a Boolean algebra, there is exactly one ultrafilter F containing a, namely $F = \{x : a \sqsubseteq x\}$. Now the ultrafilters of a Boolean algebra $\mathcal{A}$ can serve as atoms of another algebra.

Theorem 1.3.1 (*Stone's Representation Theorem*) Let $\mathcal{U}_\mathcal{A}$ be the set of all ultrafilters of $\mathcal{A}$. The function $\sigma : \mathcal{A} \to \mathcal{P}(\mathcal{U}_\mathcal{A})$ defined by

$$\sigma(x) \triangleq \{F \in \mathcal{U}_\mathcal{A} : x \in F\}$$

is a Boolean algebra embedding, i.e. $\sigma(x \sqcup y) = \sigma(x) \cup \sigma(y)$, $\sigma(\overline{x}) = \overline{\sigma(x)}$, $\sigma(\perp\!\!\!\perp) = \emptyset$ and σ is injective. □

Relation algebras

Here we define abstract algebraic structures that capture many of the algebraic properties of relations. They are based on Boolean algebras augmented by the operations $;$ (binary *relation composition*) and $\breve{\ }$ (unary *converse*), and a distinguished element $\mathbb{I}$ (*identity*).

An *(abstract) relation algebra* is of the form $(A, \sqcup, \overline{}, \perp\!\!\!\perp, ;, \breve{\ }, \mathbb{I})$ where

- $(A, \sqcup, \overline{}, \perp\!\!\!\perp)$ is a Boolean algebra,
- $;$ is associative and distributes over $\sqcup$,
- $\breve{\ }$ is an involution, distributes over $\sqcup$ and antidistributes over $;$,
- $\mathbb{I}$ is an identity for $;$ and
- $x ; \overline{x\breve{\ } ; \overline{y}} \sqsubseteq y$ for all $x, y \in A$.

The last-mentioned inequality states that $z = \overline{x\breve{\ } ; \overline{y}}$ is a solution of $x ; z \sqsubseteq y$. In the presence of the first three properties, this is in fact the largest solution. Moreover, the inequality is equivalent to the claim that, for all $x, y, z \in A$:

$$x ; y \sqcap z = \perp\!\!\!\perp \iff x\breve{\ } ; z \sqcap y = \perp\!\!\!\perp \iff z ; y\breve{\ } \sqcap x = \perp\!\!\!\perp.$$

[3]Since the axioms for lattices, distributivity and complementation are equational.

These are exactly the Schröder equivalences, and hence the properties listed here for abstract relations are the same as for relations in Sect. 1.2 (under the subheading "Algebraic properties of binary relations"). A standard example of a relation algebra is the set of all relations on a universe U, called the *full relation algebra over* U:

$$Rel(U) = (\mathcal{P}(U^2), \cup, \overline{}, \emptyset, ;, \breve{}, I_U) \ .$$

Boolean modules and Peirce algebras

Observe that in a relation algebra the nonboolean operations $;$, $\breve{}$ and $\mathbb{I}$ distribute in each argument over the Boolean join $\sqcup$.[4] Such operations are called operators, and algebras of this type are collectively referred to as *Boolean algebras with operators* or BAOs for short. An operator is said to be *normal* if it has value $\perp\!\!\!\perp$ whenever one of the arguments is $\perp\!\!\!\perp$. A *normal* BAO is one in which every operator is normal. It is easy to show that relation algebras are normal BAOs.

Let $\mathcal{A}$ be a relation algebra. A (left) *Boolean* $\mathcal{A}$*-module* is a Boolean algebra $\mathcal{B} = (B, \sqcup, \overline{}, \perp\!\!\!\perp)$ together with a mapping $f : A \times B \to B$, where $f(r,x)$ is written $r{:}x$ and called the *Peirce product*, such that for all $r, s \in A$ and $x, y \in B$:

- $r{:}(x \sqcup y) = r{:}x \sqcup r{:}y$,
- $(r \sqcup s){:}x = r{:}x \sqcup s{:}x$,
- $(r\,;s){:}x = r{:}(s{:}x)$,
- $\mathbb{I}{:}x = x$,
- $\perp\!\!\!\perp{:}x = \perp\!\!\!\perp$,
- $r^{\smile}{:}\overline{r{:}x} \sqsubseteq \overline{x}$.

To see that Boolean modules are a type of BAO, one can define for each $r \in A$ a unary operation $f_r{:}B \to B$ by $f_r(x) = r{:}x$. Then the first equation above states that each f_r is an operator. For a concrete example of a Boolean module, consider $\mathcal{A} = Rel(U)$, $\mathcal{B} = (\mathcal{P}(U), \cup, \overline{}, \emptyset)$ and $R{:}X$ is the Peirce product defined in Sect. 1.2. Thus, whereas relation algebras are an algebraic version of relations acting on each other, Boolean modules are an algebraic version of relations acting on sets.

It is occasionally useful to consider also relation-forming operations on sets. One such is the *cylindrification*, which associates with any set X contained in some universe U the relation $X^c = \{(x,u) : x \in X \text{ and } u \in U\}$. A *Peirce algebra* $(\mathcal{A}, \mathcal{B})$ is a Boolean $\mathcal{A}$-module $\mathcal{B}$ with an additional unary (postfix) operation $^c : \mathcal{B} \to \mathcal{A}$, such that for all $r \in \mathcal{A}$ and $x \in \mathcal{B}$

$$x^c{:}\top\!\!\!\top = x \quad \text{and} \quad (r{:}\top\!\!\!\top)^c = r\,;\top\!\!\!\top.$$

The concrete example above can be expanded to a Peirce algebra by defining X^c to be the relation $X \times U$.

[4]The constant $\mathbb{I}$ does so vacuously, since it has no arguments.

Residuation and conjugation

When dealing with operations on partial orders, there are several recurring concepts that are simple, but important.

Let $(X, \sqsubseteq)$ and $(Y, \sqsubseteq)$ be two partially ordered sets. A map $f : X \to Y$ is

- *order-preserving* (or *isotonic*) if $x \sqsubseteq x'$ implies $f(x) \sqsubseteq f(x')$,
- *join-preserving* if $f(x \sqcup y) = f(x) \sqcup f(y)$ whenever $x \sqcup y$ exists in X,
- *completely join-preserving* if $f(\bigsqcup S) = \bigsqcup\{f(s) : s \in S\}$ whenever $\bigsqcup S$ exists in X,
- *residuated* if there is a map $g : Y \to X$, called the *residual of* f, such that

$$f(x) \sqsubseteq y \iff x \sqsubseteq g(y) \qquad \text{for all } x \in X,\ y \in Y.$$

A residuated map f and its residual g are also referred to as a *Galois connection* or *adjunction*. In the latter case f is the *lower adjoint* (of g), and g is the *upper adjoint* (of f)[5].

A simple example of a residuated map is the function $f(x) = x^{\smile}$, defined on any relation algebra. The residual of f is in fact f itself, since the statements $x^{\smile} \sqsubseteq y$ and $x \sqsubseteq y^{\smile}$ are equivalent. (For residuals of relations see below.)

The notion of a *(completely) meet-preserving* map is defined dually by interchanging $\sqcup$ with $\sqcap$, and $\bigsqcup$ with $\bigsqcap$. The following result implies that the properties above are listed in increasing order of strength.

Theorem 1.3.2 Let $(X, \sqsubseteq)$ and $(Y, \sqsubseteq)$ be posets. If $f : X \to Y$ is residuated, then it is completely join-preserving and hence order-preserving. Furthermore, the residual g is unique, completely meet-preserving, and is given by

$$g(y) = \bigsqcup\{x \in X : f(x) \sqsubseteq y\}.$$

For a partial converse, if $(X, \sqsubseteq)$ is a complete lattice and f is completely join-preserving then f is residuated. □

For a map $f : A \to B$, where $(A, \sqcup, \overline{}, \perp\!\!\!\perp)$ and $(B, \sqcup, \overline{}, \perp\!\!\!\perp)$ are Boolean algebras, the *dual of* f is defined by $f^d(x) = \overline{f(\overline{x})}$. Furthermore, f is said to be *conjugated* if there is a map $h : B \to A$, called the *conjugate of* f, such that

$$f(x) \sqcap y = \perp\!\!\!\perp \iff x \sqcap h(y) = \perp\!\!\!\perp \qquad \text{for all } x \in X,\ y \in Y.$$

(Or, equivalently,

$$f(x) \sqsubseteq \overline{y} \iff x \sqsubseteq \overline{h(y)} \qquad \text{for all } x \in X,\ y \in Y.$$

In this form it may be compared to the definition above of a residuated map.) A natural example of a conjugated map is the function $f_a(x) = a;x$, defined for any element a in a relation algebra. The conjugate map is $h_a(y) = a^{\smile};y$, since the expressions $a;x \sqcap y = \perp\!\!\!\perp$ and $a^{\smile};y \sqcap x = \perp\!\!\!\perp$ are equivalent. In fact, in this case the conjugate map is usually called the *right* conjugate. If we reverse the order of the composition, and define the map $(f')_a(x) = x;a$, then the conjugate

[5] In categorical terms f is the left adjoint (of g), and g is the right adjoint (of f). To avoid confusion with left and right residuals of relations, we refrain from this terminology.

map (which will now be called the *left* conjugate) is $(h')_a(y) = y ; a^{\smile}$, since the expressions $x ; a \sqcap y = \perp\!\!\!\perp$ and $y ; a^{\smile} \sqcap x = \perp\!\!\!\perp$ are equivalent. This explains the use of the word "conjugate" in connection with the Schröder equivalences (under "algebraic properties of relation operations" in Sect. 1.2).

For maps between Boolean algebras, residuals and conjugates are duals of each other (i.e. $g = h^d$), so a map is conjugated if and only if it is residuated. (However, conjugation has the advantage of being a symmetric property: if h is the conjugate of f, then f is the conjugate of h.) In particular the residual g_a of the function f_a above is given by $(h_a)^d(y) = \overline{h_a(\overline{y})} = \overline{a^{\smile} ; \overline{y}}$. It is easy to check that this satisfies the definition of a residual: the expressions $f_a(x) \sqsubseteq y$ and $x \sqsubseteq g_a(y)$ are equivalent. Again, in this case the residual is usually called the *right* residual, and the *left* residual will be given by the dual of the left conjugate: $((h')_a)^d(y) = \overline{(h')_a(\overline{y})} = \overline{\overline{y} ; a^{\smile}}$. It is easy to verify that for binary relations these characterisations of right and left residual coincide with those given in Sect. 1.2.

Fixed points of order-preserving maps

Given a recursively defined function such as the factorial function g, defined by $g(0) = 1$ and $g(n+1) = (n+1) \cdot g(n)$, one can legitimately ask what object represents g. One solution is to view the recursion at a higher level: for every (partial) function $x : \mathbf{N} \to \mathbf{N}$, define a (partial) function $f(x)$ (often called a *functional* since it maps functions to functions) by the (nonrecursive) equations

$$f(x)(0) = 1 \quad \text{and} \quad f(x)(n+1) = (n+1) \cdot x(n)$$

and look for a solution of the equation $f(x) = x$. Any such solution is called a *fixed point* of f. In this example it can be shown that there is exactly one solution, namely $x(n) = n!$, but for other recursive equations there may be many or no solutions.

Recursively defined functions are prominent in computer science and, as in the above example, the meaning of a recursive definition can be viewed as a fixed point of a related function(al). So it is important to find general conditions under which a function is guaranteed to have fixed points.

Knaster and Tarski (1927) proved that, for any universe U, an inclusion-preserving function $f : \mathcal{P}(U) \to \mathcal{P}(U)$ has at least one fixed point. This result was generalized by Tarski to arbitrary complete lattices.

Theorem 1.3.3 (*Tarski's Fixed Point Theorem, 1955*) Let $(X, \sqsubseteq)$ be a complete lattice and suppose $f : X \to X$ is order-preserving. Then the set $F = \{x \in X : f(x) = x\}$ is nonempty. In fact $(F, \sqsubseteq)$ is a complete lattice[6] with least element $\bigsqcap F = \bigsqcap\{x \in X : f(x) \sqsubseteq x\}$ and largest element $\bigsqcup F = \bigsqcup\{x \in X : x \sqsubseteq f(x)\}$. □

A similar result holds for complete partial orders. The least element of $(F, \sqsubseteq)$ is (by definition) the *least fixed point* of f, denoted by μf. The *greatest fixed point* of f is $\nu f = \bigsqcup F$. When a recursive definition has several solutions, it is often the least fixed point that is of interest. The preceding result implies that μf is characterized by the conditions

[6]Note that in general $(F, \sqsubseteq)$ is not a sublattice of $(X, \sqsubseteq)$.

- $f(\mu f) = \mu f$ (computation) and
- $f(x) \sqsubseteq x \implies \mu f \sqsubseteq x$ (induction).

1.4 Relational structures and algebras

Posets and Boolean algebras are specific examples of relational structures and (universal) algebras. To define these concepts in general, we first need the notion of a *(similarity) type*, which is a function $\tau : \mathcal{F}_\tau \cup \mathcal{R}_\tau \to \{0, 1, 2, \ldots\}$ where $\mathcal{F}_\tau$ is a set of *function symbols* and $\mathcal{R}_\tau$ is a disjoint set of *relation symbols* or *predicate symbols*. For a symbol $s \in \mathcal{F}_\tau \cup \mathcal{R}_\tau$ we say that s has *arity* $\tau(s)$. A relational structure of type τ is of the form $\mathcal{U} = (U, (f^{\mathcal{U}})_{f \in \mathcal{F}_\tau}, (R^{\mathcal{U}})_{R \in \mathcal{R}_\tau})$, where

- U is a set called the *universe* of $\mathcal{U}$,
- $f^{\mathcal{U}}$ is a $\tau(f)$-ary operation on U, i.e. $f^{\mathcal{U}} : U^{\tau(f)} \to U$ and
- $R^{\mathcal{U}}$ is a $\tau(R)$-ary relation on U, i.e. $R^{\mathcal{U}} \subseteq U^{\tau(R)}$.

The distinction between the symbol $s \in \mathcal{F}_\tau \cup \mathcal{R}_\tau$ and its *interpretation* $s^{\mathcal{U}}$ is important, even though the superscript is often omitted in a context where confusion is unlikely. A 0-ary operation $c^{\mathcal{U}}$ from $U^0 = \{\emptyset\}$ to U is also called a *constant*, and $c^{\mathcal{U}}$ is usually identified with the value $c^{\mathcal{U}}(\emptyset)$.

If the set of relation symbols is empty then we say that τ is an *algebraic type*. A *(universal) algebra* is a relational structure of such a type. For example a Boolean algebra has an algebraic type $\tau = \{(\sqcup, 2), (\overline{}, 1), (\perp\!\!\!\perp, 0)\}$ and a relation algebra has an algebraic type $\tau' = \tau \cup \{(;, 2), (\breve{\ }, 1), (\mathbb{I}, 0)\}$. Some other algebras mentioned in later chapters are

- *semigroups* $(S, *)$ where $*$ is an associative binary operation,
- *monoids* $(M, *, e)$ where $(M, *)$ is a semigroup and e is an identity element,
- *groups* $(G, *, ^{-1}, e)$ where $(G, *, e)$ is a monoid and $^{-1}$ is a unary inverse operation ($x * x^{-1} = e = x^{-1} * x$),
- *semilattices* $(S, \sqcup)$, where $\sqcup$ is an associative, commutative and idempotent binary operation.

Given a set V, the set of *terms of type* τ *with variables from* V is defined as the smallest set $T_\tau(V)$ such that

- $V \subseteq T_\tau(V)$ and
- if $t_0, \ldots, t_{n-1} \in T_\tau(V)$, $f \in \mathcal{F}_\tau$ and $n = \tau(f)$ then the (uninterpreted) string of symbols $f(t_0, \ldots, t_{n-1})$ is in $T_\tau(V)$.

This set is the universe of an algebra $\mathcal{T}_\tau(V) = \mathcal{T} = (T_\tau(V), (f^{\mathcal{T}})_{f \in \mathcal{F}_\tau})$, called the *(absolutely free) term algebra of type* τ *generated by* V, with the operations $f^{\mathcal{T}}$ given by

$$f^{\mathcal{T}}(t_0, \ldots, t_{n-1}) = f(t_0, \ldots, t_{n-1}) \quad \text{for } t_i \in T_\tau(V),\ i < n = \tau(f).$$

The terms in $T_\tau(\emptyset)$ are known as *initial (or ground) terms* and $\mathcal{T}_\tau(\emptyset)$ is called the *initial algebra of type* τ.

For example the initial algebra of type $\{(z,0),(s,1)\}$ is the Peano algebra on which the arithmetic of the natural numbers is based ($0 = z$, $1 = s(z)$, $2 = s(s(z)),\ldots$).

Let $\mathcal{A}, \mathcal{B}, \mathcal{B}_i$ $(i \in I)$ be algebras of type τ.

- $\mathcal{A}$ is a *subalgebra* of $\mathcal{B}$ if $A \subseteq B$ and $f^{\mathcal{A}}(a_0,\ldots,a_{n-1}) = f^{\mathcal{B}}(a_0,\ldots,a_{n-1})$ for all $a_j \in A$, $j < n = \tau(f)$ and all $f \in \mathcal{F}_\tau$.
- $h : \mathcal{A} \to \mathcal{B}$ is a *homomorphism* if h is a function from A to B and $h(f^{\mathcal{A}}(a_0,\ldots,a_{n-1})) = f^{\mathcal{B}}(h(a_0),\ldots,h(a_{n-1}))$ for all $a_j \in A$, $j < n = \tau(f)$ and all $f \in \mathcal{F}_\tau$.
- $\mathcal{B}$ is a *homomorphic image* of $\mathcal{A}$ if there exists a surjective homomorphism from $\mathcal{A}$ to $\mathcal{B}$.
- $\mathcal{A}$ is *isomorphic* to $\mathcal{B}$, in symbols $\mathcal{A} \cong \mathcal{B}$, if there exists a bijective homomorphism from $\mathcal{A}$ to $\mathcal{B}$.
- $\mathcal{A} = \prod_{i \in I} \mathcal{B}_i$, the *product* of algebras $\mathcal{B}_i$, if $A = \prod_{i \in I} B_i$ and $f^{\mathcal{A}}$ is defined coordinatewise by

  $$\pi_i(f^{\mathcal{A}}(a_0,\ldots,a_{n-1})) = f^{\mathcal{B}_i}(\pi_i(a_0),\ldots,\pi_i(a_{n-1}))$$

  for all $a_j \in A$, $j < n = \tau(f)$ and all $f \in \mathcal{F}_\tau$.
- $\mathcal{A}$ is a *subdirect product* of algebras $\mathcal{B}_i$ $(i \in I)$ if $\mathcal{A}$ is a subalgebra of $\prod_{i \in I} \mathcal{B}_i$ and for each $i \in I$ and each $b \in B_i$ there is an $a \in A$ such that the i-th coordinate of a is b (indicated by writing $a_i = b$).
- $\mathcal{A}$ is *subdirectly irreducible* if whenever $\mathcal{A}$ is isomorphic to a subdirect product of algebras $\mathcal{B}_i$ $(i \in I)$ then $\mathcal{A} \cong \mathcal{B}_i$ for some $i \in I$.[7]

The last two notions are used in the following important result.

Theorem 1.4.1 (*Birkhoff's Subdirect Product Theorem, 1944*) Every algebra is isomorphic to a subdirect product of subdirectly irreducible algebras. □

A *congruence* on an algebra $\mathcal{A}$ is an equivalence relation R on A that is compatible with the operations of $\mathcal{A}$ in the sense that

$$a_0 R b_0 \text{ and } \ldots \text{ and } a_{n-1} R b_{n-1} \text{ implies } f^{\mathcal{A}}(a_0,\ldots,a_{n-1}) R f^{\mathcal{A}}(b_0,\ldots,b_{n-1})$$

for all $a_i, b_i \in A$, $i < n = \tau(f)$ and all $f \in \mathcal{F}_\tau$. The set of all congruences on an algebra $\mathcal{A}$ is denoted by $Con(\mathcal{A})$; it is closed under arbitrary intersections and hence is a complete lattice[8]. Congruences provide a way of describing all homomorphic images of an algebra. Given a congruence R on $\mathcal{A}$, the *quotient algebra* $\mathcal{A}/R$ is defined on the set A/R of equivalence classes ("$[a]_R$" denotes the equivalence class of a) by

$$f^{\mathcal{A}/R}([a_0]_R,\ldots,[a_{n-1}]_R) = [f^{\mathcal{A}}(a_0,\ldots,a_{n-1})]_R$$

for all $a_0,\ldots,a_{n-1} \in A$, $i < n = \tau(f)$ and all $f \in \mathcal{F}_\tau$. The compatibility condition above guarantees that these operations are well-defined. The so-called *canonical*

[7]Note that if $I = \emptyset$ then $\prod_{i \in \emptyset} \mathcal{B}_i$ is a one-element algebra, hence one-element algebras are not subdirectly irreducible.

[8]In fact it is an *algebraic lattice* since every congruence is the join of the compact (= finitely generated) congruences that it contains.

map $\pi : \mathcal{A} \to \mathcal{A}/R$ defined by $\pi(a) = [a]_R$ is a surjective homomorphism, so $\mathcal{A}/R$ is a homomorphic image of $\mathcal{A}$. Conversely, given any surjective homomorphism $h : \mathcal{A} \to \mathcal{B}$, the *kernel* of h, defined by

$$\ker(h) = \{(x, y) \in A^2 : h(x) = h(y)\},$$

is a congruence on $\mathcal{A}$, and $\mathcal{A}/\ker(h) \cong \mathcal{B}$. (This is usually called the *Homomorphism Theorem*, or sometimes the *First Isomorphism Theorem*).

Some algebras, particularly those we are interested in, such as Boolean algebras and relation algebras, are entirely defined by equations. We postpone a discussion of these till after we have dealt with equational logic in Sect. 1.6.

1.5 Categories

The language of categories is often used to present a topic at a high level of abstraction. A *category* **C** consists of a class[9] of *objects* $\mathsf{Obj}_{\mathbf{C}}$ and a class of *morphisms* $\mathsf{Mor}_{\mathbf{C}}$ that satisfy the following conditions.

- For each morphism f there is an object $\mathsf{source}\, f$ from which the morphism originates and an object $\mathsf{target}\, f$ where it ends. If $\mathsf{source}\, f = A$ and $\mathsf{target}\, f = B$ then we write $f : A \to B$.
- For all morphisms $f : A \to B$ and $g : B \to C$ there exists a *composition morphism*[10] $g \circ f : A \to C$, and for any $h : C \to D$ we have associativity $h \circ (g \circ f) = (h \circ g) \circ f$.
- For each object A there is an *identity morphism* $\mathsf{id}_A : A \to A$, and for any $f : A \to B$ and $g : B \to A$ we have $\mathsf{id}_A \circ g = g$ and $f \circ \mathsf{id}_A = f$.

The collection of all morphisms between objects $A, B \in \mathsf{Obj}_{\mathbf{C}}$ is denoted by $\mathsf{Mor}_{\mathbf{C}}[A, B]$. Some categories of interest are

- **Set** with sets as objects and functions as morphisms,
- **Rel** with sets as objects and relations as morphisms, and
- **Alg** with algebras as objects and homomorphisms as morphisms.

The language of category theory is well-suited to describing relationships between different areas of mathematics. *Functors* are structure-preserving maps between categories that make such relationships explicit and provide a reliable means for transferring results between different fields. More precisely, a *covariant functor* $\mathbf{F} : \mathbf{C} \to \mathbf{D}$ maps objects to objects and morphisms to morphisms (from **C** to **D**) such that

- $\mathbf{F} f : \mathbf{F} A \to \mathbf{F} B$ for any morphism $f : A \to B$,
- $\mathbf{F}(g \circ f) = (\mathbf{F} g) \circ (\mathbf{F} f)$ for any morphisms $f : A \to B$, $g : B \to C$ and
- $\mathbf{F}(\mathsf{id}_A) = \mathsf{id}_{\mathbf{F} A}$ for any object A.

[9] The distinction between *set* and *class* is not relevant here, so readers unfamiliar with classes may just think of them as sets.

[10] Recall that we have used the small circle rather than the semicolon to indicate functional composition; we do the same for morphisms. It is useful to keep in mind that by convention this inverts the order: $g \circ f = f ; g$.

For a *contravariant functor* the third condition is kept, but the first two conditions are changed to

- $\mathbf{F}f : \mathbf{F}B \to \mathbf{F}A$ for any morphism $f : A \to B$,
- $\mathbf{F}(g \circ f) = (\mathbf{F}f) \circ (\mathbf{F}g)$ for any morphisms $f : A \to B$, $g : B \to C$.

A functor $\mathbf{F} : \mathbf{C} \to \mathbf{D}$ is

- *full* if for all $A, B \in \mathsf{Obj}_{\mathbf{C}}$ and every $g : \mathbf{F}A \to \mathbf{F}B$ there exists $f : A \to B$ such that $\mathbf{F}f = g$, i.e. the restriction of $\mathbf{F}$ to $\mathsf{Mor}_{\mathbf{C}}[A, B]$ is *surjective* (onto $\mathsf{Mor}_{\mathbf{D}}[\mathbf{F}A, \mathbf{F}B]$).
- *faithful* if the restriction of $\mathbf{F}$ to $\mathsf{Mor}_{\mathbf{C}}[A, B]$ is *injective* (one-one) for all $A, B \in \mathsf{Obj}_{\mathbf{C}}$,
- *dense* if for every $D \in \mathsf{Obj}_{\mathbf{D}}$ there exists an object $C \in \mathsf{Obj}_{\mathbf{C}}$ such that $\mathbf{F}C = D$,
- an *equivalence* if $\mathbf{F}$ is covariant, full, faithful and dense,
- a *duality* if $\mathbf{F}$ is contravariant, full, faithful and dense.

Two categories $\mathbf{C}$ and $\mathbf{D}$ are *equivalent* if there exists an equivalence $\mathbf{F} : \mathbf{C} \to \mathbf{D}$, and they are *dual* if there exists a duality $\mathbf{F} : \mathbf{C} \to \mathbf{D}$.

1.6 Logics

There are many general perspectives on what constitutes a logic. Here we take the view that a logic consists of a collection of "similar" theories, defined by the following three items.

Syntax, specifying the *symbols* (variables, relation symbols, function symbols, connectives and/or quantifiers) of the theory, and how to combine them to obtain the properly formed expressions one wants to reason about. These expressions are traditionally called (*well-formed*) *formulae* (or *sentences*) of the logic.

Semantics, defined by a collection of models (i.e. mathematical structures, such as in Sect. 1.4) in which the symbols are interpreted, and a notion of truth of a sentence in a model. For a model M and a sentence φ, the expression 'φ is true in M' is abbreviated $M \models \varphi$.[11] A sentence φ is called a *logical consequence* of a set of sentences Σ, in symbols $\Sigma \models \varphi$ if in every model in which all sentences of Σ are true, φ is also true. Sentences that are logical consequences of the empty set (satisfied by all models under consideration) are called *valid.*

A proof system, which specifies an effective method by which a sentence φ can be "proved" from a set of sentences Σ, in symbols $\Sigma \vdash \varphi$. Sentences deducible from the empty set are called *theorems.*

For a given theory, the proof system is said to be *sound* with respect to the semantics if every proof produces only logical consequences, i.e.

$$\Sigma \vdash \varphi \quad \text{implies} \quad \Sigma \models \varphi$$

[11] $\models$ is called the *satisfaction symbol* and denotes semantical truth.

for all sentences φ and all sets of sentences Σ. The proof system is said to be *complete* if the converse holds, i.e. if every logical consequence has a proof within the logic.

A particular theory is normally identified with its set of logical truths, which coincides with its set of theorems if the proof system is sound and complete. The method for deriving proofs is usually based on

- *axioms:* a set of sentences that are considered theorems (by definition) and
- *proof rules:* a collection of rules specifying how a sentence of a given form can be deduced from finitely many other sentences.

In this case the set of all theorems is the smallest set that contains all the axioms and is closed under application of the proof rules. There are many different styles of proof systems such as natural deduction, Gentzen sequent calculus, tableaux method, resolution, connection method, Hilbert style, Fitch style, etc. Some of these systems are very close to the style of proof used in everyday mathematics, while others are more suitable for automated theorem proving. We will not consider any of these systems in general since, from a logical point of view, the exact nature of the proof system is not important, as long as it is sound and complete. (The situation is similar to a program written in different programming languages, some more suitable for humans, others for machines, but from the user's point of view the correctness of the program should be the primary concern.) Related theories of the same logic are obtained by modifying the semantics (restricting the interpretations) or the proof system (extending the set of axioms and/or proof rules). We now describe some particular logics.

Classical propositional logic

The most basic classical reasoning system restricts itself to a language $\mathcal{L}$ consisting of sentences φ built up from *propositions* $p_0, p_1, p_2, \ldots$ combined with a unary connective $\neg$ (negation) and a binary connective $\vee$ (logical or). The set of all *propositional sentences* is the smallest set $\mathcal{S}$ that contains all the propositions, as well as $\neg\varphi$ and $(\varphi \vee \psi)$ for all $\varphi, \psi \in \mathcal{S}$. The following abbreviations are used to aid readability:

$$(\varphi \wedge \psi) \triangleq \neg(\neg\varphi \vee \neg\psi),$$
$$\varphi \to \psi \triangleq \neg\varphi \vee \psi,$$
$$\varphi \leftrightarrow \psi \triangleq (\varphi \to \psi) \wedge (\psi \to \varphi).$$

To avoid excessively many parentheses, $\neg$ is given highest priority, followed by $\vee$, $\wedge$ with equal priority and finally $\to$ and $\leftrightarrow$ with lowest priority.

The semantics of a propositional theory is given by a collection of valuations, where a *valuation* is a function from $\{p_0, p_1, p_2, \ldots\}$ to $\{\mathsf{true}, \mathsf{false}\}$. For a sentence φ and a valuation v, the satisfaction relation $\mathsf{v} \models \varphi$ is defined inductively:

$$\mathsf{v} \models p_i \text{ if } \mathsf{v}(p_i) = \mathsf{true},$$
$$\mathsf{v} \models (\varphi \vee \psi) \text{ if at least one of } \mathsf{v} \models \varphi \text{ and } \mathsf{v} \models \psi \text{ holds},$$
$$\mathsf{v} \models \neg\varphi \text{ if } \mathsf{v} \not\models \varphi, \quad \text{(i.e. } \mathsf{v} \models \varphi \text{ does not hold).}$$

A set Σ of sentences *logically implies* a sentence φ, in symbols $\Sigma \models \varphi$, if every valuation v that satisfies all sentences in Σ also satisfies φ. A sentence φ is called a *logical truth* or *tautology* if $\emptyset \models \varphi$, i.e. if it is satisfied by all valuations. Two sentences φ, ψ are said to be *logically equivalent* ($\varphi \equiv \psi$) if $\varphi \leftrightarrow \psi$ is a logical truth.

There are many different proof systems for propositional logic, in many different styles. Almost any introductory textbook in logic will contain an axiomatisation of propositional logic with a proof of soundness and completeness. For any such system there is a natural Boolean algebra associated with the logic, obtained as follows. Define two formulae φ and ψ to be *provably equivalent*, written $\varphi \approx \psi$, iff $\varphi \leftrightarrow \psi$ is a theorem of the logic, i.e. $\vdash \varphi \leftrightarrow \psi$. This is an equivalence relation, and so we can form an equivalence class $[\varphi] = \{\theta : \varphi \approx \theta\}$ for every formula φ. Denote the set of all these equivalence classes by $\mathcal{L}/\approx$, then define over this *quotient set* operations arising from the logical connectives as follows:

$$[\varphi] \sqcap [\psi] \triangleq [\varphi \wedge \psi],$$
$$[\varphi] \sqcup [\psi] \triangleq [\varphi \vee \psi],$$
$$\overline{[\varphi]} \triangleq [\neg\varphi].$$

This method yields, for any logic, what is called its *Lindenbaum-Tarski algebra*. In the present case it is easy to check that the Lindenbaum-Tarski algebra of (classical) propositional logic is in fact a Boolean algebra. Since we have assumed the formalisation to be sound and complete, it could be proved further that the equivalence class which is the maximum element in $\mathcal{L}/\approx$ contains all and only the tautologies, and that two formulae φ and ψ are provably equivalent iff they are logically equivalent.

Classical first-order logic

In first-order logic we extend propositional logic by considering also the internal structure of propositions. For a fixed type τ and a set V of variables, an *atomic formula* is a string of symbols of the form $R(t_0, \ldots, t_{n-1})$ such that $R \in \mathcal{R}_\tau$ and $t_i \in T_\tau(V)$ $(i < n = \tau(R))$, where $\mathcal{R}_\tau$ and $T_\tau(V)$ are as in Sect. 1.4. The set Frm_τ of all *first-order formulae of type* τ is the smallest set that contains all atomic formulae as well as the formulae $\exists v\,(\varphi)$, $\neg\varphi$ and $(\varphi \vee \psi)$ for all $v \in V$ and $\varphi, \psi \in Frm_\tau$. The symbol $\exists$ is the *existential quantifier*. The *universally* quantified formula $\forall v\,(\varphi)$ is an abbreviation for $\neg\exists v\,(\neg\varphi)$, and the connectives $\wedge, \rightarrow, \leftrightarrow$ are defined as for propositional logic. $\exists v$ and $\forall v$ are considered unary connectives with higher priority than the binary connectives. Associated with each formula φ is a set $\mathsf{free}(\varphi)$ of *free variables*. For an atomic formula, $\mathsf{free}(\varphi)$ is the set of variables that occur in φ, and this is extended inductively by

$$\mathsf{free}(\exists v\,(\varphi)) = \mathsf{free}(\varphi) - \{v\},$$
$$\mathsf{free}(\neg\varphi) = \mathsf{free}(\varphi),$$
$$\mathsf{free}(\varphi \vee \psi) = \mathsf{free}(\varphi) \cup \mathsf{free}(\psi).$$

A *first-order sentence* is a formula φ that has no free variables ($\mathsf{free}(\varphi) = \emptyset$).

The semantics of a first-order theory of type τ is given by a collection of relational structures of type τ. Each relational structure $\mathcal{U}$ defines an *interpretation* $s \mapsto s^{\mathcal{U}}$ for the symbols $s \in \mathcal{F}_\tau \cup \mathcal{R}_\tau$. This map gives meaning to the function and relation symbols of the language. Observe that a valuation $\mathsf{v} : V \to U$ can be extended to $\bar{\mathsf{v}} : T_\tau(V) \to U$ by defining $\bar{\mathsf{v}}(f(t_0, \ldots, t_{n-1})) = f^{\mathcal{U}}(\bar{\mathsf{v}}(t_0), \ldots, \bar{\mathsf{v}}(t_{n-1}))$. We say that a formula φ is *true in* $\mathcal{U}$ *under the valuation* v, in symbols $\mathcal{U}, \mathsf{v} \models \varphi$ if one of the following conditions holds:

φ is $R(t_0, \ldots, t_{n-1})$ and $(\bar{\mathsf{v}}(t_0), \ldots, \bar{\mathsf{v}}(t_{n-1})) \in R^{\mathcal{U}}$,

φ is $\neg\psi$ and $\mathcal{U}, \mathsf{v} \not\models \psi$,

φ is $\psi \vee \psi'$ and at least one of $\mathcal{U}, \mathsf{v} \models \psi$ or $\mathcal{U}, \mathsf{v} \models \psi'$ holds,

φ is $\exists x\,(\psi)$ and $\mathcal{U}, \mathsf{v}' \models \psi$ for some $\mathsf{v}' : V \to U$ such that v and v' agree on $V - \{x\}$.

Finally, a formula φ is said to be *true in* $\mathcal{U}$ if $\mathcal{U}, \mathsf{v} \models \varphi$ for all valuations v. In this case $\mathcal{U}$ is also called a *model* of φ, in symbols $\mathcal{U} \models \varphi$.

As with syntax, the proof theory of first-order logic extends that of propositional logic, typically by giving some axioms and/or rules for dealing with the quantifiers. The algebraisation, however, of first-order logic is a considerably more complicated matter than for propositional logic. Relation algebras may be considered as an attempt to present in equational form first-order logic with no function symbols and only binary relation symbols. However, as mentioned in Chapt. 2, relation algebras only capture a certain fragment of this logic. Other attempts at presenting algebraic versions of first-order logic are the *cylindric algebras* of [Henkin, Monk+ 1971] and [Henkin, Monk+ 1985], and (earlier) the *polyadic algebras* of [Halmos 1962].

Equational logic

Equational logic can be considered as that fragment of first-order logic in which there is only one relational symbol, "=", to be interpreted as equality. It can also be added to a formalisation of first-order logic; the result is then *first-order logic with equality.* However, we consider equational logic separately because it is of interest in its own right: one of the central themes of algebraic logic is to reduce the reasoning of other proof systems to the elegant proof system of equational logic.

The syntax of an equational theory is based on an algebraic type τ and a set V of variables from which one obtains the set of terms $T_\tau(V)$ (Sect. 1.4). The collection of "formulae" is simply the set $T_\tau(V) \times T_\tau(V)$. A pair of terms (s, t) is called an *equation* and is written $s = t$.

The semantics of an equational theory of type τ is given by a collection of universal algebras of type τ. Each algebra $\mathcal{A}$ in this collection determines an interpretation which maps any $f \in \mathcal{F}_\tau$ to an operation $f^{\mathcal{A}}$. This interpretation can be extended inductively from $\mathcal{F}_\tau$ to all terms by considering a term $t \in T_\tau(V)$ as a template for composing the functions $f^{\mathcal{A}}$ that correspond to the function symbols f in t. The composite function defined by a term t is called the *induced term function* and is denoted by $t^{\mathcal{A}}$. The arity of this function is the number of distinct variables that appear in the term. An equation $s = t$ is satisfied in $\mathcal{A}$ if the induced term functions $s^{\mathcal{A}}$ and $t^{\mathcal{A}}$ are identical.

Proof systems for equational theories differ in their choice of axioms, but they all have the following in common:

- *reflexivity axiom:* $\vdash t = t$,
- *symmetry rule:* $s = t \vdash t = s$,
- *transitivity rule:* $r = s,\ s = t \vdash r = t$,
- *congruence rule:* $s_0 = t_0, \ldots, s_{n-1} = t_{n-1} \vdash f(s_0, \ldots, s_{n-1}) = f(t_0, \ldots, t_{n-1})$,
- *substitution rule:* $s = t \vdash s[v/r] = t[v/r]$ (where $s[v/r]$ is derived from s by substituting the variable v with the term r in all instances),

for all terms $r, s, s_0, \ldots, s_{n-1}, t, t_0, \ldots, t_{n-1}$, all $f \in \mathcal{F}_\tau$ and $n = \tau(f)$. The smallest equational theory (of type τ) is given by the set of reflexivity axioms since this set is closed under the above rules. The set of all equations is the largest equational theory, obtained by adding an axiom like $\vdash v = v'$ for distinct variables v, v'.

For a specific example, consider the equational theory of relation algebras, based on the type $\tau = \{(\sqcup, 2), (\overline{\ \ }, 1), (\perp\!\!\!\perp, 0), (;, 2), (\breve{\ }, 1), (\mathbb{I}, 0)\}$, and axiomatized by the equations (implicit) on Page 8. More generally, given a class $\mathcal{K}$ of algebras of type τ, we denote by $\mathbf{H}(\mathcal{K})$, $\mathbf{S}(\mathcal{K})$, $\mathbf{P}(\mathcal{K})$, $\mathbf{Ps}(\mathcal{K})$, $\mathbf{Si}(\mathcal{K})$ the class of all homomorphic images, all subalgebras, all products, all subdirect products and all subdirectly irreducibles of (members of) $\mathcal{K}$ respectively. For a set Σ of first-order formulae of type τ, let $\mathbf{Mod}(\Sigma) = \{\mathcal{M} : \mathcal{M} \models \Sigma\}$ be the class of all models of type τ that satisfy all formulae in Σ. If Σ contains only equations, $\mathbf{Mod}(\Sigma)$ is called an *equational class* or *variety.* The following is a fundamental result of universal algebra.

Theorem 1.6.1 (*Birkhoff's Preservation Theorem, 1935*) A class $\mathcal{K}$ of algebras of type τ is a variety if and only if $\mathcal{K}$ is closed under $\mathbf{H}$, $\mathbf{S}$ and $\mathbf{P}$ (i.e. $\mathbf{H}(\mathcal{K}) \subseteq \mathcal{K}$, $\mathbf{S}(\mathcal{K}) \subseteq \mathcal{K}$ and $\mathbf{P}(\mathcal{K}) \subseteq \mathcal{K}$).

Since the intersection of any collection of varieties is again a variety, every class of algebras $\mathcal{K}$ is contained in a smallest variety $\mathbf{V}(\mathcal{K})$, called the variety *generated* by $\mathcal{K}$. Based on Birkhoff's result, Tarski (1946) showed that $\mathbf{V}(\mathcal{K}) = \mathbf{HSP}(\mathcal{K})$.

Second-order logic

Many important concepts in computer science like well-foundedness cannot be formalized within the framework of first-order logic. Therefore in second-order logic the set of variables is extended by *relation variables*, which can be used when building terms, and quantification is allowed on these relation variables.

Validity of a formula φ in a relational structure $\mathcal{U}$ is defined relative to a valuation v of individual variables to elements of the universe as well as a valuation u of n-ary relation variables to n-ary relations on the universe. Similar to the first-order case we have

- $\mathcal{U}, \mathsf{u}, \mathsf{v} \models \exists R\,(\varphi)$ if $\mathcal{U}, \mathsf{u}', \mathsf{v} \models \varphi$ for some u' agreeing with u on all relation variables except R.

Again, a formula φ (possibly containing free individual or relation variables) is true in $\mathcal{U}$ if $\mathcal{U}, \mathsf{u}, \mathsf{v} \models \varphi$ for all valuations u and v. It is *valid* if it is true in every structure $\mathcal{U}$.

In contrast to first-order logic, second-order logic is incomplete: there is no proof system enumerating all valid second-order sentences. Therefore, attention is often restricted to certain fragments and/or certain classes of structures. For example *monadic second-order logic* allows only monadic (i.e. unary) relation variables (which are interpreted as subsets of a relational structure).

Modal logic

It is often helpful to formalize an area without the full generality of the quantifiers, but with more expressive power than plain propositional logic. In this case one can add further connectives to propositional logic and formalize their intended meanings. For example, when reasoning about a property Q of a computer program, statements like "after the next step Q" and "from now on Q" can be viewed as unary connectives applied to the proposition Q. Such unary connectives, often called *modalities*, are usually not truthfunctional: the truth value of a compound formula does not depend only on the truth values of its constituent formulae.

There are many modal logics; by way of example we consider a basic one called **K**. This logic is built from propositions $\{p_0, p_1, \ldots\}$ and Boolean connectives $\neg, \vee$, just like ordinary propositional logic, but it also has an additional unary connective $\Diamond$, called *possibility*. For any propositional formula φ, the formula $\Diamond\varphi$ is read "possibly φ". A *necessity* operator $\Box$ can be defined as its dual: $\Box\varphi \triangleq \neg\Diamond\neg\varphi$.

The semantics of **K**, as for other modal logics, is given in terms of a so-called *Kripke structure*: a relational structure $\mathcal{W} = (W, R)$ where W is a nonempty set of "possible worlds"[12] and R is a binary relation on W, usually called the *accessibility relation*. A valuation v is a function from $\{p_0, p_1, \ldots\}$ to $\mathcal{P}(W)$. A Kripke structure $\mathcal{W}$ together with a valuation is called a *Kripke model based on* $\mathcal{W}$. Satisfaction of a sentence φ in a Kripke model $\mathcal{M} = (W, R, \mathsf{v})$ at a world $w \in W$ is defined inductively:

$\mathcal{M}, w \models p_i$ if $w \in \mathsf{v}(p_i)$.

$\mathcal{M}, w \models (\varphi \vee \psi)$ if $\mathcal{M}, w \models \varphi$ or $\mathcal{M}, w \models \psi$.

$\mathcal{M}, w \models \neg\varphi$ if $\mathcal{M}, w \not\models \varphi$.

$\mathcal{M}, w \models \Diamond\varphi$ if $\mathcal{M}, w' \models \varphi$ for some $w' \in W$ with wRw'.

A sentence φ is valid in a Kripke model $\mathcal{M}$, if $\mathcal{M}, w \models \varphi$ for all $w \in W$.

To get a complete formalisation of the modal logic **K**, start with any complete formalisation of propositional logic, and add the following axiom and rules:

- *monotonicity*: $\Box(\varphi \to \psi) \to (\Box\varphi \to \Box\psi)$
- *necessitation*: $\varphi \vdash \Box\varphi$
- *modus ponens*: φ, $(\varphi \to \psi) \vdash \psi$ (if not already included).

[12] This traditional terminology is motivated by philosophy; in particular by the Leibnizian idea that "necessarily true" means "true in all possible worlds".

Many further modal logics can be obtained by adding further axioms to those of **K**. For example, a well-known logic called **S4** is obtained by adding to **K** the axioms

$$\Box\varphi \to \varphi$$
$$\Box\varphi \to \Box\Box\varphi.$$

In order for this formalisation of **S4** to be complete we need to add to its Kripke semantics the stipulation that the accessibility relation should be reflexive and transitive – i.e. a quasi-order. This correspondence between logical conditions on the modal operators and first-order conditions on the accessibility relation is typical of modal logic. The same idea also applies to operators of higher arity than just unary, and to the simultaneous use of many different modal operators (multi-modal logic).

1.7 Conclusion

The calculus of relations is pervasive in mathematics and logic, in the same sense of pervasiveness as the calculus of sets. In this chapter we have highlighted some aspects of it; these serve as background to the remaining chapters, in which we address the role and applications of relational methods in computer science.

The basic connection is that we may think of a program as an input-output relation: from a given initial state, it terminates (if at all) in another state, where a "state" is thought of as a snapshot of the current values of all the program variables. If we consider programs to be nondeterministic, then it is natural to think of a program simply as a binary relation over the set of all possible states. The *relational model* of program semantics, then, in its simplest form, is a relational structure $(\mathcal{S}, \{R_i\}_{i\in I})$, where $\mathcal{S}$ is the "state space" (the set S of states, usually structured in some way, e.g. as an ordered set), and $\{R_i\}_{i\in I}$ is a suitably indexed set of binary relations over the state space, representing programs. This can be thought of as a Kripke structure: the "possible worlds" now being states, and each program acting as an "accessibility relation" in the sense that from a given initial state the program can access certain other (terminal) states. A logical characterisation of program semantics would then aim to find a "program logic", with a modality for each of the programs, which is complete with respect to this semantic structure. Alternatively, each of these binary relations over S can give rise to a unary operator over $\mathcal{P}(S)$ (Peirce product, for example), and since sets of states may be thought of extensionally as properties (or predicates) of states, this would lead to the realm of so-called "predicate transformer semantics". It is this close interaction between a relational structure, a logic, and an algebra, that allows us to make use of the calculus of relations in a variety of ways.

Chapter 2

Relation Algebras

Roger D. Maddux

The contemporary theory of relation algebras is a direct outgrowth of the nineteenth century calculus of relations. After a few examples illustrating the calculus of relations (the most widely applied part of the subject), this chapter touches upon some topics in the algebraic theory of relation algebras: basic definitions, examples, constructions, elementary arithmetical theory, general algebraic results, representation theorems with applications, and connections with logic, including Tarski's formalization of set theory without variables.

2.1 The calculus of binary relations

The calculus of relations was invented and developed by De Morgan, Peirce, and Schröder. The study of binary relations and various binary and unary operations on binary relations was started by Augustus De Morgan in [De Morgan 1856] and [De Morgan 1864], who did so in order to expand the compass of classical Aristotelian syllogistic reasoning. Following the example set by George Boole in [Boole 1847], Charles Sanders Peirce created algebraic formalisms out of De Morgan's ideas in [Peirce 1870], [Peirce 1880], [Peirce 1883], [Peirce 1885], and other papers, including some first published only recently in [Peirce 1984]. He invented calculi for properties, binary relations, ternary relations, and relations of higher ranks, including many operations (and notations for operations) on relations of various ranks. Peirce's "Note B" of 1883 [Peirce 1883] lays out a single formalism, based exclusively on binary relations, with a few dozen of its laws. This algebraic system was extensively developed and applied by F. W. K. Ernst Schröder in his classic 649-page book [Schröder 1895], *Vorlesungen über die Algebra der Logik (exacte Logik), Volume 3, "Algebra und Logik der Relative", part I*, published in Leipzig in 1895.

A fragment of the Peirce-Schröder calculus of binary relations was axiomatized by Alfred Tarski [Tarski 1941]. One of Tarski's axioms is not an equation, and the class of algebras satisfying his axioms is not closed under the formation of direct products, so it is not a variety. The variety of relation algebras is the class of algebras that satisfy all but the offending axiom. A definition of relation algebras first appeared in [Jónsson, Tarski 1948], followed shortly thereafter by equivalent definitions in [Chin, Tarski 1951] and [Jónsson, Tarski 1952]. An equational axiomatization of relation algebras is given in the next section. In this section we give a few examples illustrating uses of the calculus of binary relations.

Consider a universe of discourse U with typical elements a, b, c, etc. Following De Morgan [De Morgan 1864], we let the capital letters L, M, N, etc., denote binary relations over U. We signify that a is in the relation L to b by writing $a\ L\ b$, and that the two relations L and M are the same by writing $L = M$. Let I be the identity relation on U and let $V = U \times U$ be the universal relation on U, according to Sect. 1.2. We use "iff" to abbreviate "if and only if".

Example 2.1.1 (*People*) Suppose everything in the universe of discourse is a person, Alice and Bert are in this universe, a is Alice, and b is Bert. Let L be the relation "loves". Then Alice loves Bert iff $a\ L\ b$. Alice is loved by Bert iff $a\ L^{\smile}\ b$. Alice does not love anyone who loves Bert iff $a\ \overline{L;L}\ b$. Alice loves Bert and no one but Bert iff $a\ L \cap \overline{L;\overline{I}}\ b$. Alice loves all and only those who love Bert iff $a\ \overline{L;\overline{L}} \cap \overline{\overline{L};L}\ b$. Alice is not the same as Bert, so $a\ \overline{I}\ b$, and there is a third person in the universe iff $a\ \overline{I} \cap \overline{I};\overline{I}\ b$.

Example 2.1.2 (*Numbers*) Suppose the universe is the set of real numbers, L is the relation "is strictly less than", and a and b are real numbers. Then $a\ L\ b$ iff a is less than (and not equal to) b. Some number is larger than both a and b iff $a\ L;L^{\smile}\ b$. No number is larger than both a and b iff $a\ \overline{L;L^{\smile}}\ b$. No number is larger than a iff $a\ \overline{L;V}\ b$. Two numbers are either distinct, or one of them is less than the other, or the other is less than the one. Therefore $V = L \cup L^{\smile} \cup I$. Peirce observed that for any relation M, the relation $\overline{\overline{M};M^{\smile}}$ is transitive and includes the identity relation. Hence $\overline{\overline{L};L^{\smile}} = L \cup I$.

Example 2.1.3 (*Sets*) Let the universe of discourse U be the universe of a model of set theory. Let L be the membership relation of this model, and let a and b be elements of the model. The objects in the universe of discourse are called "sets". So the set a is an element of set b iff $a\ L\ b$. We say that a is a subset of b if every element of a is an element of b, or, equivalently, there is no element of a that is not an element of b. The relation "is a subset of" is therefore $\overline{L^{\smile};\overline{L}}$, i.e, a is a subset of b iff $a\ \overline{L^{\smile};\overline{L}}\ b$. The converse of "is a subset of" is "is a superset of", namely $\overline{\overline{L^{\smile}};L}$. The Extensionality Axiom asserts that two sets are equal if one is both a subset and superset of the other. The relation between sets of being both a subset and superset of the other is $\overline{L^{\smile};\overline{L}} \cap \overline{\overline{L^{\smile}};L}$ and can also be read "has the same elements as". The Extensionality Axiom is therefore equivalent to $\overline{L^{\smile};\overline{L}} \cap \overline{\overline{L^{\smile}};L} = I$. The Union Axiom asserts that every set has a union, that is, a set whose elements are all those sets that are elements of elements of the given set. The relation "is the union of all the sets in" is $\overline{L^{\smile};\overline{L;L} \cup \overline{L^{\smile}};L;L}$. The Union Axiom is therefore equivalent to $V = V;L^{\smile};\overline{L;L} \cup \overline{L^{\smile}};L;L$.
It is easy to show that every relation obtained from L by repeated use of relative multiplication, union, conversion, and complementation are first-order definable in the model (U, L). Tarski has proved the converse, that all first-order definable

binary relations can be thus obtained from L if (U, L) is a model of the Pair Axiom $\forall a \forall b \exists c \forall x (xLc \leftrightarrow xIa \vee xIb)$. This axiom is equivalent to equations in the calculus of relations. One is given below; for others see [Tarski 1953] and [Tarski, Givant 1987, pp. 129, 137]. The *functional part* of a relation L is defined to be $L \cap \overline{\overline{L;\overline{I}}}$. The functional part of a relation is always a function. Let T be the functional part of $L^{\smile}$. We say that a set is a *singleton* if it has exactly one element. The relation T translates to "is a singleton whose only element is", so $a\ T\ b$ iff a is a singleton whose only element is b. Let R be the functional part of $L^{\smile} \cap V;T^{\smile}$ and let S be the functional part of $L^{\smile} \cap R;\overline{I}$. Then $a\ R\ b$ iff exactly one of the elements of a is a singleton and b is that unique singleton. Also, $a\ S\ b$ iff a contains exactly one singleton, b is an element of a, b is the only element of a that is distinct from the unique singleton in a. Let $P = R;T$ and let $Q = T;T \cup S;L \cap M$, where M is the functional part of $L^{\smile};L^{\smile} \cap \overline{P}$. Then (1) $c\ P\ a$ iff c contains exactly one singleton, and the sole element of that singleton is a, (2) $c\ Q\ b$ iff either (a) c is a singleton whose sole element is a singleton whose sole element is b or (b) b is an element of the only element of c that is distinct from the unique singleton in c, and b is the unique element of an element of c that is not the sole element of the unique singleton in c. To see better what this means, assume that the Extensionality Axiom happens to hold. Define the ordered pair with first component a and second component b to be $\{\{a\}, \{a, b\}\}$. Then P is the function that maps each ordered pair to its first component, and Q is the function that maps each ordered pair to its second component. Furthermore, $c\ P\ a$ and $c\ Q\ b$ iff $c = \{\{a\}, \{a, b\}\}$. Even in the absence of Extensionality, P and Q are functions and the Pair Axiom is equivalent to the equational axiom $V = P^{\smile};Q$. This equation says that, for any a and b, there is some c such that P maps c to a, Q maps c to b, c contains exactly one singleton, the sole element of that singleton is a, and either c is a singleton whose sole element is a singleton whose sole element is b (in which case $a = b$), or else b is an element of the only element of c that is distinct from the unique singleton in c and b is the unique element of an element of c that is not the sole element of the unique singleton in c (hence $b \neq a$). The equation does not imply that c is unique, so the functions P and Q are called *quasiprojections* (rather than "projections"). An equational axiom that does guarantee uniqueness for c (and hence is a weak form of Extensionality) is $P;P^{\smile} \cap Q;Q^{\smile} \subseteq I$.

2.2 Equational axioms for relation algebras

A *relation algebra* $\mathfrak{A}$ may be defined as a Boolean algebra (called the *Boolean part* or *Boolean reduct* of $\mathfrak{A}$) that has been augmented with

- an associative binary operation ; (called *relative multiplication*) that distributes from both sides over the join operation of the Boolean reduct,
- a distinguished *identity element* $\mathbb{I}$ for ;, and
- an involution $^{\smile}$ (called *conversion*) that distributes over the Boolean join and satisfies $x^{\smile};\overline{x;y} \sqsubseteq \overline{y}$.

If x and y are elements of a relation algebra, then $x;y$ is the *relative product* of x and y, and $x^{\smile}$ is the *converse* of x. An additional binary operation $\dagger$, called *relative addition*, is defined by

$$x \dagger y = \overline{\overline{x};\overline{y}} \ ,$$

and $x \dagger y$ is the *relative sum* of x and y. An additional distinguished element, called the *diversity element*, is defined as the complement of the identity element. The defining conditions can be translated directly into equational axioms for relation algebras. Just choose axioms for Boolean algebras and add equations expressing the remaining conditions. The axioms obtained this way are somewhat redundant. One of the distributive laws and one of the identity laws can be omitted. Doing this with Edward V. Huntington's axioms for Boolean algebras [Huntington 1933b] [Huntington 1933a] yields Tarski's equational axiom set for relation algebras [Tarski, Givant 1987]:

1. $(x \sqcup y) \sqcup z = x \sqcup (y \sqcup z)$,
2. $x \sqcup y = y \sqcup x$,
3. $x = \overline{\overline{x} \sqcup y} \sqcup \overline{\overline{x} \sqcup \overline{y}}$,
4. $x;(y;z) = (x;y);z$,
5. $x;\mathbb{I} = x$,
6. $(x \sqcup y);z = x;z \sqcup y;z$,
7. $x^{\smile\smile} = x$,
8. $(x \sqcup y)^{\smile} = x^{\smile} \sqcup y^{\smile}$,
9. $(x;y)^{\smile} = y^{\smile};x^{\smile}$,
10. $x^{\smile};\overline{x;y} \sqcup \overline{y} = \overline{y}$.

On the basis of these axioms, one may derive all the basic arithmetical and equational laws of relation algebras by using the rules of equational logic. The first immediate consequence of Huntington's axioms is that $x \sqcup \overline{x} = y \sqcup \overline{y}$ for every x and y. Therefore there is a single object $\top$ such that $x \sqcup \overline{x} = \top$ for all x. Furthermore, $\overline{x \sqcup \overline{x}} = \overline{y \sqcup \overline{y}}$, so there is an object $\bot$ such that $\bot = \overline{x \sqcup \overline{x}}$ for all x. Define $x \sqcap y$ for every x and y by $x \sqcap y = \overline{\overline{x} \sqcup \overline{y}}$. All the familiar Boolean algebraic identities may now be derived, and it is a challenging exercise to do so. Axioms (4)–(5) assert that $(A, ;, \mathbb{I})$ is a semigroup with right identity. From the other axioms it is possible to deduce that $\mathbb{I}$ must also be a left identity, so $(A, ;, \mathbb{I})$ is a monoid. $(A, \dagger, \overline{\mathbb{I}})$ is also a monoid. Two other equivalent versions of axiom (10) are $x^{\smile};\overline{x;y} \sqsubseteq \overline{y}$ and $x^{\smile};\overline{x;y} \sqcap y = \bot$. Another nice (but not equational) axiom set is obtained by replacing axioms (5)–(10) with De Morgan's equivalences [De Morgan 1864], which De Morgan called "Theorem K":

$$x;y \sqsubseteq z \leftrightarrow x^{\smile};\overline{z} \sqsubseteq \overline{y} \leftrightarrow \overline{z};y^{\smile} \sqsubseteq \overline{x}.$$

Peirce[Peirce 1984, Vol. 4, "1879–1884", p. 341.] worked out a more elaborate version of De Morgan's equivalences. He said,

> "Hence the rule is that having a formula of the form $x;y \sqsubseteq z$, the three letters may be cyclically advanced one place in the order of writing, those which are

carried from one side of the copula to the other being both negatived and converted.

We have, then, the following twelve propositions, all equivalent.

$$
\begin{array}{lll}
ls \prec b & s\breve{\bar{b}} \prec \breve{\bar{l}} & \breve{\bar{b}}l \prec \breve{\bar{s}} \\
\breve{s}\breve{l} \prec \breve{b} & \bar{b}\breve{s} \prec \bar{l} & \breve{l}\bar{b} \prec \bar{s} \\
\bar{b} \prec \bar{l} \dagger \bar{s} & \breve{l} \prec \bar{s} \dagger \breve{b} & \breve{s} \prec \breve{b} \dagger \bar{l} \\
\breve{\bar{b}} \prec \breve{\bar{s}} \dagger \breve{\bar{l}} & l \prec b \dagger \breve{\bar{s}} & s \prec \breve{\bar{l}} \dagger b
\end{array}
$$

There are in all 64 such sets of 12 equivalent propositions".

Schröder included four of Peirce's 64 sets of equivalent statements in his book. Here are a few other examples of laws in the calculus of relations that can be proved from the axioms for relation algebras.

- $\mathbb{0}$, $\mathbb{1}$, $\mathbb{I}$, and $\bar{\mathbb{I}}$ are fixed points of $\breve{\ }$.
- $\mathbb{I}$ is a left and right identity for $;$, and $\bar{\mathbb{I}}$ is a left and right identity for $\dagger$.
- $\mathbb{0} ; x = \mathbb{0} = x ; \mathbb{0}$ and $\mathbb{1} \dagger x = \mathbb{1} = x \dagger \mathbb{1}$.
- Conversion $\breve{\ }$ is monotonic, self-conjugate, distributes over arbitrary joins and meets, and commutes with complementation.
- The functions $x ;(\text{-})$ and $x^{\smile} ;(\text{-})$ are conjugate, and the functions $(\text{-}) ; x$ and $(\text{-}) ; x^{\smile}$ are conjugate, so relative multiplication $;$ is monotonic and distributes over arbitrary joins, i.e., the functions $x ;(\text{-})$ and $(\text{-}) ; x$ are completely join-preserving.
- $x ; y \sqcap z = (x \sqcap z ; y^{\smile}) ; y \sqcap z$,
- $x ; y \sqcap z = x ;(y \sqcap x^{\smile} ; z) \sqcap z$,
- $x ; y \sqcap z = (x \sqcap z ; y^{\smile}) ;(y \sqcap x^{\smile} ; z) \sqcap z$,
- $x \sqsubseteq x ; \mathbb{1}$, $x \sqsubseteq \mathbb{1} ; x$, and $\mathbb{1} ; \mathbb{1} = \mathbb{1}$,
- $x ;(y \dagger z) \sqsubseteq (x ; y) \dagger z$ and $(x \dagger y) ; z \sqsubseteq x \dagger (y ; z)$, which Peirce called "two formulæ so constantly used that hardly anything can be done without them,"
- $\mathbb{I} \sqsubseteq \overline{\bar{x} ; x^{\smile}}$ and $\mathbb{I} \sqsubseteq \overline{x^{\smile} ; \bar{x}}$.

2.3 Square relation algebras

A standard example of a relation algebra is the *square relation algebra on a set* U, namely $\mathfrak{Re}(U) = (\mathcal{P}(U \times U), \cup, \bar{\ }, ;, \breve{\ }, I_U)$. Every square relation algebra on a set is complete and atomic (i.e., has a complete and atomic Boolean reduct). If U is empty, then $\mathfrak{Re}(U)$ is an algebra with just one relation in it. If U contains exactly one element, then $\mathfrak{Re}(U)$ is *Boolean*, that is, it satisfies the identity $\mathbb{I} = \mathbb{1}$. On the other hand, if $\mathfrak{Re}(U)$ satisfies $\mathbb{I} = \mathbb{1}$, then U is either empty or has exactly one element. If U is finite then so is the algebra $\mathfrak{Re}(U)$. If U has exactly n elements, then the number of relations in $\mathfrak{Re}(U)$ is 2^{n^2}. Even for small sets the square algebras are large. If U has exactly 3 elements, then $\mathfrak{Re}(U)$ has 9 atoms and 512 elements. If U is a countably infinite set, then $\mathfrak{Re}(U)$ is an uncountable algebra. A *representation* of a relation algebra $\mathfrak{A}$ is an embedding of $\mathfrak{A}$ into a direct product of square relation algebras. $\mathfrak{A}$ is *representable* if there is a representation of $\mathfrak{A}$, and $\mathfrak{A}$ is *completely representable* if there is a representation of $\mathfrak{A}$ that preserves all meets and joins. If $\mathfrak{A}$ is representable, then there must be an index set I and sets U_i for every $i \in I$, such that $\mathfrak{A}$ is isomorphic to a subalgebra of $\prod_{i\in I} \mathfrak{Re}\,(U_i)$. If the sets U_i are pairwise disjoint, then they are the equivalence classes of the equivalence relation $E = \bigcup_{i\in I}(U_i \times U_i)$. The map which takes an element x of $\prod_{i\in I} \mathfrak{Re}\,(U_i)$ to $\bigcup_{i\in I} x_i$ is an isomorphism of the relation algebra $\prod_{i\in I} \mathfrak{Re}\,(U_i)$ onto the relation algebra $\mathfrak{Sb}E$, whose universe consists of all relations that are included in the equivalence relation E. A relation algebra is *proper* if it is isomorphic to a subalgebra of $\mathfrak{Sb}E$ for some equivalence relation E.

2.4 Special kinds of elements

Many interesting kinds of elements in a relation algebra can be defined and studied. Their definitions are based on the many familiar kinds of relations used throughout mathematics. We define some here and state a few of their properties; for more information see [Chin, Tarski 1951]. Let x be an element of an arbitrary relation algebra $\mathfrak{A}$. We say that x is a *domain element* or *right-ideal element* if $x\,;\mathbb{T} = x$, a *range element* or *left-ideal element* if $\mathbb{T}\,;x = x$, and an *ideal element* if x is both a left and right ideal element. For every set U, the domain elements of $\mathfrak{Re}\,(U)$ are the relations of the form $W \times U$ for some $W \subseteq U$, and the range elements are the relations of the form $U \times W$ for some $W \subseteq U$. Let *Dom*, *Ran*, and *Id* be the sets of domain elements, range elements, and ideal elements of a relation algebra $\mathfrak{A}$. Then

- $x\,;\mathbb{T}$ and $x \dagger \perp\!\!\!\perp$ are domain elements.
- $\mathbb{T}\,;x$ and $\perp\!\!\!\perp \dagger x$ are range elements.
- $\perp\!\!\!\perp \dagger x \dagger \perp\!\!\!\perp$, $(\perp\!\!\!\perp \dagger x)\,;\mathbb{T}$, $\perp\!\!\!\perp \dagger x\,;\mathbb{T}$, and $\mathbb{T}\,;x\,;\mathbb{T}$ are ideal elements.
- *Dom*, *Ran*, and *Id* are closed under $\overline{}$, $\sqcup$, and $\sqcap$,
- If $\mathfrak{A}$ is complete then *Dom*, *Ran*, and *Id* are closed under arbitrary meets and joins.
- *Dom* is closed under $x\,;(\text{-})$ and *Ran* is closed under $(\text{-})\,;x$.

The set of domain (or range) elements in a complete relation algebra always forms a complete Boolean algebra. In particular, the Boolean algebra of all subsets of U is isomorphic to the Boolean algebra of domain elements of $\mathfrak{Re}\,(U)$ and isomorphic to the Boolean algebra of range elements of $\mathfrak{Re}\,(U)$.

An element x of a relation algebra $\mathfrak{A}$ is a *functional element* if $x^{\smile}\,;x \sqsubseteq \mathbb{I}$. The functional elements of $\mathfrak{Re}\,(U)$ are the partial functions from U to U. In every relation algebra,

- $x \sqcap \overline{x\,;\overline{\mathbb{I}}}$ is a functional element,
- if $x \sqsubseteq \mathbb{I}$ then x is a functional element,
- if x and y are functional elements, then so is $x\,;y$,
- if x is a functional element then $x\,;(y \sqcap z) = x\,;y \sqcap x\,;z$ for all $y, z \in A$, and

- x is a functional element iff $x;y \sqcap x;\overline{y} = \mathbb{⊥}$ for every element $y \in A$.

We say that x is a *symmetric element* if $x^{\smile} = x$, a *transitive element* if $x;x \sqsubseteq x$, an *equivalence element* if x is a symmetric and transitive element, a *reflexive element* if $x = (\mathbb{I} \sqcap x);x;(\mathbb{I} \sqcap x)$, and a *partial ordering element* if x is a symmetric and transitive element. All of these types correspond to similarly named relations on U. For example, a relation on U is a reflexive element of $\mathfrak{Re}(U)$ iff it is reflexive over its field and it is an equivalence element of $\mathfrak{Re}(U)$ iff it is an equivalence relation on its field. Many familiar properties of relations can be shown to hold in arbitrary relation algebras. For example, every symmetric and transitive element is also reflexive.

Some relations on U contain exactly one ordered pair. There is no equational characterization of such elements, but one can come close. An element x of a relation algebra $\mathfrak{A}$ is said to be a *singleton* iff $x \neq \emptyset$ and $x^{\smile};\mathbb{T};x \sqcup x;\mathbb{T};x^{\smile} \sqsubseteq \mathbb{I}$, a *point* iff $x \neq \emptyset$ and $x;\mathbb{T};x \sqsubseteq \mathbb{I}$, and a *pair* iff $x \neq \emptyset$ and $x;\overline{\mathbb{I}};x;\overline{\mathbb{I}};x \sqsubseteq \mathbb{I}$. Then a nonzero element x in $\mathfrak{Re}(U)$ is a singleton iff $x = \{(a,b)\}$ for some $a, b \in U$, a point iff $x = \{(a,a)\}$ for some $a \in U$, and a pair iff $x = \{(a,a),(b,b)\}$ for some $a, b, \in U$. Every point is both a singleton and a pair. Furthermore, in an arbitrary simple relation algebra every singleton, point, and pair must be an atom. (An algebra is simple if it has at least two elements and every homomorphism is either one-to-one or collapses the algebra onto the one-element algebra.)

2.5 The variety of relation algebras

RA is defined to be the class of all relation algebras. Because this class has an equational axiomatization, it is an equational class or variety, and is therefore closed under the formation of subalgebras, homomorphic images, and direct products. Peirce [Peirce 1883] noticed that each of the terms $0 \dagger l \dagger 0$, $(0 \dagger l)\infty$, $0 \dagger l\infty$, and $\infty l \infty$ "have the remarkable property that each is either 0 of ∞". In current notation this is the

Peirce property (1883) $\quad \{\mathbb{⊥} \dagger l \dagger \mathbb{⊥},\ (\mathbb{⊥} \dagger l);\mathbb{T},\ \mathbb{⊥} \dagger (l;\mathbb{T}),\ \mathbb{T};l;\mathbb{T}\} \subseteq \{\mathbb{⊥}, \mathbb{T}\}$

Not every relation algebra has the Peirce property. The relation algebras that have the Peirce property are exactly the ones that are simple or have just one element.

Theorem 2.5.1 For every relation algebra $\mathfrak{A}$ the following are equivalent:

- $\mathfrak{A}$ is simple,
- $\mathfrak{A}$ is subdirectly irreducible (has no nontrivial representation as a subdirect product),
- $\mathfrak{A}$ is directly irreducible (is not isomorphic to a nontrivial direct product),
- $\mathfrak{A}$ has exactly two ideal elements, namely $\mathbb{⊥}$ and $\mathbb{T}$,
- for every $x \in A$, $x \neq \mathbb{⊥}$ iff $\mathbb{T};x;\mathbb{T} = \mathbb{T}$,
- for every $x \in A$, $x \neq \mathbb{T}$ iff $\mathbb{⊥} \dagger x \dagger \mathbb{⊥} = \mathbb{⊥}$,
- for all $x, y \in A$, $x = \mathbb{⊥}$ or $y = \mathbb{⊥}$ iff $x;\mathbb{T};y = \mathbb{⊥}$,
- $\mathfrak{A}$ is nontrivial and has the Peirce property.

The simple relation algebras are remarkable for another reason, due essentially to Schröder [Schröder 1895, pp. 150–153].

Theorem 2.5.2 For every Boolean combination φ of relation-algebraic equations there is a correlated relation-algebraic term $\varphi^\star$ such that φ and the equation $\varphi^\star = \perp\!\!\!\perp$ are equivalent in every simple relation algebra. One such mapping * can be defined recursively as follows:

- $(x = y)^\star = (x \sqcap \overline{y}) \sqcup (\overline{x} \sqcap y)$,
- $(\varphi \vee \psi)^\star = \top\!\!\!\top ; \varphi^\star ; \top\!\!\!\top ; \psi^\star ; \top\!\!\!\top$,
- $(\varphi \wedge \psi)^\star = \varphi^\star \sqcup \psi^\star$,
- $(\neg\varphi)^\star = \perp\!\!\!\perp \dagger \overline{\varphi^\star} \dagger \perp\!\!\!\perp$.

Homogenous relation algebras.

A relation algebra is said to be *homogeneous* (see Chapt. 3) if it complete, atomic, and simple. The class of homogeneous relation algebras is not a variety, i.e. it has no equational axiomatization. In fact, it is not even an elementary class. It is not closed under subalgebras, nor under homomorphisms, nor under direct products. Many relation algebras are not homogeneous, such as every proper relation algebra whose unit element has two or more equivalence classes.

2.6 Representable relation algebras

The class RRA of representable relation algebras is defined as the class of all subalgebras of direct products of square relation algebras. Square relation algebras are simple. Direct products of square relation algebras are representable but not necessarily simple. For example, $\mathfrak{Re}\,(U_0) \times \mathfrak{Re}\,(U_1)$ is not simple if U_0 and U_1 are nonempty. It has two simple homomorphic images, $\mathfrak{Re}\,(U_0)$ and $\mathfrak{Re}\,(U_1)$.

Not all relation algebras are representable. A nonrepresentable relation algebra was first discovered by Lyndon [Lyndon 1950], who presented a finite one with 56 atoms. Jónsson [Jónsson 1959] constructed another one with infinitely many atoms. An infinite family of finite nonrepresentable relation algebras, based on Jónsson's construction and on the nonexistence of certain projective planes, were found by Lyndon [Lyndon 1961]. The smallest nonrepresentable relation algebra obtained from Lyndon's construction has 8 atoms.

It is clear from its definition that RRA is closed under the formation of subalgebras and direct products. Tarski [Tarski 1955] proved that RRA is also closed under the formation of homomorphic images, and is therefore a variety by Birkhoff's Theorem (see Sect. 1.4). On the other hand, the construction of various nonrepresentable relation algebras in [Lyndon 1950; Lyndon 1961; Jónsson 1959] led to Monk's proof [Monk 1964] that RRA is not finitely axiomatizable. The finite axiomatization of RA is therefore incomplete in the sense that it does not axiomatize the calculus of relations. Complete (and necessarily infinite) axiomatizations have been described by Lyndon [Lyndon 1956] and McKenzie (see [Monk 1969]).

Lyndon [Lyndon 1956, p. 307, footnote 13] noted that every finite relation algebra with 3 or fewer atoms is representable, so the number of atoms in the smallest possible nonrepresentable relation algebras is at least 4. McKenzie [McKenzie 1970] was the first to find a finite nonrepresentable relation algebra with 4 atoms. It turns out that there are 31 nonrepresentable relation algebras with 4 atoms. For an example, let $U = \{e, a, b, c\}$ with $|U| = 4$ and let

$$\begin{aligned} T = \; & \{(e,x,x) : x \in U\} \cup \{(x,e,x) : x \in U\} \cup \{(x,x,e) : x \in U\} \cup \\ & \cup \{(x,y,z) : x,y,z \in \{a,b,c\} \wedge |\{x,y,z\}| = 2\}. \end{aligned}$$

The inclusion

$$t \sqcap (u ; v \sqcap w) ; (x \sqcap y ; z) \sqsubseteq u ; [(u^{\smile} ; t \sqcap v ; x) ; z^{\smile} \sqcap v ; y \sqcap u^{\smile} ; (t ; z^{\smile} \sqcap w ; y)] ; z$$

is true in every $\mathfrak{Re}(U)$, and therefore holds in every representable relation algebra. However, this equation fails in so-called complex algebras constructed from T, which are thus not representable [Maddux 1976]. Finite nonrepresentable relation algebras are quite numerous. The number of nonrepresentable relation algebras with n atoms is roughly 2^{n^3} when n is large [Maddux 1985].

2.7 Representation theorems

Since there are nonrepresentable relation algebras, it is interesting to ask how one can recognize whether a relation algebra is representable. In particular, is there an algorithm for determining whether a finite relation algebra is representable? This problem is still unsolved, and may present considerable difficulties. In spite of this, there are many general results that guarantee representability. Here are two of the earliest.

Theorem 2.7.1 ([Jónsson, Tarski 1952]) A relation algebra is representable if it is atomic and every one of its atoms is a functional element.

Theorem 2.7.2 ([Jónsson, Tarski 1952]) A relation algebra is representable if its unit element is the join of a finite set of functional elements.

These two theorems were generalized in the next one.

Theorem 2.7.3 ([Maddux, Tarski 1976]) A relation algebra is representable if every nonzero element contains a nonzero functional element.

Tarski's QRA-theorem requires some terminology. If x and y are functional elements of a relation algebra and $\mathbb{T} = x^{\smile} ; y$, then x and y are called *conjugated quasiprojections* [Tarski, Givant 1987, 8.4(i)].

Theorem 2.7.4 ([Tarski 1953]) A relation algebra is representable if is a QRA, i.e. it contains a pair of conjugated quasiprojections.

All of the representation theorems given so far are generalized by the next one.

Theorem 2.7.5 ([Maddux 1978b], [Maddux 1978a]) A relation algebra is representable if for every nonzero element z there are functional elements x and y such that $z \sqcap x^{\smile} ; y$ is nonzero.

Here is one similar in form but not a special case of the preceding one. A relation algebra $\mathfrak{A}$ is said to be *pair-dense* (or *point-dense*) if every nonzero element below the identity contains a pair (or point), and $\mathfrak{A}$ is *singleton-dense* if every nonzero element contains a singleton.

Theorem 2.7.6 ([Maddux 1991c]) Every pair-dense (point-dense, or singleton-dense) relation algebra is completely representable. $\mathfrak{Re}(U)$ has the following characterization: $\mathfrak{A} \cong \mathfrak{Re}(U)$ iff $\mathfrak{A}$ is a simple complete point-dense relation algebra and $|U|$ is the number of points in $\mathfrak{A}$.

Here are a couple representation results concerning algebras generated by equivalence elements.

Theorem 2.7.7 ([Jónsson 1988]) A relation algebra is representable if it is generated by an equivalence element.

Theorem 2.7.8 ([Givant 1994]) A relation algebra is representable if it is generated by a chain of equivalence elements, a set of disjoint equivalence elements, or, generalizing both these conditions, a tree of equivalence elements.

The hypothesis that the equivalence elements form a chain or tree is necessary, since there are nonrepresentable relation algebras generated by their equivalence elements.

Pairing and fork algebras

Here are some applications of Tarski's theorem to pairing algebras [Maddux 1989] and fork algebras (see Chapt. 4 of this book). A *pairing algebra* is an algebra of the form $(\mathfrak{A}, p, q)$, where $\mathfrak{A}$ is a relation algebra (called the relation algebra reduct) and p and q are conjugated quasiprojections. An equational axiomatization for this variety can be obtained from any equational axiom set for relation algebras by adding three more equations (see also Sect. 3.6):

(P1) $p^{\smile} ; p = \mathbb{I}$,

(P2) $q^{\smile} ; q = \mathbb{I}$, and

(P3) $\mathbb{T} = p^{\smile} ; q$,

or just one, such as $\mathbb{T} = p^{\smile} ; q \sqcap (\overline{p}^{\smile} \dagger \overline{p} \sqcup \mathbb{I}) \sqcap (\overline{q}^{\smile} \dagger \overline{q} \sqcup \mathbb{I})$. If $(\mathfrak{A}, p, q)$ is a pairing algebra then $\mathfrak{A}$ is a representable relation algebra by Tarski's theorem, so we may assume that $\mathfrak{A}$ is a proper relation algebra whose unit is an equivalence relation E. Then (P1)–(P3) assert that p and q are functions and for every $(a, b) \in E$ there is some c such that $p(c) = a$ and $q(b) = c$. If $(\mathfrak{A}, p, q)$ also satisfies the "unique-pair" identity

(P4) $p ; p^{\smile} \sqcap q ; q^{\smile} \sqsubseteq \mathbb{I}$,

then this c is unique, and we may denote it by $\star(a, b)$. Then we have $p(\star(a, b)) = a$, $q(\star(a, b)) = b$, and $\star(a, b) = \star(a', b')$ implies $a = a'$ and $b = b'$. In every pairing algebra define a binary operation ∇ (called "fork") by

(D) $x \nabla y = x ; p^{\smile} \sqcap y ; q^{\smile}$.

In a proper pairing algebra that satisfies the "unique-pairs" identity (P4), this definition is equivalent to

$$x \nabla y = \{(c, \star(a, b)) : c\ x\ a, c\ y\ b\}.$$

In every pairing algebra, the operation ∇ defined by (D) satisfies two identities:

(F1) $x \nabla y = x ;(\mathbb{I} \nabla \mathbb{T}) \sqcap y ;(\mathbb{T} \nabla \mathbb{I})$,

(F2) $(v \nabla w) ;(x \nabla y)^{\smile} = v ; x^{\smile} \sqcap w ; y^{\smile}$.

It is easy to prove (F1) from the axioms of relation algebra together with (D) alone. (F2) requires (P1)–(P3) and is not so easy to prove. For a direct proof of (F2), see [Tarski, Givant 1987, pp.97–98]. For an indirect proof, recall that the relation algebra reduct of a pairing algebra is representable, so one need only check that (F2) holds in all representable relation algebras. Every pairing algebra satisfies the identity $p ; p^{\smile} \sqcap q ; q^{\smile} = (\mathbb{I} \nabla \mathbb{T})^{\smile} \nabla (\mathbb{T} \nabla \mathbb{I})^{\smile}$. This also does not require (P1)–(P3). It follows that (P4) is equivalent to:

(F3) $(\mathbb{I} \nabla \mathbb{T})^{\smile} \nabla (\mathbb{T} \nabla \mathbb{I})^{\smile} \sqsubseteq \mathbb{I}$.

A fork algebra is an algebra of the form $(\mathfrak{A}, \nabla)$, where $\mathfrak{A}$ is a relation algebra and ∇ is a binary operation on $\mathfrak{A}$ satisfying (F1)–(F3). Define p and q by $p = (\mathbb{I} \nabla \mathbb{T})^{\smile}$ and $q = (\mathbb{T} \nabla \mathbb{I})^{\smile}$. Then (P1)–(P3) follow from (F2) alone, so $(\mathfrak{A}, p, q)$ is a pairing relation algebra whose fork operation coincides with ∇ by (F1), and which also satisfies (P4) because of (F3). By the reasoning given above, $(\mathfrak{A}, \nabla)$ is isomorphic to a proper relation algebra with a binary operation ∇ that arises from some $\star$ by $x \nabla y = \{(c, \star(a, b)) : cRa, cSb\}$.

2.8 Relation algebras and logic

The definition of relative multiplication states that two objects are in the relative product of two relations iff there is a third object related to the first two in certain ways. The other operations can be defined by referring to only two objects, and every equation in the theory of relation algebras is equivalent to a sentence that uses only three variables. Conversely, every such sentence is equivalent to a relation-algebraic equation. This opens the possibility of constructing a correspondence between relations algebras and a 3-variable fragment of first-order predicate calculus. The completeness theorem makes it possible to ignore the distinction between means of expression and means of proof, but for languages with only finitely many variables this theorem does not hold. In fact, Monk proved that "it is impossible to write down finitely many schemata which will give a notion of proof which is sound and complete" [Monk 1971], so if we choose any standard axiomatization of first-order logic, and define a formula φ to be 3-*provable* if it has a proof whose formulas use only v_0, v_1, and v_2, then 3-provability is incomplete and some logically valid 3-variable sentences are not 3-provable. Assume that the equivalence of two formulas is 3-provable whenever one of them can be obtained from the other by respelling bound variables. (If necessary, this can be arranged by strengthening the axiom system with finitely many schemata.) It turns out that the equational theory of relation algebras is equipollent in means of proof with the resulting 3-variable fragment of 4-variable logic, i.e., an equation is true in every relation algebra iff its translation into a 3-variable sentence can be proved using at most 4 variables. This section includes more precise formulations of these results. We take the approach of Tarski [Tarski, Givant 1987] (the need for which can be found in the warning on page 42 about the nontransitivity of equipollence), incorporating the first-order logic of binary relations with equality and the equational logic of relation algebras into a single formalism that contains these two types as subformalisms. Proofs can be found by consulting [Tarski, Givant 1987], [Maddux 1978b], [Maddux 1983], [Maddux 1989], [Maddux 1994b].

The first-order language $\mathcal{L}$

Let $\mathcal{L}$ be a first-order predicate language with equality symbol 1' and a countable set of binary relation symbols R_0, R_1, R_2, $R_3, \ldots$, but no function symbols or constants. The set of variables of $\mathcal{L}$ is $\mathsf{var} = \{ v_k : k \in \mathbb{N} \} = \{v_0, v_1, v_2, \ldots\}$ and let $\mathbf{in}(v_i) = i$. The logical constants of $\mathcal{L}$ are the implication symbol $\rightarrow$, the negation symbol $\neg$, the universal quantifier $\forall$, and the equality symbol 1'. (The relation symbols R_n are nonlogical constants.) The existential quantifier $\exists$ and the other sentential connectives are introduced by abbreviations, *e.g.*, $\exists x\psi = \neg\forall x\neg\psi$. The atomic formulas of $\mathcal{L}$ are xR_ny and $x1\text{'}y$ for $n \in \mathbb{N}$ and $x, y \in \mathsf{var}$. The set of all formulas of $\mathcal{L}$ is form. For any $\psi \in \mathsf{form}$, $\mathsf{free}(\psi)$ is the set of variables which occur freely in ψ, so ψ is a sentence iff $\mathsf{free}(\psi) = \emptyset$. The set of all sentences of $\mathcal{L}$ is sent. For every $\psi \in \mathsf{form}$, $\forall[\psi]$ is a sentence, called the **closure** of ψ, defined by $\forall[\psi] = \psi$ if $\mathsf{free}(\psi) = \emptyset$, and $\forall[\psi] = \forall x_0 \ldots \forall x_m\psi$ if $\mathsf{free}(\psi) = \{x_0, \ldots, x_m\}$ and $\mathbf{in}(x_k) < \mathbf{in}(x_{k+1})$ for all $k < m$. The set ax of logical axioms of $\mathcal{L}$ is the set of all instances of the following schemata, in which $\psi, \varphi, \xi \in \mathsf{form}$ and $x, y \in \mathsf{var}$:

(A_1) $\forall[(\psi \rightarrow \varphi) \rightarrow ((\varphi \rightarrow \xi) \rightarrow (\psi \rightarrow \xi))]$,

(A_2) $\forall[(\neg\psi \rightarrow \psi) \rightarrow \psi]$,

(A_3) $\forall[\psi \rightarrow (\neg\psi \rightarrow \varphi)]$,

(A_4) $\forall[\forall x\forall y\psi \rightarrow \forall y\forall x\psi]$,

(A_5) $\forall[\forall x(\psi \rightarrow \varphi) \rightarrow (\forall x\psi \rightarrow \forall x\varphi)]$,

(A_6) $\forall[\forall x\psi \rightarrow \psi]$,

(A_7) $\forall[\psi \rightarrow \forall x\psi]$ where $x \notin \mathsf{free}(\psi)$,

(A_8) $\forall[\exists x(x1\text{'}y)]$,

(A_9) $\forall[x1\text{'}y \rightarrow (\varphi \rightarrow s_{xy}\varphi)]$, where $s_{xy}\varphi$ is the formula obtained from φ by simultaneously replacing every (free or bound) occurrence of x by y and y by x.

The only rule of inference is *modus ponens*, *i.e.*, infer φ from ψ and $\psi \rightarrow \varphi$. For every $\Psi \subseteq \mathsf{sent}$ and every $\psi \in \mathsf{sent}$, $\Psi \vdash \psi$ if ψ belongs to every set of sentences of $\mathcal{L}$ which contains $\Psi \cup \mathsf{ax}$ and is closed under *modus ponens*.

The definitional extension $\mathcal{L}^+$

$\mathcal{L}^+$ is obtained from $\mathcal{L}$ by the addition of operators on predicates and atomic formulas expressing the equality of predicates. Let pred be a set with the following properties:

- $1', R_0, R_1, R_2, R_3, \ldots \in \mathsf{pred}$,
- if $A, B \in \mathsf{pred}$ then $A + B, \overline{A}, A\,;B, A^{\smile} \in \mathsf{pred}$,
- $\mathfrak{P} = (\mathsf{pred}, +, ^-, ;, ^{\smile}, 1')$ is an absolutely free algebra of similarity type $\{(+,2), (^-,1), (;,2), (^{\smile},1), (1',0))\}$ absolutely freely generated by $\{1', R_0, R_1, \ldots\}$.

Let $1 = 1' + \overline{1'}$, $0 = \overline{1' + \overline{1'}}$, and, for any $A, B \in \mathsf{pred}$, $A \cdot B = \overline{\overline{A} + \overline{B}}$ and $A \dagger B = \overline{\overline{A}\,; \overline{B}}$. The atomic formulas of $\mathcal{L}^+$ are xAy and $A = B$, where $A, B \in \mathsf{pred}$ and $x, y \in \mathsf{var}$. The set of formulas form^+ and the set of sentences sent^+ of $\mathcal{L}^+$ are otherwise defined the same as in $\mathcal{L}$. The set of logical axioms ax^+ of $\mathcal{L}^+$ contains all instances of (A_1)–(A_9) above (with $\psi, \varphi, \xi \in \mathsf{form}^+$ and $x, y \in \mathsf{var}$) and all instances of (D_1)–(D_5) below, where $A, B \in \mathsf{pred}$:

(D_1) $\forall v_0 \forall v_1 (v_0 A + B v_1 \leftrightarrow (v_0 A v_1 \vee v_0 B v_1))$,

(D_2) $\forall v_0 \forall v_1 (v_0 \overline{A} v_1 \leftrightarrow \neg v_0 A v_1)$,

(D_3) $\forall v_0 \forall v_1 (v_0 A\,; B v_1 \leftrightarrow \exists v_2 (v_0 A v_2 \wedge v_2 B v_1))$,

(D_4) $\forall v_0 \forall v_1 (v_0 A^{\smile} v_1 \leftrightarrow v_1 A v_0)$,

(D_5) $A = B \leftrightarrow \forall v_0 \forall v_1 (v_0 A v_1 \leftrightarrow v_0 B v_1)$.

As in $\mathcal{L}$, the only rule of inference in $\mathcal{L}^+$ is *modus ponens.* For every $\Psi \subseteq \mathsf{sent}^+$ and every $\psi \in \mathsf{sent}^+$, $\Psi \vdash^+ \psi$ if ψ belongs to every set of sentences in $\mathcal{L}^+$ which contains $\Psi \cup \mathsf{ax}^+$ and is closed under *modus ponens.*

Theorem 2.8.1 ([Tarski, Givant 1987, 2.3(i)(ii)]) $\mathcal{L}$ is a subformalism of $\mathcal{L}^+$, in the sense that

- $\mathsf{form} \subseteq \mathsf{form}^+$ and $\mathsf{sent} \subseteq \mathsf{sent}^+$,
- for every $\Psi \subseteq \mathsf{sent}$ and every $\varphi \in \mathsf{sent}$, if $\Psi \vdash \varphi$ then $\Psi \vdash^+ \varphi$.

To show that the passage from $\mathcal{L}$ to $\mathcal{L}^+$ does not increase the means of expression or proof, let G be the *elimination mapping* from form^+ to form determined by the following recursive definition (where $n \in \mathrm{N}$, $x, y \in \mathsf{var}$, $A, B \in \mathsf{pred}$, $\varphi, \psi \in \mathsf{form}^+$).

- $G(xR_n y) = xR_n y$,
- $G(x1'y) = x1'y$,
- $G(xA + By) = G(xAy) \vee G(xBy)$,
- $G(x\overline{A}y) = \neg G(xAy)$,
- $G(xA\,; By) = \exists z (xAz \wedge zBy)$ where z is the variable with least index that is distinct from x and y,
- $G(xA^{\smile}y) = G(yAx)$,
- $G(A = B) = \forall v_0 \forall v_1 (G(v_0 A v_1) \leftrightarrow G(v_0 B v_1))$,
- $G(\varphi \rightarrow \psi) = G(\varphi) \rightarrow G(\psi)$,
- $G(\neg\varphi) = \neg G(\varphi)$,
- $G(\forall x(\varphi)) = \forall x(G(\varphi))$.

Theorem 2.8.2 ([Tarski, Givant 1987, 2.3(iv)(v)]) $\mathcal{L}$ and $\mathcal{L}^+$ are equipollent in means of expression and proof, in that the mapping G has the following properties.

- G is a recursive function.
- G maps form^+ onto form and sent^+ onto sent.
- If $\varphi \in \mathsf{form}$ then $G(\varphi) = \varphi$.
- If $\varphi \in \mathsf{form}^+$ then $\vdash^+ \varphi \leftrightarrow G(\varphi)$.
- If $\Psi \subseteq \mathsf{sent}^+$ and $\varphi \in \mathsf{sent}^+$ then $\Psi \vdash^+ \varphi$ iff $\{G(\psi) : \psi \in \Psi\} \vdash \varphi$.
- If $\Psi \subseteq \mathsf{sent}$ and $\varphi \in \mathsf{sent}$ then $\Psi \vdash^+ \varphi$ iff $\Psi \vdash \varphi$.

Theorem 2.8.3 ([Tarski, Givant 1987, 8.3(viii)]) Define the relation $\equiv$ on predicates $A, B \in \mathsf{pred}$ by $A \equiv B$ iff $\vdash^+ A = B$. Then $\equiv$ is a congruence relation on the free predicate algebra $\mathfrak{P}$, and the quotient algebra $\mathfrak{P}/\equiv$ is a free representable relation algebra generated by $\{R_0/\equiv, R_1/\equiv, \ldots\}$.

The formalism $\mathcal{L}^\times$

$\mathcal{L}^\times$ is the equational logic of relation algebras. The set $\mathsf{sent}^\times$ of sentences of $\mathcal{L}^\times$ is $\{A = B : A, B \in \mathsf{pred}\}$. The set $\mathsf{ax}^\times$ of logical axioms of $\mathcal{L}^\times$ consists of all equations of the following forms, where $A, B, C \in \mathsf{pred}$:

(R_1) $A + B = B + A$,

(R_2) $A + (B + C) = (A + B) + C$,

(R_3) $A = \overline{\overline{A} + B} + \overline{\overline{A} + \overline{B}}$,

(R_4) $A;(B;C) = (A;B);C$,

(R_5) $(A + B);C = (A;C) + (B;C)$,

(R_6) $A;1' = A$,

(R_7) $\left(A^{\smile}\right)^{\smile} = A$,

(R_8) $(A + B)^{\smile} = A^{\smile} + B^{\smile}$,

(R_9) $(A;B)^{\smile} = B^{\smile};A^{\smile}$,

(R_{10}) $A^{\smile};\overline{A;B} + \overline{B} = \overline{B}$. The only rule of inference in $\mathcal{L}^\times$ is the *rule of replacement*: infer $A' = B'$ from $A = B$ and $C = D$, whenever C occurs as a part of A or B, and $A' = B'$ is the equation obtained from $A = B$ by replacing some occurrence of C by D. Let $\Psi \subseteq \mathsf{sent}^\times$ and $\psi \in \mathsf{sent}^\times$. Then $\Psi \vdash^\times \psi$ if ψ belongs to every set of sentences of $\mathcal{L}^\times$ which contains $\Psi \cup \mathsf{ax}^\times$ and is closed under the *rule of replacement*.

Theorem 2.8.4 ([Tarski, Givant 1987, 3.4(i)(ii)]) $\mathcal{L}^\times$ is a subformalism of $\mathcal{L}^+$, i.e.,

- $\mathsf{form}^\times \subseteq \mathsf{form}^+$ and $\mathsf{sent}^\times \subseteq \mathsf{sent}^+$,
- for every $\Psi \subseteq \mathsf{sent}^\times$ and every $\varphi \in \mathsf{sent}^\times$, if $\Psi \vdash^\times \varphi$ then $\Psi \vdash^+ \varphi$.

Theorem 2.8.5 ([Tarski, Givant 1987, 8.2(x)]) Define a relation $\equiv^\times$ on predicates $A, B \in \mathsf{pred}$ by $A \equiv^\times B$ iff $\vdash^\times A = B$. Then $\equiv^\times$ is a congruence relation on the free predicate algebra $\mathfrak{P}$, and the quotient algebra $\mathfrak{P}/\equiv$ is a free relation algebra generated by $\{R_0/\equiv^\times, R_1/\equiv^\times, \ldots\}$.

Theorem 2.8.6 ([Tarski, Givant 1987, 3.4(iv)(vi)])
$\mathcal{L}^{\times}$ is strictly weaker than $\mathcal{L}^{+}$ in means of expression and proof.

- There is a sentence in $\mathcal{L}$ not $^{+}$-equivalent to any equation $A = B$ of $\mathcal{L}^{\times}$.
- There is an equation $A = B$ of $\mathcal{L}^{\times}$ such that $\vdash^{+} A = B$ but not $\vdash^{\times} A = B$.

Already in 1915 Löwenheim [Löwenheim 1915] presented a proof (taken from a letter by Korselt) that the sentence saying "there are at least four elements", namely

$$\exists v_0 \exists v_1 \exists v_2 \exists v_3 (\neg v_0 1' v_1 \wedge \neg v_0 1' v_2 \wedge \neg v_1 1' v_2 \wedge \neg v_0 1' v_3 \wedge \neg v_1 1' v_3 \wedge \neg v_2 1' v_3)$$

is not equivalent to any relation-algebraic equation. The existence of a $\mathcal{L}^{+}$-provable but not $\mathcal{L}^{\times}$-provable equation $A = B$ was established by Lyndon [Lyndon 1950]. An example of such an equation is

$$\begin{aligned} & R_0 \cdot (R_1 ; R_2 \cdot R_3) ;(R_4 \cdot R_5 ; R_6) \\ & \quad \leq R_1 ;[(R_1^{\smile} ; R_0 \cdot R_2 ; R_4) ; R_6^{\smile} \cdot R_2 ; R_5 \cdot R_1^{\smile} ;(R_0 ; R_6^{\smile} \cdot R_3 ; R_5)] ; R_6 \end{aligned}$$

(where $A \leq B$ is $A + B = B$ and $A \cdot B = \overline{\overline{A} + \overline{B}}$ for all $A, B \in \mathsf{pred}$). This equation holds in every representable relation algebra and is $\mathcal{L}^{+}$-provable, but it fails in the 4-atom nonrepresentable relation algebra given earlier, so it is not $\mathcal{L}^{\times}$-provable.

The formalisms $\mathcal{L}_n$ and $\mathcal{L}_n^{+}$

For all $A, B \in \mathsf{pred}$, the only variables that can occur (either free or bound) in $G(A = B)$ are v_0, v_1, and v_2. These three variables are the only ones needed in (D_1)–(D_5). Tarski proved that if φ is a sentence in which the only variables that occur are v_0, v_1, and v_2, then φ is equivalent to $G(A = 1)$ for some predicate $A \in \mathsf{pred}$. It is therefore natural to seek a formalism equipollent with $\mathcal{L}^{\times}$ among those obtainable from $\mathcal{L}$ and $\mathcal{L}^{+}$ by restricting the number of available variables to 3 or more. Let $3 \leq n \in \mathbb{N}$. Obtain $\mathcal{L}_n$ and $\mathcal{L}_n^{+}$ from $\mathcal{L}$ and $\mathcal{L}^{+}$ by using an atomic formula φ only if $\mathsf{free}(\varphi) \subseteq \{v_0, \ldots, v_{n-1}\}$. The set of variables for $\mathcal{L}_n$ and $\mathcal{L}_n^{+}$ is $\mathsf{var}_n = \{ v_k : k < n \} = \{v_0, v_1, v_2, \ldots, v_{n-1}\}$. The atomic formulas of $\mathcal{L}_n$ are xR_ky and $x1'y$, where $k \in \mathbb{N}$ and $x, y \in \mathsf{var}_n$. The atomic formulas of $\mathcal{L}_n^{+}$ are xAy and $A = B$, where $A, B \in \mathsf{pred}$ and $x, y \in \mathsf{var}_n$. The definitions of $\mathcal{L}_n$ and $\mathcal{L}_n^{+}$ are otherwise the same as $\mathcal{L}$ and $\mathcal{L}^{+}$. Notations for formulas, sentences, axioms, etc., of $\mathcal{L}_n$ and $\mathcal{L}_n^{+}$ are obtained by adding a subscript n, as in form_n, form_n^{+}, sent_n, sent_n^{+}, ax_n, and ax_n^{+}. Let G_n is the function obtained by restricting the domain of the elimination mapping G to form_n.

Theorem 2.8.7 $\mathcal{L}_n$ and $\mathcal{L}_n^{+}$ are equipollent in means of expression and proof, in that all parts of Theorem 2.8.2 remain true when the notions occurring there are referred to $\mathcal{L}_n$ and $\mathcal{L}_n^{+}$ by the addition of appropriate subscripts.

Theorem 2.8.8 ([Tarski, Givant 1987, 3.9(iii)]) $\mathcal{L}_3^{+}$ and $\mathcal{L}^{\times}$ are equipollent in means of expression, since there is a recursive function H mapping sent_3^{+} onto $\mathsf{sent}^{\times}$ such that $H(\varphi) = \varphi$ whenever $\varphi \in \mathsf{sent}^{\times}$ and $\vdash_3^{+} \varphi \leftrightarrow H(\varphi)$ for every $\varphi \in \mathsf{sent}_3^{+}$.

$\mathcal{L}^\times$ is a subformalism of $\mathcal{L}_3^+$. If it could be shown that $\Psi \vdash_3^+ \varphi$ iff $\{G_3(\psi) : \psi \in \Psi\} \vdash \varphi$, whenever $\Psi \subseteq \mathsf{sent}^\times$ and $\varphi \in \mathsf{sent}^\times$, it would follow that $\mathcal{L}^\times$ is equipollent with $\mathcal{L}_3^+$ in means of proof. But this cannot be done because some instances of the associative law (R_4) are not $\mathcal{L}_3^+$-provable. For instance, it is not the case that $\vdash_3^+ R_0;(R_1;R_2) = (R_0;R_1);R_2$. Tarski proved this by using an algebra, constructed by J.C.C. McKinsey, that satisfies all the axioms for relation algebras except the associativity of relative multiplication. For other proofs, see [Henkin 1973] or [Maddux 1983]. A formalism equipollent with $\mathcal{L}^\times$ is constructed in [Tarski, Givant 1987, 3.7, 3.8] by adding an axiom schema to $\mathcal{L}_3^+$ to remedy this defect. However, it turns out that a formalism equipollent with $\mathcal{L}_3^+$ can be constructed by altering the associative law.

Theorem 2.8.9 Let $\mathcal{L}w^\times$ be the formalism obtained from $\mathcal{L}^\times$ by replacing (R_4) with $1;(1;A) = 1;A$, where $A \in \mathsf{pred}$. Then H (from Theorem 2.8.8) has all the properties required to establish that $\mathcal{L}w^\times$ is equipollent with $\mathcal{L}_3^+$ in means of expression and proof.

All instances of the associative law are $\mathcal{L}_4^+$-provable. It follows that if $\vdash^\times A = B$ then $\vdash_4^+ A = B$. It is a remarkable fact that the converse happens to hold, so that $\mathcal{L}^\times$ is complete with respect to $\mathcal{L}_4^+$-provability.

Theorem 2.8.10 For all $A, B \in \mathsf{pred}$, we have $\vdash_4^+ A = B$ iff $\vdash^\times A = B$.

If the associative law for relative multiplication in the axioms for relation algebras is replaced by the equation $\mathbb{T};(\mathbb{T};x) = \mathbb{T};x$, then the resulting axiom set defines the variety of semiassociative relation algebras.

Theorem 2.8.11 Define relations $\equiv_3$ and $\equiv_4$ on predicates $A, B \in \mathsf{pred}$ by $A \equiv_3 B$ iff $\vdash_3^+ A = B$ and $A \equiv_4 B$ iff $\vdash_4^+ A = B$. Then

- $\equiv_3$ and $\equiv_4$ are congruence relations on the free predicate algebra $\mathfrak{P}$.
- The quotient algebra $\mathfrak{P}/{\equiv_3}$ is a free semiassociative relation algebra freely generated by $\{R_0/{\equiv_3}, R_1/{\equiv_3}, \ldots\}$.
- The quotient algebra $\mathfrak{P}/{\equiv_4}$ is a free relation algebra freely generated by $\{R_0/{\equiv_4}, R_1/{\equiv_4}, \ldots\}$.

For more details concerning the last three results see [Tarski, Givant 1987, 3.10], [Maddux 1978b], [Maddux 1989], [Maddux 1994b].

Formalizing set theory in $\mathcal{L}^\times$

Despite the inequipollence of $\mathcal{L}^\times$ with $\mathcal{L}$ and $\mathcal{L}^+$, it is possible to formalize many first-order theories within $\mathcal{L}^\times$, including every theory that proves the pairing axiom for R_0: $\forall v_0 \forall v_1 \exists v_2 \forall v_3 (v_3 R_0 v_2 \leftrightarrow v_3 1' v_0 \vee v_3 1' v_1)$. For any $A, B \in \mathsf{pred}$, define

$$Q_{AB} = \{A^\smile;A = 1', B^\smile;B = 1', A^\smile;B = 1\} \subseteq \mathsf{sent}^\times \ .$$

Let $P^0_{AB} = B$ and $P^{n+1}_{AB} = A;P^n_{AB}$ whenever $n \in \mathrm{N}$. If φ is a formula and the variables occurring free in φ are $v_{i_0}, v_{i_1}, \ldots, v_{i_n}$, where $i_0 < i_1 < \ldots < i_n$, let $U(\varphi) = \overline{P^{i_0}_{AB}} \dagger \left(\overline{P^{i_0}_{AB}}\right)^\smile + \ldots + \overline{P^{i_n}_{AB}} \dagger \left(\overline{P^{i_n}_{AB}}\right)^\smile$. Define an auxiliary mapping M_{AB}, from form^+ to pred, as follows.

- $M_{AB}(v_i 1' v_j) = (P^i_{AB} \cdot P^j_{AB}) ; 1$,
- $M_{AB}(v_i C v_j) = (P^i_{AB} ; C \cdot P^j_{AB}) ; 1$,
- $M_{AB}(C = D) = 0 \dagger (C \cdot D + \overline{C} \cdot \overline{D}) \dagger 0$,
- $M_{AB}(\neg\varphi) = \overline{M_{AB}(\varphi)}$,
- $M_{AB}(\varphi \to \psi) = \overline{M_{AB}(\varphi)} + M_{AB}(\psi)$,
- $M_{AB}(\forall v_i \varphi) = U(\forall v_i \varphi) \dagger M_{AB}(\varphi)$.

Define the mapping K_{AB} from sent^+ to $\mathsf{sent}^\times$ as follows. If $\varphi \in \mathsf{sent}^\times$ then $K_{AB}(\varphi)$ is φ. If $\varphi \in \mathsf{sent}^+$ but $\varphi \notin \mathsf{sent}^\times$ then $K_{AB}(\varphi)$ is $M_{AB}(\varphi) = 1$.

Theorem 2.8.12 ([Tarski, Givant 1987, 4.4(xxxiii)(xxxiv)])
Let $A, B \in \mathsf{pred}$ and $\varphi \in \mathsf{form}$.

- K_{AB} is a recursive function.
- K_{AB} maps sent^+ onto $\mathsf{sent}^\times$.
- $K_{AB}(\varphi) = \varphi$ iff $\varphi \in \mathsf{sent}^\times$.
- $Q_{AB} \vdash^+ K_{AB}(\varphi) \leftrightarrow \varphi$.
- For every $\Psi \subseteq \mathsf{sent}^+$ and every $\varphi \in \mathsf{sent}^+$,

$$Q_{AB} \cup \Psi \vdash^+ \varphi \quad \text{iff} \quad Q_{AB} \cup \{K_{AB}(\psi) : \psi \in \Psi\} \vdash^\times K_{AB}(\varphi)$$

This is the main mapping theorem of [Tarski, Givant 1987]. Using it, Tarski was able to show that most set theories are formalizable as equational theories in $\mathcal{L}^\times$. For example, if the set theory Ψ proves the pairing axiom for R_0, then predicates A, B can be constructed from R_0 (as in Example 2.1.3) such that $\Psi \vdash Q_{AB}$, hence, by the main mapping theorem,

$$\Psi \vdash^+ \varphi \quad \text{iff} \quad \{K_{AB}(\psi) : \psi \in \Psi\} \vdash^\times K_{AB}(\varphi)$$

The main mapping theorem can be proved semantically with the help of Tarski's Theorem 2.7.4; see [Tarski, Givant 1987, footnote 1, pp.242–3, and footnote 3, p.244].

2.9 Conclusion

In a chapter as short as this it is only possible to touch briefly on a few topics in the calculus of relations and theory of relation algebras. The scope of the applications of these subjects to computer science is well documented in other chapters in this book. For further general reading, deeper results, and applications of relation algebras to other areas of mathematics and computer science, consult [De Morgan 1864], [Peirce 1883] [Schröder 1895], [Tarski 1941], [Lyndon 1950], [Chin, Tarski 1951], [Jónsson, Tarski 1952], [Tarski 1955], [Lyndon 1956], [Lyndon 1961], [Jónsson 1982], [Maddux 1982], [Maddux 1983], [Henkin, Monk+ 1985], [Tarski, Givant 1987], [Maddux 1989], [Maddux 1991c], [Maddux 1991a], [Maddux 1991b], [Maddux 1994a], [Givant 1994], [Ladkin, Maddux 1994], and [Maddux 1996].

Chapter 3

Heterogeneous Relation Algebra

Gunther Schmidt, Claudia Hattensperger, Michael Winter

So far, relational algebra has been presented in its classical form. Relations are often conceived as something that might be called *quadratic* or *homogeneous*; a relation *over* a set. It is interpreted as a subset $R \subseteq U \times U$ of a Cartesian product of the universe U with itself. If relations *between* two or more sets are considered, this may easily be subsumed under this view, uniting all the sets in question into one huge set and calling this set the universe U. On the other hand, a variant of the theory has evolved that treats relations from the very beginning as *heterogeneous* or *rectangular*, i.e. as relations where the normal case is that they are relations between two different sets. The present chapter is devoted to this variant form.

3.1 Distinguishing domains

Programmers are used to defining the *types* they work with very precisely. Starting with, e.g., *integers* and *reals*, they proceed to *records*, *variant records*, *recursively defined types* and *higher order types*. Also, engineers and those working with numerical algorithms usually learn to work with $(n \times n)$- and also $(n \times m)$-matrices.

Basic practical examples of relations often start out from such examples as a set of boys and a set of girls between which a relation of attraction is given. This is then often visualized as a graph or as a boolean matrix, as in Fig. 3.1.

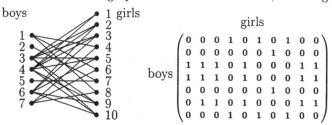

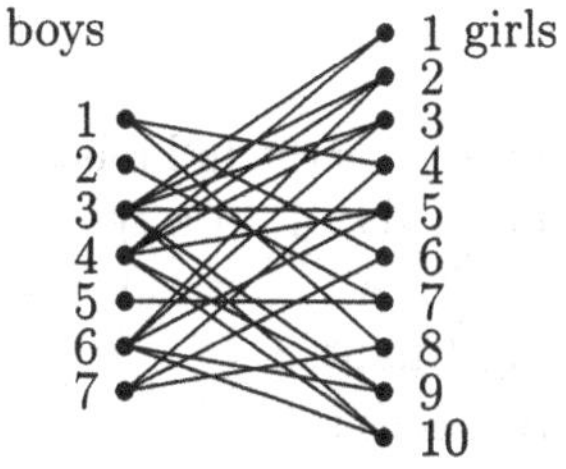

$$\text{boys} \overset{\text{girls}}{\begin{pmatrix} 0&0&0&1&0&1&0&1&0&0 \\ 0&0&0&0&0&0&1&0&0&0 \\ 1&1&1&0&1&0&0&0&1&1 \\ 1&1&1&0&1&0&0&0&1&1 \\ 0&0&0&0&0&0&1&0&0&0 \\ 0&1&1&0&1&0&0&0&1&1 \\ 0&0&0&1&0&1&0&1&0&0 \end{pmatrix}}$$

Fig. 3.1 Heterogeneous relation

Subsets of this relation might define the relation of marriage, for instance.

Therefore, we now switch to heterogeneous relations, where $\sqcup, \sqcap$ and $;$ are *partial* operations. With regard to applicability, one can let oneself be guided by the calculus of $(n \times m)$-matrices or by category theory[1].

For reasons of economy, when considering a relation algebra, we want to say as little as possible about the question of whether relations can be joined, intersected, or multiplied. In practical problems, these operations will be permitted if the right to do so can be deduced from some corresponding union, intersection, or product given in the first place.

Definition 3.1.1 A *(heterogeneous abstract) relation algebra* is a locally small category $\mathcal{R}$ consisting of a class $\mathsf{Obj}_\mathcal{R}$ of objects and a set $\mathsf{Mor}_\mathcal{R}[A,B]$ of morphisms for all $A, B \in \mathsf{Obj}_\mathcal{R}$. The morphisms are usually called *relations*. For $S \in \mathsf{Mor}_\mathcal{R}[A,B]$ we use the notation $S : A \leftrightarrow B$. Composition[2] is denoted by ";" and identities are denoted by $\mathbb{I}_A \in \mathsf{Mor}_\mathcal{R}[A,A]$. In addition, there is a totally defined unary operation $\breve{\ }_{A,B} : \mathsf{Mor}_\mathcal{R}[A,B] \longrightarrow \mathsf{Mor}_\mathcal{R}[B,A]$ between the sets of morphisms, called conversion. The operations satisfy the following rules:

i) Every set $\mathsf{Mor}_\mathcal{R}[A,B]$ carries the structure of a complete atomic boolean algebra with operations $\sqcup_{A,B}, \sqcap_{A,B}, \overline{}_{A,B}$, zero element $\perp\!\!\!\perp_{A,B}$, universal element $\top\!\!\!\top_{A,B}$ (the latter two non-equal), and inclusion ordering $\sqsubseteq_{A,B}$.

ii) The Schröder equivalences

$$Q;R \sqsubseteq_{A,C} S \iff Q\breve{\ };\overline{S} \sqsubseteq_{B,C} \overline{R} \iff \overline{S};R\breve{\ } \sqsubseteq_{A,B} \overline{Q}$$

hold for relations Q, R and S (where the definedness of one of the three formulæ implies that of the other two).

iii) The Tarski rule

$$R \neq \perp\!\!\!\perp_{A,B} \iff \top\!\!\!\top_{C,A};R;\top\!\!\!\top_{B,D} = \top\!\!\!\top_{C,D}$$

holds for all $R \in \mathsf{Mor}_\mathcal{R}[A,B]$ and $C, D \in \mathsf{Obj}_\mathcal{R}$.

All the indices of elements and operations are usually omitted for brevity and can easily be reinvented.

There is an identity, a zero and a universal relation per morphism set $\mathsf{Mor}_\mathcal{R}[A,A]$, but only a zero and a universal relation per morphism set $\mathsf{Mor}_\mathcal{R}[A,B]$ if $A \neq B$. As with sets and mappings between them, one has always to keep track of from where to where the relation goes. Of course, one can clutter up the notation with always giving the index of source and target.

A standard example of a relation algebra in our sense is the full relation algebra *Rel* where all sets are objects and a relation from a set A to a set B is just a subset of the Cartesian product $A \times B$ of A and B. All operations are defined analogously to the homogeneous case introduced in Chapt. 1.

An equivalent substitute for the Schröder equivalences is the Dedekind formula.

[1]To avoid set-theoretic problems we restrict ourselves to categories where every class of morphisms $\mathsf{Mor}_\mathcal{R}[A,B]$ is a set. Such categories are called locally small.

[2]Recall that in Chapt. 1 the small circle was used for composition of morphisms in preference to the semicolon. Here, however, we wish to emphasize that morphisms are to be thought of as relations, therefore we revert to the semicolon.

Theorem 3.1.2 (*Dedekind Formula*). Given Q, R, and S, one has

$$Q;R \sqcap S \sqsubseteq (Q \sqcap S;R^{\smile});(R \sqcap Q^{\smile};S). \qquad \square$$

The following computational rules – well-known in the homogeneous case – may also be proven from these axioms in the heterogeneous case.

$$\begin{array}{rcl} \perp\!\!\!\perp_{A,B} = R & \Longleftrightarrow & R;\top\!\!\!\top_{B,C} = \perp\!\!\!\perp_{A,C}, \\ Q;(R \sqcap S) & \sqsubseteq & Q;R \sqcap Q;S, \\ R^{\smile\smile} & = & R, \\ R \sqsubseteq S & \Longleftrightarrow & R^{\smile} \sqsubseteq S^{\smile}, \\ (R \sqcap S)^{\smile} & = & R^{\smile} \sqcap S^{\smile}, \\ \perp\!\!\!\perp^{\smile}_{A,B} & = & \perp\!\!\!\perp_{B,A}, \\ \mathbb{I}^{\smile} & = & \mathbb{I}, \end{array} \qquad \begin{array}{rcl} P \sqsubseteq Q, R \sqsubseteq S & \Longrightarrow & P;R \sqsubseteq Q;S, \\ Q;(R \sqcup S) & = & Q;R \sqcup Q;S, \\ \overline{R^{\smile}} & = & \overline{R}^{\smile}, \\ (R;S)^{\smile} & = & S^{\smile};R^{\smile}, \\ (R \sqcup S)^{\smile} & = & R^{\smile} \sqcup S^{\smile}, \\ \top\!\!\!\top^{\smile}_{A,B} & = & \top\!\!\!\top_{B,A}, \\ \top\!\!\!\top_{A,B};\top\!\!\!\top_{B,C} & = & \top\!\!\!\top_{A,C}. \end{array}$$

However, some of the proofs may differ from those in the homogeneous case. For example, the inclusion $\top\!\!\!\top \sqsubseteq \top\!\!\!\top;\top\!\!\!\top$ can in the homogeneous case be proven by

$$\top\!\!\!\top = \top\!\!\!\top;\mathbb{I} \sqsubseteq \top\!\!\!\top;\top\!\!\!\top.$$

In the heterogeneous case the result is slightly more general in as far as $\top\!\!\!\top_{A,B} \sqsubseteq \top\!\!\!\top_{A,C};\top\!\!\!\top_{C,B}$ is universally quantified over $A, B, C \in \mathsf{Obj}_{\mathcal{R}}$. Now it is obvious that the last step in the homogeneous proof, replacing the identity $\mathbb{I}_{B,B}$ by $\top\!\!\!\top_{C,B}$, can not be done if B and C are not equal.

As above, throughout this chapter most proofs are, for reasons of brevity, omitted. They can be found, for example, in [Schmidt, Ströhlein 1989] or [Schmidt, Ströhlein 1993].

There has sometimes been a discussion as to whether the heterogeneous or the homogeneous concept should be taken. Today at least, there is unanimity that everything can more or less be expressed both ways, however, on either side with certain benefits or drawbacks. So it is to a certain extent a matter of taste and style which to use.

As relation algebras are algebraic structures there is a notion of homomorphism between relation algebras.

Definition 3.1.3 Let $\mathcal{R}$ and $\mathcal{S}$ be relation algebras and $F : \mathcal{R} \to \mathcal{S}$ a functor. Then F is called a *homomorphism between relation algebras* if

i) $F(R \sqcap S) = F(R) \sqcap F(S)$,
ii) $F(\overline{R}) = \overline{F(R)}$,
iii) $F(R^{\smile}) = F(R)^{\smile}$,

hold for all relations R, S.

It is easy to show that for all objects A, B and all relations R, S we have that:

$$F(\perp\!\!\!\perp_{A,B}) = \perp\!\!\!\perp_{F(A),F(B)} \quad F(\top\!\!\!\top_{A,B}) = \top\!\!\!\top_{F(A),F(B)} \quad F(R \sqcup S) = F(R) \sqcup F(S).$$

In the heterogeneous case the question arises what representability means. A representation of a homogeneous relation algebra is a homomorphism into a proper relation algebra (cf. Chapt. 2). The heterogeneous relation algebra *Rel* can be regarded as the collection – in the sense of a subalgebra relation – of all proper heterogeneous relation algebras. So, a representation is just an embedding into *Rel*.

Definition 3.1.4 A relation algebra $\mathcal{R}$ is called *representable* if there is a faithful homomorphism $F : \mathcal{R} \to Rel$.

A one-object relation algebra is a homogeneous relation algebra. In this case the above definition of representation corresponds to the definition in Chapt. 2.

3.2 Specific formulæ and representations

Concrete heterogeneous relations can be used in an elegant way to characterize subsets of a set or – nearly equivalently – predicates on this set. To this end, associate the relation

$$m := \{(x, y) \mid x \in M \wedge y \in A\} \text{ for an arbitrary set } A$$

with a subset $M \subseteq U$. It imposes a condition on just the first component of pairs. Being a set of pairs, m is a concrete relation and can be represented by a matrix. For example:

$$m = \begin{array}{c} \\ a \\ b \\ c \\ d \\ e \end{array} \begin{array}{c} 1\ 2\ 3 \\ \begin{pmatrix} 0 & 0 & 0 \\ 1 & 1 & 1 \\ 0 & 0 & 0 \\ 1 & 1 & 1 \\ 0 & 0 & 0 \end{pmatrix} \end{array}$$

This matrix is "row-constant", and it is easy to see that such matrices are characterized by $m = m;V$ where V is the greatest relation over a suitable universe U. This motivates the definition of a subset or a vector in an abstract relation algebra.

Definition 3.2.1 A relation R satisfying $R = R;\mathbb{T}$ is called a *vector*. In a more elaborate notation: A relation $R \in \mathsf{Mor}_{\mathcal{R}}[A, B]$ satisfying $R_{A,B} = R_{A,B};\mathbb{T}_{B,B}$ is called a vector. □

For any relation $R : A \leftrightarrow B$ the construct $R;\mathbb{T}$ is a vector describing "where R is defined". In analogy to mappings, this vector is called the *domain* of R. It is of course not equal to **source** R, which is an object of the relation algebra and not a relation (see Chapt. 1 and Chapt. 2). By duality, the *range* of the relation R is the vector $R^{\smile};\mathbb{T}$.

In matrix calculus, matrices with only one column are often used. They are characterized as vectors to the one-element set. The adequate notion for objects like a one-element set is a terminal object in category theory. An object $\mathbf{1}$ is called terminal if for every object A there is a unique morphism $f : A \to \mathbf{1}$. Now, in an abstract relation algebra a vector corresponds to relations in $\mathsf{Mor}_{\mathcal{R}}[A, \mathbf{1}]$ where $\mathbf{1}$ is the terminal object in the subcategory of mappings (cf. Definition 3.2.8).

Sometimes the notion of partial identities $R \sqsubseteq \mathbb{I}$ for subsets is used. In the homogeneous case these notions correspond to each other: we can map a partial identity R to the vector $R;\mathbb{T}$ and a vector x to the partial identity $\mathbb{I} \sqcap x$. The example above shows that this is not always possible in the heterogeneous case.

Intersecting a matrix with a vector amounts to singling out certain rows and replacing the entries of all other rows by zeros.

Theorem 3.2.2 The following formulas hold for all relations $P_{A,B}, Q_{C,A}, R_{C,B}$ and $S_{A,D}$.

i) $(Q \sqcap R;\mathbb{T}_{B,A});S = Q;S \sqcap R;\mathbb{T}_{B,D}$,

ii) $(Q \sqcap (P;\mathbb{T}_{B,C})^{\smile});S = Q;(S \sqcap P;\mathbb{T}_{B,D})$. □

A one-element subset can be characterized via relations as follows:

Definition 3.2.3 A relation $x \neq \mathbb{L}$ is called a *point* if x is a vector satisfying $x;x^{\smile} \sqsubseteq \mathbb{I}$.

The one-element subsets of a set are the atoms in the powerset ordering. A similar proposition holds for points in a relation algebra.

Theorem 3.2.4 Every point is an atom among the vectors, i.e., if x is a point and y is a vector fulfilling $\mathbb{L} \neq y \sqsubseteq x$, then $x = y$. □

Considering concrete relations it is easy to prove that for every relation $R \neq \emptyset$ there exist two points such that $x;y^{\smile} \subseteq R$. A result like this cannot be proved in an abstract relation algebra. Indeed, there are relation algebras where this does not hold. So one can work with relation algebras which do, or do not, satisfy this as an independent axiom.

Axiom 3.2.5 (*Point Axiom*). For every relation $R \neq \mathbb{L}$ there exist two points x, y such that $x;y^{\smile} \sqsubseteq R$. □

Assuming the point axiom, composition of abstract relations gets the flavor of composition of concrete relations, in the following sense:

Theorem 3.2.6 (*Intermediate Point Theorem*). Let $\mathcal{R}$ be a relation algebra such that the point axiom is valid. Then for points x, y and relations R, S we have

$$x \sqsubseteq R;S;y \iff \text{There exists a point } z \text{ with } x \sqsubseteq R;z \text{ and } z \sqsubseteq S;y. \quad \square$$

This gives rise to the following representability result:

Theorem 3.2.7 A relation algebra satisfying the point axiom is representable.□

The idea of the proof is to define a functor $\mu : \mathcal{R} \to \mathit{Rel}$ as follows:

i) Define $\mathcal{Q}_{A,B}$ to be the set of points in $\mathsf{Mor}_{\mathcal{R}}[A,B]$.

ii) On objects A define $\mu(A) := \mathcal{Q}_{A,A}$.

iii) For $R \in \mathsf{Mor}_{\mathcal{R}}[A,B]$ define $\mu(R) := \{(x,y) \in \mathcal{Q}_{A,A} \times \mathcal{Q}_{B,B} \mid x;\mathbb{T}_{A,B};y^{\smile} \sqsubseteq R\}$.

For the homogeneous case the proof can be found in [Schmidt, Ströhlein 1985].

We now give the relational formulation of properties of a relation that make it a partial function or a total function. If variants are given in the definition, they may easily be proven.

Definition 3.2.8 For any relation R we say that

R is *total* $:\iff \quad \top = R;\top \iff \mathbb{I} \sqsubseteq R;R^{\smile} \iff \overline{R} \sqsubseteq R;\overline{\mathbb{I}}$,

(Or, in a more elaborate notation:

$R \in \mathsf{Mor}_{\mathcal{R}}[A,B]$ total $:\iff$ There exists a C with $\top_{A,C} = R_{A,B};\top_{B,C}$),

R is *univalent* $:\iff \quad R^{\smile};R \sqsubseteq \mathbb{I} \iff R;\overline{\mathbb{I}} \sqsubseteq \overline{R}$,

(Or: a (partial) function)

R is a *mapping* $:\iff$ R is univalent and total $\iff \overline{R} = R;\overline{\mathbb{I}}$,

R is *surjective* $:\iff$ $R^{\smile}$ total,

R is *injective* $:\iff$ $R^{\smile}$ univalent.

If R is total, it can easily been seen that $\top_{A,C} = R_{A,B};\top_{B,C}$ for all C. This condition just means that the domain of R is all of A.

The following laws, well-known from the homogeneous case, are also valid in the heterogeneous case.

Theorem 3.2.9 For relations Q, R, S (where R and S are in every case restricted to relations that can appropriately be composed) the following holds:

i) Q, R univalent $\implies$ $Q;R$ univalent,

ii) Q univalent $\iff$ $Q;(R \sqcap S) = Q;R \sqcap Q;S$ for all R, S,

iii) Q univalent $\iff$ $R;Q \sqcap S = (R \sqcap S;Q^{\smile});Q$ for all R, S,

iv) Q univalent $\iff$ $Q;\overline{R} = Q;\top \sqcap \overline{Q;R}$ for all R,

v) Q total $\iff$ $Q;\overline{R} \sqsupseteq \overline{Q;R}$ for all R,

vi) Q mapping $\iff$ $Q;\overline{R} = \overline{Q;R}$ for all R,

vii) If Q is a mapping then $R \sqsubseteq S;Q^{\smile} \iff R;Q \sqsubseteq S$. □

3.3 Homomorphy of structures

In many applications, we have structures over a set or between several sets. Structures are often compared with one another: A group may be mapped homomorphically to another group or a vector space may be linearly mapped to another vector space. In both cases we had algebraic structures, i.e., given by mappings. We can likewise map a graph to another graph, which is an example of a structure given by relations. In Fig. 3.2 we give an example of such a relation Φ being a homomorphism between two the graphs on the left and right side of it. When generalizing all this, we arrive at the concept of a homomorphism to compare the structures in question.

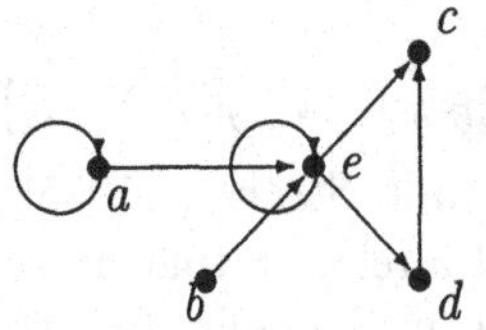

$$\Phi = \begin{array}{c} \\ a \\ b \\ c \\ d \\ e \end{array}\begin{array}{c} \begin{array}{cccc} w & x & y & z \end{array} \\ \begin{pmatrix} 0 & 0 & 0 & 1 \\ 1 & 0 & 0 & 0 \\ 0 & 1 & 0 & 0 \\ 0 & 1 & 0 & 0 \\ 0 & 0 & 1 & 0 \end{pmatrix} \end{array}$$

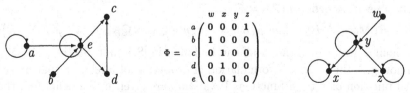

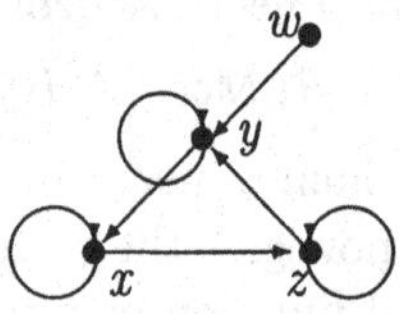

Fig. 3.2 Graph homomorphism

Now we want to introduce the concept of homomorphism *inside* a heterogeneous relation algebra (not between relation algebras). If some structure is to be mapped homomorphically into another one, so that *its structure is preserved*, one needs mappings (these mappings are relations) to relate the structures (structures are relations). The structures in question may be relational (as graphs or orderings) or algebraic (as groups or fields). They may consist of more than one relation; here however, in order to present the principle, let a structure be given by just one relation B.

Definition 3.3.1 If B and B' are relations, we call the pair Φ, Ψ a *homomorphism* from B to B', provided Φ, Ψ are mappings and $B \sqsubseteq \Phi; B'; \Psi^{\smile}$. If $\Phi = \Psi$, we simply write Φ instead of Φ, Φ.

Using the mapping properties, one immediately sees that $B \sqsubseteq \Phi; B'; \Psi^{\smile}$ is equivalent to each of the following three inclusions

$$B; \Psi \sqsubseteq \Phi; B', \qquad \Phi^{\smile}; B; \Psi \sqsubseteq B', \qquad \Phi^{\smile}; B \sqsubseteq B'; \Psi^{\smile}.$$

So the equivalent conditions for homomorphy are produced in a cyclic way by suitably multiplying by Φ, Ψ and the converses.

Definition 3.3.2 Let relations B, B', and Φ, Ψ be given. We call the pair Φ, Ψ an *isomorphism* if both Φ, Ψ and $\Phi^{\smile}, \Psi^{\smile}$ are homomorphisms.

Thus, an isomorphism is characterized by the following conditions

$$\Phi, \Psi \text{ are bijective mappings}, \qquad B; \Psi \sqsubseteq \Phi; B', \qquad B'; \Psi^{\smile} \sqsubseteq \Phi^{\smile}; B.$$

The two inclusions on the right may equally well be replaced by those, cyclically produced, equivalent conditions from above. If Φ, Ψ is an isomorphism, then each of the cyclically produced homomorphy conditions becomes an equality:

$$B = \Phi; B'; \Psi^{\smile}, \quad B; \Psi = \Phi; B', \quad \Phi^{\smile}; B; \Psi = B', \quad \Phi^{\smile}; B = B'; \Psi^{\smile}.$$

However, these four equations are no longer equivalent to one another if Φ, Ψ are not bijective.

Homomorphisms of algebraic structures

As we have seen, our homomorphisms of relational structures are not characterized by equations, in contrast to other mathematical theories, e.g., the theory of groups. Now we consider an *algebraic* structure, as opposed to a relational structure, i.e. we assume that the relations considered are mappings.

Theorem 3.3.3 Let mappings B and B' be given and let Φ, Ψ be mappings.

i) $$B; \Psi \sqsubseteq \Phi; B' \iff B; \Psi = \Phi; B'.$$

ii) *Each* of the following three equations implies $B; \Psi = \Phi; B'$:

$$B = \Phi; B'; \Psi^{\smile}, \quad \Phi^{\smile}; B; \Psi = B', \quad \Phi^{\smile}; B = B'; \Psi^{\smile}.$$

For any other two of the four equations there exist examples where one is satisfied while the other is not. □

The condition $B; \Psi \sqsubseteq \Phi; B'$ for homomorphy, therefore, may be used for algebraic as well as for relational structures. The use of the equation $B; \Psi = \Phi; B'$ is restricted to algebraic structures.

3.4 Congruences

Another application of a typically heterogeneous nature is that of forming the *quotient set* of a given set under an equivalence relation (or a congruence, in the case of a structured set). Recall that an equivalence relation is a relation $R \in \mathsf{Mor}_{\mathcal{R}}[A, A]$ fulfilling $\mathbb{I}_A \sqsubseteq R$ (reflexivity), $R;R \sqsubseteq R$ (transitivity) and $R^{\smile} \sqsubseteq R$ (symmetry). Within a given relation algebra it is in general not possible to pass on to equivalence classes as one would normally do by dividing out an equivalence relation. So we can only show that there is at most one quotient object up to isomorphism.

Theorem 3.4.1 Let Ξ be equivalence relation. Then there is at most one (up to isomorphism) surjective mapping χ usually called *natural projection* such that $\Xi = \chi;\chi^{\smile}$. □

We now consider structure-respecting equivalence relations over a structured set.

Definition 3.4.2 Let $B : C \leftrightarrow D$ and equivalence relations $\Xi : C \leftrightarrow C$ and $\Xi' : D \leftrightarrow D$ be given, then $\Xi;B \sqsubseteq B;\Xi'$ is called the *substitution property*. If $C = D$ and $\Xi = \Xi'$ then Ξ is called a B*-congruence*.

As congruences are equivalence relations, we get the following.

Theorem 3.4.3 Let B, Ξ, Ξ' be given such that the substitution property holds. Then there is at most one (up to isomorphism in the sense of Definition 3.3.2) surjective homomorphism χ, ψ such that $\Xi = \chi;\chi^{\smile}$ and $\Xi' = \psi;\psi^{\smile}$. □

If we consider $\Xi = \Xi'$ in the theorem above, the surjective homomorphism gives us the quotient structure under the B-congruence Ξ.

Congruences are often generated by specific homomorphisms.

Theorem 3.4.4 Let $B \in \mathsf{Mor}_{\mathcal{R}}[A, A]$ be a relation and χ a homomorphism from B to B' such that $\chi;B' \sqsubseteq B;\chi$ holds. Then $\Xi := \chi;\chi^{\smile}$ is a B-congruence. □

Covering of graphs

As an example of such congruences creating homomorphisms we now want to develop a theory of coverings of graphs. Motivated by the visualisation of a concrete relation as a graph, we conceive a graph as an arbitrary abstract relation $G \in \mathsf{Mor}_{\mathcal{R}}[A, A]$. This relation is also called the *associated relation* of the graph. A *graph homomorphism* is a homomorphism with respect to the associated relations of graphs.

A root of a graph is a point from which all points can be reached. This reachability condition can be expressed by the inclusion of the relation corresponding to the graph in its reflexive transitive closure G^* (cf. Chapt. 1).

Definition 3.4.5 A point a is called a *root* of a graph G if $a;\mathbb{T} \sqsubseteq G^*$ holds. If G has a distinguished root then G is called a *rooted* graph.

We want to characterize those homomorphisms of a graph that preserve to a certain extent the possibilities of traversing. It turns out that such a homomorphism is surjective and that it carries the successor relation at any point onto that at the image point.

Definition 3.4.6 A surjective homomorphism $\Phi : G \to G'$ is called a *covering*, if $\Phi ; G' \sqsubseteq G ; \Phi$ and $G^{\smile} ; G \sqcap \Phi ; \Phi^{\smile} \sqsubseteq \mathbb{I}$.

By Theorem 3.4.4 every covering induces a G-congruence on the graph. Dividing out this congruence, we get the following.

Theorem 3.4.7 Let G be a rooted graph with root a. Then there is (up to isomorphism) at most one rooted tree covering G. This tree is called the *universal covering*[3] of G. □

Universal coverings are closely connected to bisimulations (Chapt. 5). If the universal covering of the reduction graphs of two processes are equal they should be bisimilar.

In the next section we consider flow diagrams and programs. The effect of such a flow diagram is particularly easy to understand when the underlying rooted graph is in fact a rooted tree. So, the last proposition gives us the possibility to switch between rooted graphs and its universal covering.

3.5 Flow diagrams

In Fig. 3.3, we consider a flow diagram, where the arcs are labelled with letters resembling program steps. These steps are graphs again, transforming local states to local states. Matrices with coefficients in a relation algebra are adequate to describe such a situation.

Theorem 3.5.1 Let $\mathcal{R}$ be a relation algebra and define a category $\mathcal{S}$ as follows:

i) The objects are pairs (A, n) consisting of an object A in $\mathsf{Obj}_{\mathcal{R}}$ and an $n \in \mathbb{N}$.
ii) A relation $R \in \mathsf{Mor}_{\mathcal{S}}[(A,n),(B,m)]$ is a $(n \times m)$-matrix with coefficients in $\mathsf{Mor}_{\mathcal{R}}[A, B]$.
iii) The operations $\sqcap$, $\sqcup$, $\overline{}$, $\mathbb{T}$ and $\perp\!\!\!\perp$ are defined componentwise.
iv) Transposition and multiplication is defined by

$$S^{\smile}(i,k) := [S(k,i)]^{\smile} \qquad (R;S)(i,k) := \bigsqcup\{R(i,j);S(j,k) \mid 1 \le j \le n\}.$$

Then $\mathcal{S}$ is a relation algebra. □

Such relation algebras can be seen as frames (Chapt. 5). To every pair of worlds (corresponding to a column and a row of the matrix) one relation is assigned.

Using those matrices with relational coefficients, we can work as in Chapt. 10 of [Schmidt, Ströhlein 1989] or [Schmidt, Ströhlein 1993].
A program is a pair of graphs connected by a suitable graph homomorphism.

Definition 3.5.2 A quintuple $\mathcal{P} := (G, S, \Theta, e, a)$ is called a *program* or a *flow diagram*, if

i) S is a graph, called the *situation graph*,
ii) G is a graph, called the underlying *flowchart*,
iii) Θ is a surjective homomorphism of S onto G.

[3]The universal covering got its name from a similar concept in the theory of Riemann surfaces.

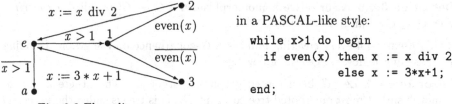

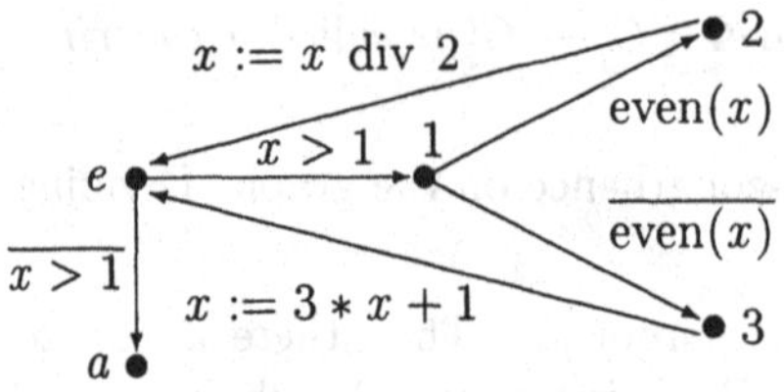

Fig. 3.3 Flow diagram

in a PASCAL-like style:

```
while x>1 do begin
   if even(x) then x := x div 2
              else x := 3*x+1;
end;
```

iv) e is a so-called *input relation*, satisfying:

$$e^{\smile};e = \mathbb{I}, \quad e;e^{\smile} \sqsubseteq \mathbb{I}, \quad e;\mathbb{T} = \Theta;\Theta^{\smile};e;\mathbb{T}, \quad \Theta^{\smile};e = \Theta^{\smile};e;\mathbb{T}.$$

v) a is a so-called *output relation*, satisfying:

$$a^{\smile};a = \mathbb{I}, \quad a;a^{\smile} \sqcap \Theta;\Theta^{\smile} \sqsubseteq \mathbb{I}, \quad a;\mathbb{T} = \Theta;\Theta^{\smile};a;\mathbb{T}.$$

If S is univalent, we call $\mathcal{P}$ *deterministic*.

In the example above we have the following associated relation of the flowchart.

$$\begin{array}{c} \\ e \\ 1 \\ 2 \\ 3 \\ a \end{array} \begin{array}{c} \begin{array}{ccccc} e & 1 & 2 & 3 & a \end{array} \\ \begin{pmatrix} 0 & 1 & 0 & 0 & 1 \\ 0 & 0 & 1 & 1 & 0 \\ 1 & 0 & 0 & 0 & 0 \\ 1 & 0 & 0 & 0 & 0 \\ 0 & 0 & 0 & 0 & 0 \end{pmatrix} \end{array}$$

It is now convenient to replace the matrix entries 0, 1 by the operations executed during the program steps. These operations correspond to concrete relations Q, R, S, T and $\emptyset$ in our example. So one obtains the associated relation of the situation graph.

$$\begin{array}{c} \\ e \\ 1 \\ 2 \\ 3 \\ a \end{array} \begin{array}{c} \begin{array}{ccccc} e & 1 & 2 & 3 & a \end{array} \\ \begin{pmatrix} \emptyset & R & \emptyset & \emptyset & \overline{R} \\ \emptyset & \emptyset & S & \overline{S} & \emptyset \\ T & \emptyset & \emptyset & \emptyset & \emptyset \\ Q & \emptyset & \emptyset & \emptyset & \emptyset \\ \emptyset & \emptyset & \emptyset & \emptyset & \emptyset \end{pmatrix} \end{array} \qquad e = \begin{array}{c} e \\ 1 \\ 2 \\ 3 \\ a \end{array} \begin{pmatrix} I \\ \emptyset \\ \emptyset \\ \emptyset \\ \emptyset \end{pmatrix} \qquad a = \begin{array}{c} e \\ 1 \\ 2 \\ 3 \\ a \end{array} \begin{pmatrix} \emptyset \\ \emptyset \\ \emptyset \\ \emptyset \\ I \end{pmatrix}$$

The input and output relation should be the relations projecting out the legal input and the output states namely the first and last row of the associated relation of the situation graph corresponding to the first and last row of the associated relation of the flowchart.

Termination of a program is a property of the situation graph. If it is possible to proceed only finitely many steps along this graph, the program will always terminate (*well-foundedness*).

Definition 3.5.3 Let G be a graph and consider vectors x. The vector of all points, from which only paths of finite length emerge,

$$J(G) := \bigsqcap\{x \mid \overline{x} \sqsubseteq G;\overline{x}\},$$

is called the *initial part* of G.

Now, the question arises, which input data are mapped by the program to which output data.

Definition 3.5.4 In a program $\mathcal{P}$ with input relation e, output relation a and situation graph S, we call

i)	S^*	the *action*,
ii)	$C := S^* \sqcap \overline{S;\mathbb{T}}^{\smile}$	the *terminating action*,
iii)	$\Sigma := e^{\smile};C;a$	the *effect*

of the program $\mathcal{P}$.

The action of a program is thus the reachability in the situation graph. The terminating action, on the other hand, associates with a given state only the last state reachable from it, which need not belong to an output position of the program. Operations e and a extract the part leading from input to output positions. The effect of a program is the relation that relates the states before entering into the program to those after exiting it. This relation can be seen as the semantics of the program. Based on this semantic we can define partial correctness as follows.

Definition 3.5.5 A program with effect Σ is called *partially correct* with respect to a precondition $v = v;\mathbb{T}$ and a postcondition $n = n;\mathbb{T}$, if $\Sigma^{\smile};v \sqsubseteq n$.

This condition reads as follows: If an output fulfils the postcondition, there is an input, which fulfils the precondition and a program run mapping this input to the given output. Clearly, one would like to check correctness of a program more directly by looking at its associated relation S rather than investigating the more complicated effect Σ. The essential ideas in this direction are contained in the Floyd-Hoare-Calculus [Floyd 1967] and [Hoare 1969]. The core of these methods is contained in the following theorem.

Theorem 3.5.6 (*Contraction Theorem*) A program $\mathcal{P}$ with input relation e and output relation a is partially correct with respect to v and n, if and only if there is a vector Q such that the following holds:

$$S^{\smile};Q \sqsubseteq Q, \quad v \sqsubseteq e^{\smile};Q, \quad a^{\smile};(Q \sqcap \overline{S;\mathbb{T}}) \sqsubseteq n. \qquad \square$$

Analogously, total correctness of a program is defined as follows.

Definition 3.5.7 A program $\mathcal{P}$ with terminating action C, initial part $J(S)$ and effect Σ is called *totally correct* with respect to a precondition $v = v;\mathbb{T}$ and a postcondition $n = n;\mathbb{T}$, if the following holds:

$$v \sqsubseteq e^{\smile};J(S) \sqcap e^{\smile};\overline{C;\overline{a;n}}.$$

So, a program $\mathcal{P}$ to be totally correct with respect to v and n means the following: Given a state satisfying v, there are no computation sequences of arbitrary length. Moreover, with the termination action C only states satisfying the postcondition n after output a can be reached.

If the program $\mathcal{P}$ is deterministic, this definition reduces to a simple inclusion.

Theorem 3.5.8 A deterministic program $\mathcal{P}$ is totally correct with respect to v and n, if and only if $v \sqsubseteq \Sigma; n$. □

Analogous to partial correctness, we ask whether total correctness can be directly read off from the program, i.e. the situation graph S. The answer is contained in a "Contraction Theorem" for total correctness.

Theorem 3.5.9 (*Complement-Expansion Theorem*) Let $\mathcal{P}$ be a program with input relation e and output relation a. $\mathcal{P}$ is totally correct with respect to a precondition v and a postcondition n, if and only if for every vector Q with $\overline{Q} \sqsubseteq S;\overline{Q} \sqcup \overline{a;n \sqcup S;\top}$ also $v \sqsubseteq e^{\smile};Q$ holds. □

The weakest precondition of a given program and a given postcondition is the greatest relation (the weakest predicate) such that the program is totally correct with respect to wp and the given postcondition.

Definition 3.5.10 Let $\mathcal{P}$ be a program and postcondition n. The *weakest precondition* of $\mathcal{P}$ with respect to n is defined by

$$\mathrm{wp}(\mathcal{P}, n) := e^{\smile};J(S) \sqcap e^{\smile};\overline{C;\overline{a;n}}.$$

In 1974 resp. 1975 Dijkstra (see [Dijkstra 1974], [Dijkstra 1975]) postulated a set of conditions for the weakest precondition of a program. The next result shows that our definition of wp satisfies these conditions.

Theorem 3.5.11 For the weakest precondition defined above we have

i)	$\mathrm{wp}(\mathrm{skip}, n) = n$,	iv)	$\mathrm{wp}(\mathcal{P}, n_1) \sqcap \mathrm{wp}(\mathcal{P}, n_2) = \mathrm{wp}(\mathcal{P}, n_1 \sqcap n_2)$,
ii)	$\mathrm{wp}(\mathrm{abort}, n) = \perp\!\!\!\perp$,	v)	$\mathrm{wp}(\mathcal{P}, n_1) \sqcup \mathrm{wp}(\mathcal{P}, n_2) \sqsubseteq \mathrm{wp}(\mathcal{P}, n_1 \sqcup n_2)$,
iii)	$\mathrm{wp}(\mathcal{P}, \perp\!\!\!\perp) = \perp\!\!\!\perp$,	vi)	$n_1 \sqsubseteq n_2 \Longrightarrow \mathrm{wp}(\mathcal{P}, n_1) \sqsubseteq \mathrm{wp}(\mathcal{P}, n_2)$. □

3.6 Direct product

Most operations occurring in real life involve several arguments and not just one. Using relational algebra, therefore, requires a means to deal with n-ary functions. One could try to work with a huge relation algebra that contains with the basic sets also *all* the product sets. It seems much simpler to demand that *sometimes* - and hopefully in those cases where we need one -- a product will exist in the relation algebra. Later in Chapt. 4, a fork operator is defined as a fundamental operation, yielding another approach.

Definition 3.6.1 We call two relations π and ρ of a heterogeneous relation algebra a *direct product* if the following conditions hold:

i)	$\pi^{\smile};\pi = \mathbb{I}$,	iii)	$\pi^{\smile};\rho = \top$,
ii)	$\rho^{\smile};\rho = \mathbb{I}$,	iv)	$\pi;\pi^{\smile} \sqcap \rho;\rho^{\smile} = \mathbb{I}$.

In particular, π, ρ are mappings; usually called *projections*. It would even suffice to require only $\pi^{\smile};\pi \sqsubseteq \mathbb{I}$; according to (3.2.9 iii) for multiplying a univalent relation as π from the right, it follows that $\pi^{\smile};\pi = \pi^{\smile};(\pi;\pi^{\smile} \sqcap \rho;\rho^{\smile});\pi = \mathbb{I} \sqcap \pi^{\smile};\rho;\rho^{\smile};\pi = \mathbb{I} \sqcap \top;\top = \mathbb{I}$.

Interpreting the condition $\pi;\pi^{\smile} \sqcap \rho;\rho^{\smile} \sqsubseteq \mathbb{I}$ in the case of two sets A, B and their Cartesian product $A \times B$, it ensures that for every $a \in A$ and $b \in B$ there is *at most* one pair c in $A \times B$ such that $\pi(c) = a$ and $\rho(c) = b$. In addition, "$= \mathbb{I}$" means that π, ρ are total, i.e., that there are no "unprojected pairs". Finally, the condition $\pi^{\smile};\rho = \mathbb{T}$ implies that for *every* element in A and *every* element in B there is a pair in $A \times B$.

The concept of projections and of some kind of product in connection with algebras of first-order logic and, especially, with relational algebras appears as early as 1946 with the work on projective algebra [Everett, Ulam 1946] and since 1951 by Tarski in the papers [Jónsson, Tarski 1951] and [Jónsson, Tarski 1952]. Later on, de Roever [de Roever, Jr. 1972], Schmidt [Schmidt 1977], Schmidt and Ströhlein [Schmidt, Ströhlein 1989], Zierer [Zierer 1988], Berghammer and Zierer [Berghammer, Zierer 1986], and Backhouse et al. [Backhouse, de Bruin+ 1991a], introduced the product and projections either as data types or as operations.

For the Cartesian product to be uniquely determined, we need to show that there can be no two essentially distinct pairs of natural projections π, ρ and π', ρ'. Being not essentially distinct means that they are equal *up to isomorphism*. We leave the question open whether projections π, ρ do at all exist – in a given heterogeneous relation algebra they need not. Assuming that two different direct products are given with isomorphic component sets, we now establish an isomorphism between them.

Theorem 3.6.2 (*Monomorphic characterization of direct products*). Let direct products (π, ρ) and (π', ρ') and bijective mappings Φ, Ψ be given. Whenever the products $\pi;\Phi;\pi'^{\smile}$, $\rho;\Psi;\rho'^{\smile}$ are defined, then $\Xi := \pi;\Phi;\pi'^{\smile} \sqcap \rho;\Psi;\rho'^{\smile}$ is a bijective mapping satisfying $\pi;\Phi = \Xi;\pi'$ and $\rho;\Psi = \Xi;\rho'$. Therefore, Ξ gives rise to an isomorphism with respect to the direct product structure: The pair Φ, Ξ is an isomorphism from π to π' and Ψ, Ξ from ρ to ρ', respectively.

Proof: Using (3.2.9 iii) and univalence of π', we obtain

$$\Xi;\pi' = (\pi;\Phi;\pi'^{\smile} \sqcap \rho;\Psi;\rho'^{\smile});\pi' = \pi;\Phi \sqcap \rho;\Psi;\rho'^{\smile};\pi' = \pi;\Phi \sqcap \rho;\Psi;\mathbb{T} = \pi;\Phi.$$

Other isomorphism formulæ for π and ρ may be derived similarly. We now show that Ξ is a mapping. Since π' and ρ' are mappings, we can use (3.2.9 vi):

$$\begin{aligned}\Xi;\overline{\mathbb{I}} = \Xi;\overline{\pi';\pi'^{\smile} \sqcap \rho';\rho'^{\smile}} &= \Xi;(\pi';\overline{\pi'^{\smile}} \sqcup \rho';\overline{\rho'^{\smile}}) = \Xi;\pi'\overline{\pi'^{\smile}} \sqcup \Xi\rho';\overline{\rho'^{\smile}} \\ &= \pi;\Phi;\overline{\pi'^{\smile}} \sqcup \rho;\Psi;\overline{\rho'^{\smile}} = \overline{\pi;\Phi;\pi'^{\smile} \sqcap \rho;\Psi;\rho'^{\smile}} = \overline{\Xi}.\end{aligned}$$

Due to symmetry, $\Xi^{\smile}$ is a mapping and Ξ a bijective mapping. □

In the subcategory of mappings the direct product establishes a product in the categorical sense. This observation leads to another proof of the proposition above.

Further details and applications on products in relation algebras can be found in [Schmidt, Ströhlein 1989] and [Zierer 1988].

3.7 Direct power

Every concrete relation between sets A and B can be transformed to a mapping between A and the powerset $\mathcal{P}(B)$. Such a powerset is determined by the "is element relation" $\varepsilon : B \leftrightarrow \mathcal{P}(B)$. Using the relation ε the required mapping $f : A \leftrightarrow \mathcal{P}(B)$ corresponding to the relation $R : A \leftrightarrow B$ can be defined as the greatest solution of $R^{\smile};f \sqsubseteq \varepsilon$ and $f;\varepsilon^{\smile} \sqsubseteq R$. Following the definitions of a left and right residual in Chapt. 1 we compute this relation as the *symmetric quotient* of $R^{\smile}$ and ε. A symmetric quotient is defined by $\mathrm{syq}(Q,S) := Q\backslash S \sqcap Q^{\smile}/S^{\smile}$. Some basic properties of the symmetric quotient are given in the next theorem.

Theorem 3.7.1 i) $\mathrm{syq}(R,S)^{\smile} = \mathrm{syq}(S,R)$,
ii) $\mathrm{syq}(\overline{R},\overline{S}) = \mathrm{syq}(R,S)$,
iii) $\mathbb{I} \sqsubseteq \mathrm{syq}(R,R)$,
iv) $R;\mathrm{syq}(R,S) = S \sqcap \mathbb{T};\mathrm{syq}(R,S)$,
v) $\mathrm{syq}(Q,R);\mathrm{syq}(R,S) = \mathrm{syq}(Q,S) \sqcap \mathrm{syq}(Q,R);\mathbb{T}$,
vi) $\mathrm{syq}(R,S) \sqsubseteq \mathrm{syq}(Q;R,Q;S)$, for arbitrary Q. □

The proof of the theorem can be found in [Schmidt, Ströhlein 1993]. Using the symmetric quotient we can now define the direct power.

Definition 3.7.2 A relation ε is called a *direct power* if it satisfies the following properties:
i) $\mathrm{syq}(\varepsilon,\varepsilon) \sqsubseteq \mathbb{I}$,
ii) $\mathrm{syq}(R,\varepsilon)$ is total for every relation R with $\mathsf{source}\ R = \mathsf{source}\ \varepsilon$.

The first condition implies that $\mathrm{syq}(R,\varepsilon)$ is univalent for all R because

$$\mathrm{syq}(R,\varepsilon)^{\smile};\mathrm{syq}(R,\varepsilon) = \mathrm{syq}(\varepsilon,R);\mathrm{syq}(R,\varepsilon) \sqsubseteq \mathrm{syq}(\varepsilon,\varepsilon) \sqsubseteq \mathbb{I}.$$

Together with the second condition $\mathrm{syq}(R,\varepsilon) : A \leftrightarrow \mathcal{P}(B)$ is a mapping for every relation $R : B \leftrightarrow A$.

Cantor showed that the collection of all sets cannot be a set. In an abstract relation algebra this fact can be re-proved by considering a homogeneous relation algebra containing an ε relation.

Theorem 3.7.3 A one-object relation algebra (i.e. a homogeneous relation algebra) with a direct power does not exist.

Proof: In a one-object relation algebra we are allowed to consider the relation $\mathbb{I} \sqcap \varepsilon$. Now, we may recreate Cantor's proof by defining $R := \perp\!\!\!\perp/(\mathbb{I} \sqcap \varepsilon)^{\smile}$. This relation is characterized by

$$S \sqsubseteq R \Longleftrightarrow S;(\mathbb{I} \sqcap \varepsilon)^{\smile} = \perp\!\!\!\perp \qquad (*)$$

which follows directly from the definition of the left residual. Let us define $T := \mathrm{syq}(R^{\smile},\varepsilon)$. As mentioned before, this relation is total. From Theorem 3.7.1 we further conclude $T;\varepsilon^{\smile} = (\varepsilon;\mathrm{syq}(\varepsilon,R^{\smile}))^{\smile} \sqsubseteq R$ and $R^{\smile};T = R^{\smile};\mathrm{syq}(R^{\smile},\varepsilon) \sqsubseteq \varepsilon$. As partial identities are idempotent, we have

$$T;(\mathbb{I} \sqcap \varepsilon)^{\smile} = T;(\mathbb{I} \sqcap \varepsilon)^{\smile};(\mathbb{I} \sqcap \varepsilon)^{\smile} \sqsubseteq T;\varepsilon^{\smile};(\mathbb{I} \sqcap \varepsilon)^{\smile} \sqsubseteq R;(\mathbb{I} \sqcap \varepsilon)^{\smile} = \perp\!\!\!\perp$$

from which we can conclude $T \sqsubseteq R$ by $(*)$. The inclusion chain

$$T \sqsubseteq T;(\mathbb{I} \sqcap T^{\smile};T)^{\smile} \sqsubseteq R;(\mathbb{I} \sqcap R^{\smile};T)^{\smile} \sqsubseteq R;(\mathbb{I} \sqcap \varepsilon)^{\smile} = \perp\!\!\!\perp$$

leads by the totality of T to $\mathbb{I} \sqsubseteq T;T^{\smile} = \perp\!\!\!\perp$. On the other hand the Tarski-rule implies by $\mathbb{T};\mathbb{I};\mathbb{T} = \mathbb{T}$ that $\mathbb{I} \neq \perp\!\!\!\perp$, which is a contradiction. □

3.8 Conclusion

Motivated by concrete relations between two different sets, by boolean $(m \times n)$-matrices, and last but not least by category theory, we introduced the concept of heterogeneous relation algebra. It differs from the homogeneous case by the relations having explicit types and is therefore an extension. All the rules can be taken over from the homogeneous relation algebra, but sometimes not with the most general typing and the proofs may not be the same when relations are typed heterogeneously. We have shown that typed relations are very often more natural, as for example in the definition of the ε-relation.

Facilitated by the small set of axioms, we developed the graphical interactive proof assistant RALF (**r**elation **al**gebraic **f**ormula manipulation system) to support proving relation algebraic theorems (see [Hattensperger, Berghammer$^+$ 1993]). It allows (only) mathematically correct modifications of subexpressions of relational formulæ according to a given set of axioms by taking into account the context in which the expression occurs. The user selects the subexpression and one of all correct replacements by mouse click and the proof can then be pretty-printed in LaTeX. Thus relation algebraic proofs can be done in a very natural and user-friendly way similar to "working with paper and pencil".

Chapter 4

Fork Algebras

Armando Haeberer, Marcelo Frias, Gabriel Baum, Paulo Veloso

In this chapter we present the class of fork algebras, an extension of relation algebras with an extra operator called *fork*. We will present results relating fork algebras both to logic and to computer science. The interpretability of first-order theories as equational theories in fork algebras will provide a tool for expressing program specifications as fork algebra equations. Furthermore, the finite axiomatizability of this class of algebras will be shown to have deep influence in the process of program development within a relational calculus based on fork algebras.

4.1 Motivation

Let us consider a simple and well-known problem such as testing whether a given string is a palindrome. This problem is known to be non-computable by a finite automaton. It can, however, be computed by a non-deterministic push-down automaton, since push-down automata are capable of storing a copy of their data. Therefore, we can suggest that programs that solve the palindrome problem must be able to duplicate data. A direct solution for our problem could be: making two copies of the input list, reversing one of them, while leaving the other untouched, and, finally, comparing both results. In order to do so, it would be convenient to have a copying operation.

In a relational framework, the desired copying operation should be regarded as a relation, receiving an object x as input, and two copies of x as output. But what do we mean by "two copies of x"? We do not want a ternary relation, for we wish to retain the input-output character of programs. One way of making the best of both worlds, so to speak, is to turn the output into a pair of form $[x, x]$. Thus, we will have the following copying relation – which will be denoted by 2 –, described by

$$2 = \{(u, [u, u]) : u \in U\}$$

where U is the underlying universe of our relations.

In order to obtain such pairs as outputs, the domain underlying our relations must be closed under the pairing operation $[\_,\_]$, having a structured universe rather than a mere set of points. The structured character of our universe U renders natural the introduction of some *structural operations* on binary relations, namely *fork* and *direct product*.

We call *fork of relations R and S* the relation

$$R \underline{\nabla} S = \{(u, [v, w]) : uRv \wedge uSw\}, \tag{4.1}$$

and define the *direct product of relations R and S* as

$$R \otimes S = \{([u, v], [w, z]) : uRw \wedge vSz\}.$$

These new operations enable us to describe some interesting behavior, especially when seen from the programming point of view. In particular, if we recall that I_U is the identity relation on the set U, we can now define our copying relation by

$$2 = I_U \underline{\nabla} I_U.$$

Furthermore, since we now have structured objects, it is natural to have operations to "unpack" them, i.e., relations π and ρ satisfying

$$\pi = \{([u, v], u) : u,\ v \in U\} \qquad \text{and} \qquad \rho = \{([u, v], v) : u,\ v \in U\}$$

If we recall that V is the greatest relation, these *projections*, which decompose objects into their components, are convenient shorthand for some special relations, since they can be defined as

$$\pi = (I_U \underline{\nabla} V)^{\smile} \qquad \text{and} \qquad \rho = (V \underline{\nabla} I_U)^{\smile}. \tag{4.2}$$

It also happens that the converse of (4.2) holds true, in fact, fork can also be defined from projections [Berghammer, Haeberer+ 1993]. It is a school exercise to show that

$$R \underline{\nabla} S = (R;\pi^{\smile}) \cap (S;\rho^{\smile}). \tag{4.3}$$

This result implies that projections could have been chosen, instead of fork, as our primitive operations. While for a category theoretician this approach could seem natural, from the programming viewpoint, fork arises as more natural, in much the same way that data types are defined from constructors, and not from their observers.

4.2 Proper and abstract fork algebras

A second reading of (4.1) shows a notational difference between pairs within relations – between parentheses – and pairs produced by the pairing operation [_, _]. Furthermore, there are conceptual differences between both kinds of objects. While the meaning of pairs (a, b) is the usual one, pairs $[a, b]$ may be different, for their sole purpose is storing data. Since set theoretical pairs fulfill this requirement, they could be natural candidates for replacing [_, _]. However, we are looking for a programming calculus, so we need an abstract counterpart of fork which is suitable for symbolic manipulations. Thus, a finite characterization of fork is desirable. When choosing set theoretical pairs for [_, _], we come to a blind alley, for, as Németi, Sain, Mikulás, and Simon [Mikulás, Sain+ 1992; Sain, Németi 1994] have proved, this class of algebras is not finitely axiomatizable.

From (4.3), it follows that relation algebras containing projection elements allow a definition of fork. One example of such class of algebras is quasiprojective relation algebras [Tarski, Givant 1987]. It happens that there are quasiprojective relation algebras on which it is not possible to define a useful pairing operation $[\_,\_]$, for the functionality of such operation cannot be guaranteed.

Fork algebras are extensions of relation algebras by a new binary operator called *fork*. In their original version, proper fork algebras were made of *strings*, i.e., the operation $[\_,\_]$ was interpreted as string concatenation [Veloso, Haeberer 1991]. The definition of projections becomes clumsy under these circumstances, since a projection is required for each position in the strings. Moreover, concatenation is not injective, and thus not useful for coding pairs. As a second-stage improvement, proper fork algebras were made of *trees* of finite height [Baum, Haeberer$^+$ 1992]. In this case, projections can be easily defined, but the class of algebras as a whole is then non finitely axiomatizable, for an isomorphism can be defined between finite trees and set-theoretical ordered pairs. In the current version of fork algebras, all that is required is that $[\_,\_]$ be injective. This characterization of fork algebras allows – as will be shown in Theorem 4.3.15 – finite axiomatizability, and it suffices for obtaining a relational programming calculus as the one to be described in Sect. 4.5.

In the rest of this section we introduce some basic definitions concerning fork algebras. We shall be interested in both proper[1] (or standard) and abstract versions of fork algebras. Much as Boolean algebras provide an abstract theory of set-theoretical operations, fork algebras provide an abstract theory of operations on structured binary relations.

In order to define the class of proper fork algebras (PFAs), we will first define the class of $\star$PFAs by

Definition 4.2.1 A $\star$PFA is a two-sorted structure with domains A and U

$$(A, U, \cup, \overline{}, \emptyset, ;, \breve{}, I_U, \underline{\nabla}, \star)$$

such that

- $(A, \cup, \overline{}, \emptyset, ;, \breve{}, I_U)$ is an algebra of concrete relations with global supremum V,
- $\star : U \times U \to U$ is an injective function when restricted to V,
- $R \underline{\nabla} S = \{(x, \star(y, z)) : x\, R\, y \text{ and } x\, S\, z\}$.

Definition 4.2.2 The class of PFAs is defined as **Rd** $\star$PFA, where **Rd** takes reducts of similarity type $(\cup, \overline{}, \emptyset, ;, \breve{}, I_U, \underline{\nabla})$, i.e., while the universe keeps unchanged, the type U and the operation $\star$ are eliminated from the similarity type.

In Defs. 4.2.1 and 4.2.2, the function $\star$ performs the roles of pairing and encoding of pairs of objects into single objects. As we have mentioned previously, there may be $\star$ functions which are far from being set-theoretical pair formation.

[1]It is important to remark that proper fork algebras are quasi-concrete structures, since, as was pointed out by Andréka and Németi in a private communication, concrete structures must be fully characterized by their underlying domain, which does not happen with proper fork algebras because of the (hidden) operation $\star$.

For instance, let us consider a PFA whose universe U is the set $\{a\}$. Let us take as V the equivalence relation $\{(a,a)\}$. If we define the function $\star : V \to U$ by

$$\star(a,a) = a,$$

it follows that $\star$ is an injective function. We will also see that $(\{a\},\star)$ cannot be isomorphic to a structure $(A, (\_,\_))$ for any set A. If such a structure exists, let us have $i : \{a\} \to A$, establishing an isomorphism between both structures. Since i is a homomorphism, it must be $i(\star(a,a)) = (i(a), i(a))$. From the definition of $\star$, we conclude $i(a) = (i(a), i(a))$, which, accordingly to the set theoretical definition of $(\_,\_)$, is forbidden by the regularity axiom.

If by FullPFA we denote the subclass of proper fork algebras with universe $\mathcal{P}(U \times U)$ for some set U, a first characterization of the algebraic structure of PFAs is given by the following theorem, whose proof follows easily from [Jónsson, Tarski 1951][2].

Theorem 4.2.3 PFA = **SP**FullPFA.

Since first-order definable classes are closed under isomorphism, i.e. the language of first-order logic does not allow the distinction between isomorphic structures, a further treatment of the class of PFAs is necessary in order to obtain finite axiomatizability.

Definition 4.2.4 We define the class of representable fork algebras (RFAs), the candidate for finite axiomatization, as **I** PFA.

In the following definition we will present the class of abstract fork algebras (AFAs for short). We will show in Theorem 4.3.15 that this class provides an abstract characterization of RFAs (and thus of PFAs).

Definition 4.2.5 An abstract fork algebra is an algebraic structure

$$(R, \sqcup, \overline{}, \perp\!\!\!\perp, ;, \breve{}, \mathbb{I}, \nabla)$$

satisfying the following axioms:

1. Axioms stating that $(R, \sqcup, \overline{}, \perp\!\!\!\perp, ;, \breve{}, \mathbb{I})$ is a relation algebra,
2. $R \nabla S = (R;(\mathbb{I} \nabla \mathbb{T})) \sqcap (S;(\mathbb{T} \nabla \mathbb{I}))$,
3. $(R \nabla S);(T \nabla Q)\breve{} = (R;T\breve{}) \sqcap (S;Q\breve{})$,
4. $(\mathbb{I} \nabla \mathbb{T})\breve{} \nabla (\mathbb{T} \nabla \mathbb{I})\breve{} \sqsubseteq \mathbb{I}$.

As a definitional extension of AFAs, we introduce relations π and ρ by

$$\pi = (\mathbb{I} \nabla \mathbb{T})\breve{} \quad \text{and} \quad \rho = (\mathbb{T} \nabla \mathbb{I})\breve{}.$$

In Theorem 4.3.10 we will show that π and ρ characterize functional relations. Furthermore, it will be shown in Theorem 4.3.15 that axioms 2 to 4 in Definition 4.2.5 induce the existence of an injective function $\star$ satisfying the relationship

$$\pi(\star(x,y)) = x \quad \text{and} \quad \rho(\star(x,y)) = y.$$

[2]Along the next theorems, by **S** we denote the operation of taking subalgebras of a given class of algebras. **P** takes direct products of algebras in a given class, and **I** takes isomorphic copies.

4.3 Algebraic and metalogical properties of fork algebras

It is well known that relation algebras have weak expressive power, since their logical counterpart - denoted by $\mathcal{L}^\times$ in [Tarski, Givant 1987] (much of which is summarised in Sect. 2.8) - is equipollent in means of expression and proof with a three-variable fragment of first-order predicate logic (see [Tarski, Givant 1987, pp. 76–89] for a detailed proof of this). It is shown in [Tarski, Givant 1987], for instance, that the formula

$$\forall x, y, z\, (\exists u\, (\neg x = u \ \wedge \ \neg y = u \ \wedge \ \neg z = u))^3$$

from Tarski's elementary theory of binary relations [Tarski 1941] cannot be expressed by any formula from Tarski's calculus of binary relations. Such lack of expressiveness has an impact in programming, for first-order specifications of programs are not likely to have faithful translations into the calculus of relations. Thus, some modification is necessary. If we examine the formula above and analyze the reasons why it is not expressible, we notice that some information which should be kept about the individual variables is lost. In order to keep that information, an attempt was made at having relations over a structured universe, instead of relations on a plain set. As a consequence of considering a structured universe, this extension of relation algebras has a greater expressive power than that of relation algebras.

The construction of programs from specifications requires a specification language, a programming language, and a set of rules that allow us to obtain programs from specifications. The framework we propose in Sect. 4.5 uses first-order logic with equality as its specification language, and the program construction process is carried out within the framework of fork algebras. We immediately notice that both settings are incompatible from the point of view of their languages, hence some theoretical results are required to bridge the gap between them.

In [Tarski, Givant 1987, 2.4 and 2.5], Tarski and Givant present a way to prove such relationship, when they define the notion of *equipollence of two formalisms relative to a common equipollent extension.*

Definition 4.3.1 We say that two formalisms F_1 and F_2 (F_2 an extension of F_1) are

equipollent in means of expression if, for every sentence X of F_1, there exists a sentence Y of F_2 such that X and Y are equivalent in the formalism F_2, and for every sentence Y of F_2 there exists a sentence X of F_1 such that X and Y are equivalent in the formalism F_2,

equipollent in means of proof if, for every set of sentences Ψ of F_1 and for every sentence X of F_1 we deduce X from Ψ in F_2 iff we deduce X from Ψ in F_1.

Since equipollence is defined between a formalism and an extension of it, for the task of comparing two formalisms where none of them is an extension of the other, a further treatment is required. In order to overcome this difficulty, we will introduce the notion of *equipollence of two formal systems relative to a common extension.*

[3]Notice that this formula has four variables ranging over individuals.

Definition 4.3.2 We say that two formalisms F_1 and F_2 are

equipollent in means of expression relative to a formalism F_3 if F_3 is a common extension of F_1 and F_2 (i.e., the sets of sentences of F_1 and F_2 are subsets of the set of sentences of F_3) which is equipollent, in means of expression, with each one of them.

equipollent in means of proof relative to a formalism F_3 if F_3 is a common extension of F_1 and F_2 that is equipollent with each one of them in means of proof.

Once the notion of equipollence is defined, Theorem 4.3.6 below establishes a close relationship between the equational theory of fork algebras and a large class of theories of first-order logic with equality [Frias, Baum 1995]. This class of first-order theories will be of particular interest, since first-order logic with equality provides an adequate basis for program specification.

Let us call $\mathcal{L}^\star$ the first-order logic of one binary injective function, i.e., the first-order logic language has only one non logical constant which is a binary function symbol standing for an injective function.

Let us call $\mathcal{L}^\nabla = (S, \vdash_\nabla)$ the formal system whose set of formulas S is the set of equations in the language of fork algebras, and inferences are performed by replacing equals by equals according to the axioms of abstract fork algebras.

Let us call $\mathcal{L}^+ = (S, \vdash_+)$ the extension of first-order logic whose set of formulas S is defined by:

atomic formulas: $t_1 A t_2$ where A is a fork algebra term and t_1 and t_2 are terms built from variables and the binary function symbol $\star$, or equations $A = B$ where both A and B are fork algebra terms.

formulas: Atomic formulas are formulas, and first-order formulas built up from atomic formulas are formulas.

Besides the logical axioms, we add the following extralogical ones:

A sentence stating that $\star$ is injective, $\forall x, y\,(x(A \sqcup B)y \leftrightarrow xAy \vee xBy)$,
$\forall x, y\,(x(A \sqcap B)y \leftrightarrow xAy \wedge xBy)$, $\forall x, y\,(xA;By \leftrightarrow \exists z\,(xAz \wedge zBy))$,
$\forall x, y\,(xA\breve{}\,y \leftrightarrow yAx)$, $\forall x, y\,(x\overline{A}y \leftrightarrow \neg xAy)$,
$\forall x, y\,(x \perp\!\!\!\perp y \leftrightarrow \mathsf{false})$, $\forall x, y\,(x \mathbb{T} y \leftrightarrow \mathsf{true})$,

$\forall x, y\,(x\mathbb{I}y \leftrightarrow x = y)$,
$\forall x, y\,(xA\nabla By \leftrightarrow \exists u, v\,(y = \star(u, v) \wedge xAu \wedge xBv))$.

Once we have defined the three formalisms, we will show some theorems relating them.

Theorem 4.3.3 $\mathcal{L}^+$ is a definitional extension of $\mathcal{L}^\star$, and thus, from [Tarski, Givant 1987] it follows that $\mathcal{L}^+$ and $\mathcal{L}^\star$ are equipollent in means of expression and proof.

Theorem 4.3.3 can be generalized to a language with predicate and function symbols besides the function symbol $\star$. If we simultaneously add new symbols $S_1, S_2, \ldots, S_n, \ldots$ to $\mathcal{L}^\star$, and the same amount of binary relation symbols $R_1, R_2, \ldots, R_n, \ldots$ in $\mathcal{L}^+$, the following result holds:

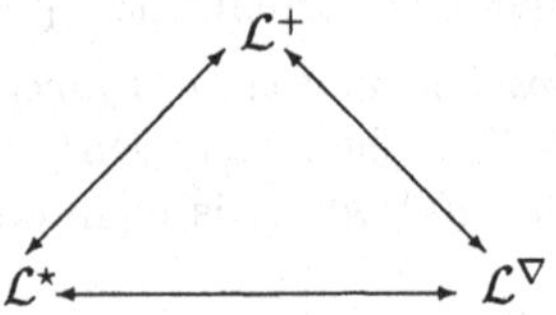

Fig. 4.1 Equipollence of $\mathcal{L}^\star$ and $\mathcal{L}^\nabla$ relative to $\mathcal{L}^+$.

Theorem 4.3.4 Given a formula $\varphi(x_1, \ldots, x_k)$ in $\mathcal{L}^\star$, there exists a fork term t_φ made from the relational constants $R_1, R_2, \ldots, R_n, \ldots$ such that

$$\vdash \exists x_1, \ldots, x_k\, (\varphi(x_1, \ldots, x_k)) \iff \vdash_+ \exists x_1, \ldots, x_k\, (\star(x_1, \star(x_2, \ldots)) t_\varphi \star(x_1, \star(x_2, \ldots))).$$

Theorem 4.3.4 is essential in order to show the adequacy of fork algebras as a framework suitable for program specification, since it shows that first-order specifications can be faithfully reflected by means of fork algebra terms.

Theorem 4.3.5 The formalisms $\mathcal{L}^+$ and $\mathcal{L}^\nabla$ are equipollent in means of expression and proof. As a consequence of the equipollence in means of proof, given a formula $\varphi(x_1, \ldots, x_k)$ in $\mathcal{L}^+$ there exists a fork algebra term t_φ such that

$$\vdash_+ \forall x_1, \ldots, x_k\, (\varphi(x_1, \ldots, x_k)) \qquad \iff \qquad \vdash_\nabla t_\varphi = \mathbb{T}.$$

Joining Theorems 4.3.3 and 4.3.5, we can prove Theorem 4.3.6.

Theorem 4.3.6 $\mathcal{L}^\star$ and $\mathcal{L}^\nabla$ are equipollent in means of expression and proof relative to their common extension $\mathcal{L}^+$ [Frias, Baum 1995].

The equipollence in means of expression stated in Theorem 4.3.6 establishes that every first-order formula has a relational counterpart in fork algebras. The equipollence in means of proof says that consequences of first-order specifications are also captured in fork algebras, and thus, when a property follows from a specification, it can be proved by using only equational reasoning in fork algebras.

Our proof of Theorem 4.3.5 (and thus of Theorem 4.3.6), uses a translation mapping from first-order logic into fork algebra terms, which yields a constructive way of finding the term t_φ. Its definition proceeds in two steps, since first-order terms, as well as formulas, also need to be translated.

Given a first-order language whose non-logical symbols are: constant symbols $c_1, \ldots, c_i$, function symbols $f_1, \ldots, f_j$ and predicate symbols $p_1, \ldots, p_k$, we will work over an extension of the language of fork algebras over the language $C_1, \ldots, C_i, F_1, \ldots, F_j, P_1, \ldots, P_k$ (where the intentional semantics of each symbol, will now be a binary relation). The mappings δ and τ translating respectively first-order terms and formulas are defined by:

Definition 4.3.7
$\delta(v_n) = \mathbb{I}, \qquad \delta(c_n) = C_n,$
$\delta(f_n(t_1[x_{1,1}, \ldots, x_{1,i_1}], \ldots, t_m[x_{m,1}, \ldots, x_{m,i_m}])) = (\Pi_1;\delta(t_1) \nabla \cdots \nabla \Pi_m;\delta(t_m));F_n.$

Definition 4.3.8

$\tau(t_1[x_{1,1},\ldots,x_{1,i}] \equiv t_2[x_{2,1},\ldots,x_{2,j}]) = ((\Pi_1;\delta(t_1))\,\nabla\,(\Pi_2;\delta(t_2)));2^{\smile};\mathbb{T}$,
$\tau(p_n(t_1[x_{1,1},\ldots,x_{1,i_1}],\ldots,t_m[x_{m,1},\ldots,x_{m,i_m}])) = (\Pi_1;\delta(t_1)\,\nabla\cdots\nabla\,\Pi_m;\delta(t_m));P_n$,
$\tau(\neg\alpha[x_1,\ldots,x_n]) = \overline{\tau(\alpha)}$,
$\tau(\alpha[x_{1,1},\ldots,x_{1,i}] \vee \beta[x_{2,1},\ldots,x_{2,j}]) = \Pi_1;\tau(\alpha) \sqcup \Pi_2;\tau(\beta)$,
$\tau(\alpha[x_{1,1},\ldots,x_{1,i}] \wedge \beta[x_{2,1},\ldots,x_{2,j}]) = \Pi_1;\tau(\alpha) \sqcap \Pi_2;\tau(\beta)$,
$\tau(\exists x\,(\alpha[x,x_1,\ldots,x_n])) = \Pi_1^{\smile};\tau(\alpha)$,
$\tau(\forall x\,(\alpha[x,x_1,\ldots,x_n])) = \Pi_1\backslash\tau(\alpha)$.

In the definitions 4.3.7 and 4.3.8, the symbols Π_i stand for terms based on the relations π and ρ, and are meant to project the values for the variables in each subformula. In order to clarify this point, let us consider the following example.

Example 4.3.9 Let us have some finite axiomatization of the theory of lists, i.e., a finite set of first-order formulas T characterizing the relevant properties of lists. Let φ be any first-order definable property deducible from T. Then the equipollence in means of proof implies that

$$\{\tau(t) = \mathbb{T} : t \in T\} \vdash_{\nabla} \tau(\varphi) = \mathbb{T}.$$

As an example let us consider the formula

$$\varphi = \forall L_1 : \mathsf{List}, L_2 : \mathsf{List}\,(\mathit{Sort}(\mathit{Conc}(L_1, L_2)) = \mathit{Sort}(\mathit{Conc}(L_2, L_1))) \;,$$

where *Sort* is a function that orders the elements of a list, and *Conc* concatenates two lists, the one in the first argument before the one in the second. If we apply the mapping τ to the formula φ, we obtain

$$\tau(\varphi) = \pi\backslash\left(\pi\backslash\left(\begin{array}{c}((\pi;\rho)\nabla\rho);\mathit{Conc};\mathit{Sort}\\ \nabla\\ (\rho\nabla(\pi;\rho));\mathit{Conc};\mathit{Sort}\end{array};2^{\smile};\mathbb{T}\right)\right).$$

The question of whether or not PFAs are finitely axiomatizable, was immediately raised, since frameworks with a few simple rules are avidly sought for for programming. Moreover, such finite axiomatization should have variables ranging only over relations, in the same way that most functional calculi have variables ranging only over functions [Bird 1990], and thus allowing a clear expression of their properties.

Fork algebras' expressiveness theorems establish that the specifications and the properties of the application domain, which may be expressed in first-order logic, can also be expressed in the *equational* theory of fork algebras. However, this expressibility is insufficient for one to formulate, within the theory, many of the fundamental aspects of the program construction process. As we have already remarked, program constructing calculations require more than the possibility to express the specification of requirements; it is necessary to be able to check their correctness and termination, supply general rules, strategies and heuristics, and demonstrate their validity.

The finitization theorem [Frias, Haeberer$^+$ 1995b; Frias, Baum$^+$ 1995; Frias, Haeberer$^+$ 1995a], to be presented in Theorem 4.3.15, provides important arguments for overcoming the limitations of the equational theory of fork algebras. In order to prove the main finitization result, we will firstly present some theorems.

Theorem 4.3.10 The objects $(\mathbb{I}\nabla\mathbb{T})^{\smile}$ and $(\mathbb{T}\nabla\mathbb{I})^{\smile}$ are functional. Moreover, the equation $(\mathbb{I}\nabla\mathbb{T})^{\smile\smile};(\mathbb{T}\nabla\mathbb{I})^{\smile} = \mathbb{T}$ holds in every AFA.

This theorem establishes that the relations $(\mathbb{I}\nabla\mathbb{T})^{\smile}$ and $(\mathbb{T}\nabla\mathbb{I})^{\smile}$ satisfy the definition of quasiprojections as given in [Tarski, Givant 1987, p. 96]. We will call $(\mathbb{I}\nabla\mathbb{T})^{\smile}$ the first quasiprojection and $(\mathbb{T}\nabla\mathbb{I})^{\smile}$ the second quasiprojection of an AFA. We will denote the first quasiprojection by π and the second by ρ. If we recall that quasiprojective relation algebras (QRA for short) are relation algebras with a couple of functional relations A and B satisfying $A^{\smile};B = \mathbb{T}$, from Theorem 4.3.10 we immediately obtain the following result.

Theorem 4.3.11 The relation algebra reduct of any AFA is a QRA.

If we want to show that the axioms characterizing AFAs offer an axiomatization of PFAs, we can proceed as follows. It is a standard exercise to show that all AFAs' axioms hold in the class of PFAs. Now, all we have to do is prove a representation theorem, asserting that every AFA is isomorphic to a PFA. Both results together guarantee that AFAs and RFAs are the same class of algebras.

By checking that the axioms characterizing AFAs are satisfied in any PFA, we establish the following theorem.

Theorem 4.3.12 RFA $\subseteq$ AFA.

Following with the proof of the finitization theorem, it remains to prove the other inclusion, i.e., every AFA is a PFA. This inclusion will be obtained as an instance of the following theorem.

In order to complete the proof of the finitization theorem, we must now prove the other inclusion, i.e., every AFA is a RFA. This inclusion will be obtained as an instance of the theorem below.

Theorem 4.3.13 Consider $A \in$ AFA with RA reduct A'. Given an RA homomorphism $\mathcal{H} : A' \to B$ into a proper relation algebra B, there is an expansion of B to $B^* \in$ PFA so that $\mathcal{H} : A \to B^*$ is an AFA homomorphism.

Proof: The algebra B consists of relations over a set U, with unit element $V \subseteq U \times U$.

Since the quasiprojections π, $\rho \in A'$, we have $\mathcal{H}(\pi)$, $\mathcal{H}(\rho) \in B$.

According to Theorem 4.3.10, A' satisfies

$$\pi^{\smile};\rho = \mathbb{T},\ \pi^{\smile};\pi \sqsubseteq \mathbb{I} \text{ and } \rho^{\smile};\rho \sqsubseteq \mathbb{I}.$$

So

$$\mathcal{H}(\pi)^{\smile};\mathcal{H}(\rho) = \mathcal{H}(\mathbb{T}) = V, \text{ and } \mathcal{H}(\pi) \text{ and } \mathcal{H}(\rho) \text{ are functional.}$$

We define the relation $F \subseteq (U \times U) \times U$ by

$$((a,b),c) \in F \iff (c,a) \in \mathcal{H}(\pi) \text{ and } (c,b) \in \mathcal{H}(\rho).$$

By making some easy calculations with concrete relations, we prove that $V \subseteq domF$, F is functional on V, and F is injective on V. Hence, the restriction of F to V renders a well defined injective function $\star : V \to U$ such that

$$\star(a,b) = c \iff (c,a) \in \mathcal{H}(\pi) \text{ and } (c,b) \in \mathcal{H}(\rho).$$

We expand B to a PFA B^* by

$$R \underline{\nabla} S = \{(x, \star(y,z)) : (x,y) \in R,\ (x,z) \in S, \text{ and } (y,z) \in V\}.$$

In order to show that $\underline{\nabla}$ is well defined, we leave the proving that

$$\mathcal{H}(R) \underline{\nabla} \mathcal{H}(S) = (\mathcal{H}(R) ; \mathcal{H}(\pi)^{\smile}) \sqcap (\mathcal{H}(S) ; \mathcal{H}(\rho)^{\smile}) \tag{4.4}$$

as an exercise.

We now wish to show that $\mathcal{H}$ is an AFA homomorphism from A into B^*, i.e., that $\mathcal{H}$ preserves fork.

Consider $R, S \in A$. By Ax. 2 in Definition 4.2.5 $R \nabla S = (R ; \pi^{\smile}) \sqcap (S ; \rho^{\smile})$. Therefore,

$$\mathcal{H}(R \nabla S) = \mathcal{H}((R ; \pi^{\smile}) \sqcap (S ; \rho^{\smile})).$$

Since $\mathcal{H} : A' \to B$ is an RA homomorphism, we have

$$\mathcal{H}((R ; \pi^{\smile}) \sqcap (S ; \rho^{\smile})) = (\mathcal{H}(R) ; \mathcal{H}(\pi)^{\smile}) \sqcap (\mathcal{H}(S) ; \mathcal{H}(\rho)^{\smile})$$

which by 4.4 equals $\mathcal{H}(R) \underline{\nabla} \mathcal{H}(S)$. □

Theorem 4.3.14 Every AFA is isomorphic to some PFA.

Proof: Given $A \in$ AFA, its RA reduct A' is a QRA by Theorem 4.3.11. Since, by [Tarski, Givant 1987, p. 242], every QRA is a representable relation algebra, we have an RA isomorphism $\mathcal{H} : A' \to B$ onto a proper relation algebra B.

By Theorem 4.3.13, B can be expanded to a PFA B^*, so that $\mathcal{H} : A \to B^*$ is an AFA isomorphism. □

Summarizing the results presented in theorems 4.3.12 and 4.3.14, we prove the main result concerning the finitization problem.

Theorem 4.3.15 AFA = RFA.

Proof: By Theorem 4.3.14, AFA$\subseteq$ **I** PFA= RFA, and by Theorem 4.3.12, RFA $\subseteq$ AFA. □

Figure 4.2, inspired by the finitization theorem, shows that its *algebraic* portion (AFA $\overset{\cong}{\longleftrightarrow}$ PFA), which reflects the theorem's thesis, induces its *logic* portion (*Th*(AFA) $\overset{\equiv}{\longleftrightarrow}$ *Th*(PFA)), which reflects the elementary equivalence between both theories. Moreover, the arrow AFA $\longrightarrow$ *Th*(PFA) shows the soundness of our axiomatization, and the arrow PFA $\longrightarrow$ *Th*(AFA) reflects its completeness.

Theorem 4.3.16 *Th*(AFA) = *Th*(PFA), i.e., AFAs and PFAs are elementarily equivalent classes of algebras.

It is not difficult to show that the theory of AFAs is incomplete. This can be proved by taking two AFA's A_1 and A_2 such that $Th(A_1) \neq Th(A_2)$. Let us consider A_1 to be an atomless PFA, and A_2 one with atoms. The formula

$$\exists x\, (\emptyset \neq x \wedge \forall y\, (\emptyset \subseteq y \subseteq x \to y = x))$$

clearly holds for A_2, while it is false for A_1, thus stating the next theorem.

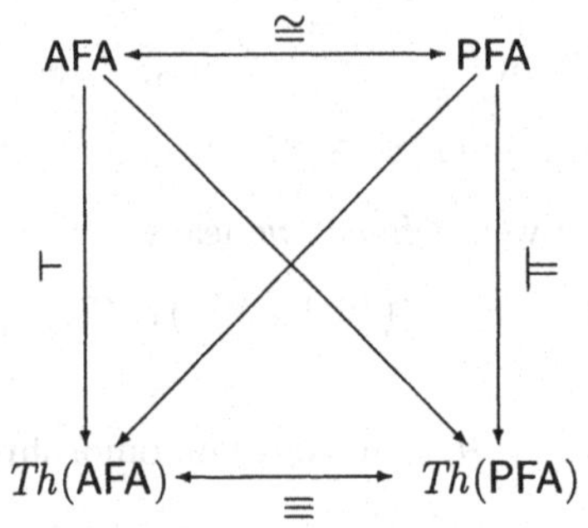

Fig. 4.2 Consequences of the finitization theorem.

Theorem 4.3.17 *Th*(AFA) is incomplete, i.e., there are sentences in *Th*(AFA) which are neither true nor false.

The next result we will present shows that the equational theory of AFAs is undecidable, i.e., there is no finitary method or algorithm informing wether or not a given equation is provable from the axioms of AFAs. From the programming calculus viewpoint, this result shows that program derivations require the use of heuristics and *rationales* (experiences from previous derivations) to be applied by a user with knowledge about the methodology.

Theorem 4.3.18 The equational and first-order theories of AFA's are undecidable.

Proof: Firstly, we will prove that the equational theory of AFA's is undecidable. As a consequence of the equipollence in means of proof between $\mathcal{L}^{\star}$ and $\mathcal{L}^{\nabla}$, for every sentence φ we have

$$\vdash \varphi \quad \Longleftrightarrow \quad \vdash_{\nabla} \tau(\varphi) = \mathbb{T}.$$

If there is a decision algorithm for the equational theory of AFAs, being the mapping τ recursive, the composition of both mappings would yield a decision method for the logic $\mathcal{L}^{\star}$, which is known to be undecidable. Therefore, no decision algorithm exists for the equational theory of AFA's.

Now, let us consider the first-order theory of AFAs. Since the testing wether a given AFA sentence is an equation is decidable, a decision method for the first-order theory of AFAs would yield a decision method for the equational theory of AFAs, which is not possible by the previous reasoning. □

In the rest of this section we shall demonstrate more results showing the close relationship between first-order logic with equality and AFAs. Unlike the previous results in this direction, we will use semantical arguments instead of proof theoretical ones [Frias, Haeberer+ 1995a].

Definition 4.3.19 By a (*proper*) *fork model* of a set T of AFA terms we mean a (proper) fork algebra $\mathcal{A}$ such that $\mathcal{A} \models t \neq \perp\!\!\!\perp$ for every $t \in T$.

We shall see that this notion of fork model is directly related to the notion of model in first-order logic. We know from Theorem 4.3.4 that every satisfiable first-order formula can be translated as an AFA term that admits its own fork model. Now we will see that the converse of this proposition also holds.

Theorem 4.3.20 If φ is a first-order formula s.t. $\tau(\varphi)$ has a fork model, then there is a first-order structure $\mathcal{U}$ and an assignment of values s to the free variables of φ such that $\mathcal{U} \models \varphi[s]$.

Proof: Let us suppose that for every structure $\mathcal{U}$ and every assignment s we have $\mathcal{U} \not\models \varphi[s]$. Hence, for each structure $\mathcal{U}$ and each assignment s, we have $\mathcal{U} \models \neg\varphi[s]$. Due to the completeness of first-order logic it follows that $\vdash \neg\varphi$. Therefore, as a consequence of the equipollence in means of proof between $\mathcal{L}^\star$ and $\mathcal{L}^\nabla$, we conclude that the equation $\tau(\neg\varphi) = \mathbb{T}$ is valid in AFA. Because of the definition of τ, we have $\overline{\tau(\varphi)} = \mathbb{T}$, and thus $\tau(\varphi) = \mathbb{L}$ contradicting the hypothesis of the theorem. □

It is known that one of the characteristics of classic first-order logic is the fact that it satisfies the compactness theorem and the Löwenheim-Skolem theorem. We will now show that relational versions of these theorems can be proved for fork algebras, thus showing that fork algebras provide an adequate framework for algebraization of first-order logic with equality [Veloso, Haeberer+ 1995].

Theorem 4.3.21 Let Γ be a possibly infinite set of first-order sentences. If every finite subset of $\tau_\Gamma \triangleq \{\tau(\gamma) : \gamma \in \Gamma\}$ has a proper fork model, then so does τ_Γ.

Proof: Let $\Gamma_0 \subseteq \Gamma$ be finite. Let us suppose $\{\tau(\gamma) : \gamma \in \Gamma_0\}$ has a fork model which we shall call $\mathcal{A}_{\Gamma_0}$. From Theorem 4.3.20, there is a first-order structure $\mathcal{U}_{\Gamma_0}$ such that $\mathcal{U}_{\Gamma_0} \models \Gamma_0$. Since the same holds for every finite subset of Γ, by first-order logic compactness there is a structure $\mathcal{U}_\Gamma$ such that $\mathcal{U}_\Gamma \models \Gamma$. Finally, from $\mathcal{U}_\Gamma$ we construct a proper fork algebra $\mathcal{A}_{\mathcal{U}_\Gamma}$ which is a proper fork model for $\{\tau(\gamma) : \gamma \in \Gamma\}$ [Frias, Haeberer+ 1995a]. □

Theorem 4.3.22 Let Γ be a set of sentences, s.t. $\{\tau(\gamma) : \gamma \in \Gamma\}$ has a fork model whose underlying set U is infinite. There is then a denumerable proper fork model for $\{\tau(\gamma) : \gamma \in \Gamma\}$.

Proof: Suppose $\{\tau(\gamma) : \gamma \in \Gamma\}$ has an infinite fork model $\mathcal{A}$. Then, by Theorem 4.3.20, there exists a structure $\mathcal{U}_\mathcal{A}$ s.t. $\mathcal{U}_\mathcal{A} \models \Gamma$. By first-order Löwenheim-Skolem, there exists a denumerable structure $\mathcal{U}_0$ s.t. $\mathcal{U}_0 \models \Gamma$. Finally, from $\mathcal{U}_0$ we construct a denumerable proper fork model for $\{\tau(\gamma) : \gamma \in \Gamma\}$ [Frias, Haeberer+ 1995a]. □

4.4 Arithmetic issues

In this section we present – without proof – several results which are extremely useful when proving new properties in the arithmetic of fork algebras. These results certainly will be very useful in the construction of algorithms from specifications, since arithmetical manipulation of fork terms is a crucial step in our methodology.

1. $(R \nabla S);\pi = R \sqcap (S;\mathbb{T})$,
2. $(R \nabla S);\rho = S \sqcap (R;\mathbb{T})$,
3. $((R \nabla S);\pi)\nabla((R \nabla S);\rho) = R \nabla S$,
4. $dom\,R = (\mathbb{I} \nabla R);\pi$,
5. $(R \nabla S);\pi = dom\,S\,;R$,
6. $(R \nabla S);\rho = dom\,R\,;S$,
7. $U \sqsubseteq R$ and $V \sqsubseteq S$ implies $U \nabla V \sqsubseteq R \nabla S$,
8. $R;(S \nabla T) \sqsubseteq (R;S)\nabla(R;T)$;
9. $2^{\smile};(\mathbb{T} \nabla \mathbb{T}) \sqsubseteq \mathbb{T} \otimes \mathbb{T}$,
10. If F is a functional relation, then $F;(R \nabla S) = (F;R)\nabla(F;S)$,
11. $dom(R \nabla S) = dom\,R \sqcap dom\,S$,
12. If $T = T;(\mathbb{I}\otimes\mathbb{I})$, $T;\pi \sqsubseteq (R \nabla S);\pi$ and $T;\rho \sqsubseteq (R \nabla S);\rho$ then $T \sqsubseteq R \nabla S$,
13. $2;\pi = 2;\rho = \mathbb{I}$,
14. $dom\,\pi = dom\,\rho = \mathbb{I}\otimes\mathbb{I}$,
15. 2 is a functional relation,
16. $2;(R \otimes S) = R \nabla S$,
17. 2 is an injective relation,
18. If $R \neq \mathbb{L}$, and $R \sqsubseteq \mathbb{T} \nabla \mathbb{T}$ then $R;\pi \neq \mathbb{L}$, and $r;\rho \neq \mathbb{L}$,
19. If $R \nabla S \neq \mathbb{L}$, then $R \neq \mathbb{L}$ and $S \neq \mathbb{L}$,
20. $(R \nabla S);(T \otimes Q) = (R;T)\nabla(S;Q)$,
21. $(R \otimes S);(U \otimes V) = (R;U)\otimes(S;V)$,
22. $(R \otimes S)^{\smile} = (R^{\smile} \otimes S^{\smile})$,
23. $ran(R \nabla S) \sqsubseteq ran(R \otimes S) = ran\,R \otimes ran\,S$,
24. $(R \nabla S)\sqcap(U \nabla V) = (R \sqcap U)\nabla(S \sqcap V)$,
25. $(R \otimes S)\sqcap(U \otimes V) = (R \sqcap U)\otimes(S \sqcap V)$,
26. $(R \otimes S)\sqcap\mathbb{I} = (R \otimes S)\sqcap(\mathbb{I}\otimes\mathbb{I})$,
27. If R, S, T, U are distinct of $\mathbb{L}$, and $R \otimes S = T \otimes U$, then $R = T$ and $S = U$,
28. $(R \nabla S);\pi;T = ((R;T)\nabla S);\pi$ and $(R \nabla S);\rho;T = (R \nabla(S;T));\rho$,
29. $dom\,R \sqsubseteq dom\,S \implies (R \nabla S);\pi = R$ and $dom\,S \sqsubseteq dom\,R \implies (R \nabla S);\rho = S$,
30. If $R \sqsubseteq S$ and $T \sqsubseteq U$ then $(R \otimes T) \sqsubseteq (S \otimes U)$.

4.5 A fork-algebraic programming calculus

A programming calculus can be viewed as a set of rules to obtain programs from specifications in a systematic way.

A very interesting and popular approach is the one based on functional programming languages [Bird 1990]. In these functional frameworks, specifications and programs are expressed in the same language, and transformation rules are defined in a suitable, frequently ad-hoc, metalanguage. A drawback of functional settings, however, is the lack of expressiveness of their specification languages, which are confined to functional expressions. Since these functional expressions can be viewed as programs (probably inefficient ones), in order to specify a problem in a functional framework, we must have in advance at least one algorithm that solves it.

On the other hand, relational calculi have a more expressive specification language (due to the existence of the converse and the complement of relations), allowing for more declarative specifications. However, choosing a relational framework does not guarantee that a calculus is totally adequate. These frameworks, as the one proposed by Möller in [Möller 1991b], despite having a powerful specification language, also have some methodological drawbacks. The process of program derivation seeks to use only abstract properties of relations, on which variables ranging over individuals (often called dummy variables) are avoided. Nevertheless, since there is no complete set of abstract rules capturing all the information of the relational (semantical) framework, the process goes back and forth between abstract and concrete properties of relations.

When using fork algebras as a programming calculus [Baum, Frias+ 1996; Frias, Aguayo 1994; Frias, Aguayo+ 1993; Frias, Gordillo 1995; Haeberer, Baum+ 1994; Haeberer, Veloso 1991], we have (as shown in Theorem 4.3.6), the expressiveness of first-order logic. Furthermore, as the arrow $\mathsf{PFA} \longrightarrow Th(\mathsf{AFA})$ from Fig. 4.2 shows, the axioms of AFA's provide a *complete* characterization of PFA's. These results enable us to use first-order logic as a specification language and certain fork algebra equations as programs, and to reason about the properties of specifications and programs within the theory. Moreover, Theorem 4.3.16, which establishes the elementary equivalence between $Th(\mathsf{AFA})$ and $Th(\mathsf{PFA})$, allows the formulation of strategies and heuristics of the program construction process in the shape of first-order formulas about relations.

A very popular strategy used in the solution of problems and in the design of programs is *Divide and Conquer* (*D&C*). Its basic idea is to decompose a problem into two or more subproblems and then combine their solutions to solve the original problem. Divide and Conquer has been extensively studied and several variants to it have been proposed, most of them regarding the type of subproblems into which the original problem is decomposed. Many of the best known algorithms used to solve classic programming problems (sorting, searching, etc.) stem from the application of recursive Divide and Conquer principle (i.e., the original problem itself becomes a subproblem for even simpler data). Some examples of this are QuickSort and MergeSort for the problem of sorting, and BinarySearch for the problem of searching. Actually, Divide and Conquer is a combination of two simpler strategies, namely, *Case Analysis* (*C_A*) and *Recomposition*. The former's idea is to express a problem as a sum of other problems, usually by decomposing its domain. Whereas the technique of Recomposition decomposes complex data into simpler pieces, where each piece will be processed by some subalgorithm. The results arising from those subalgorithms are later composed into a single output.

In the context of our theory, strategies can be expressed as first-order formulas over relations. In that way, for example, a formula that characterizes the application of Case Analysis will state that it is possible to decompose a problem into a number of subproblems $P_1, \ldots, P_k$ over mutually disjoint domains. Formally, the strategy can be represented by the formula

$$C\_A(P, P_1, \ldots, P_k) \leftrightarrow \bigwedge_{1 \le i < j \le k} dom P_i \sqcap dom P_j = \mathbb{L} \wedge P = \bigsqcup_{1 \le i \le k} P_i.$$

As a particular instance of Case Analysis, we have the strategy of *Trivialization* which requires one of the subproblems to be *Easy* (i.e., there must be a solution at hand for that problem). We can formally model this strategy by the formula

$$Trivialization(P, P_0, P_1, \ldots, P_k) \leftrightarrow C\_A(P, P_0, P_1, \ldots, P_k) \wedge Easy(P_0).$$

This strategy is widely used when constructing recursive algorithms. In this case, usually one of the subalgorithms is required not to contain a call to the original one, so that it can be an end point for the recursion. The predicate *Easy* can thus be formalized as

$$Easy(Q, P, P_0) \leftrightarrow Q = P_0 \wedge \forall X\,(P_0[X/P] = P_0).$$

The previous formula characterizes a relation P_0 as being an *easy* representation of Q, a subproblem of the problem P. This means that P_0 might have occurrences of P within it, but these occurrences are innocuous, since they do not influence the behavior of P_0. Under this formal version of *Easy*, Trivialization becomes

$$Trivialization(P, P_0, P_1, \ldots, P_k) \leftrightarrow \exists Q\,(C\_A(P, Q, P_1, \ldots, P_k) \wedge Easy(Q, P, P_0)).$$

Similarly, the following formula over relations characterizes the strategy of Recomposition.

$$Recomposition(P, Split, Q_1, \ldots, Q_k, Join) \leftrightarrow P = Split\,;(Q_1 \otimes \cdots \otimes Q_k)\,; Join,$$

where the relations *Split* and *Join* stand for programs so that the former effectively decomposes the data, and the latter combines the results of $Q_1, \ldots, Q_k$ in order to provide a solution for P.

By joining the strategies of *Recomposition* and *Trivialization*, we obtain the formalization of Divide and Conquer

$$D\&C(P, P_0, Split, Q_1, \ldots, Q_k, Join) \leftrightarrow$$
$$\exists Q\,(Trivialization(P, P_0, Q) \wedge Recomposition(Q, Split, Q_1, \ldots, Q_k, Join)),$$

where the original program P may appear inside some of the terms $Q_1, \ldots, Q_k$, but does not affect the term P_0.

Each strategy comes with an explanation about how to construct a program solving the original problem. It is easy to see how the previously given strategies induce the structure of the programs. For example, it is clear from the definition of Case Analysis that whenever $C\_A(P, P_0, P_1)$ holds, we can infer that $P = P_0 \sqcup P_1$, thus rendering a program (equation) of the desired shape, and solving the problem

P. In the same way, when $D\&C(P, P_0, Split, Q_1, \ldots, Q_k, Join)$ holds, we have a program of the shape $P = P_0 \sqcup Split\,;(Q_1 \otimes \cdots \otimes Q_k)\,; Join$ solving P.

Once we have shown how to express strategies as first-order relational formulas, we are ready to explain the process of program construction. The derivation of a program (ground equation with algorithmic operators), using strategies as the ones shown above, consists essentially in:

1. Providing a specification for P in the form of equations or first-order formulas. These specifications define the theory of P ($Th(P)$). A possible way to obtain this specification could be using the translation τ.
2. Selecting a strategy. This is a design decision which will define the general form of the solution we are seeking. A strategy for solving a problem P, would be a formula $S(P, R_1, \ldots, R_k)$ as the ones shown above, together with the explanation about how to construct a program $P = T$ from it.
3. Performing calculations in fork algebras, so that we find ground terms $Q_1, \ldots, Q_k$ such that $Th(\mathsf{AFA}), Th(P) \vdash S(P, Q_1, \ldots, Q_k)$.
4. Iterating steps 2 and 3, choosing new strategies, not only for P, but also for any subterm of $Q_1, \ldots, Q_k$.

When we decide to stop applying the steps above, the methodology guarantees we have found a term T (according to the strategies applied) satisfying

$$Th(\mathsf{AFA}), Th(P), S(P, Q_1, \ldots, Q_k) \vdash P = T.$$

If we now interpret the equation $P = T$ in a programming language (for example, the relational product ; is viewed as sequential composition of programs, and recursive equations as loops), we will obtain an algorithm for solving the problem according to the strategies chosen.

4.6 Conclusion

In this chapter we have shown some of the applications of fork algebras in logic and computer science, as well as theoretical results. We have also presented a programming calculus based on fork algebras. This calculus, because of the finite axiomatizability of fork algebras, and the interpretability of first-order theories as equational theories in fork algebras has clear advantages over many other functional and relational calculi. These advantages stem from the fact that problem specifications have a natural representation in fork algebras, and the fact that a finite characterization of the calculi simplifies program manipulation.

Chapter 5

Relation Algebra and Modal Logics

Holger Schlingloff, Wolfgang Heinle

This chapter gives an introduction to modal logics as seen from the context of relation algebra. We illustrate this viewpoint with an example application: verification of reactive systems. In the design of safety-critical software systems formal semantics and proofs are mandatory. Whereas for *functional systems* (computing a certain function) usually denotational semantics and Hoare-style reasoning is employed, *reactive systems* (reacting to an environment) mostly are modelled in an automata-theoretic framework, with a modal or temporal logic proof system. Much of the success of these logics in the specification of reactive systems is due to their ability to express properties without explicit use of first-order variables. For example, consider a program defined by the following transition system:

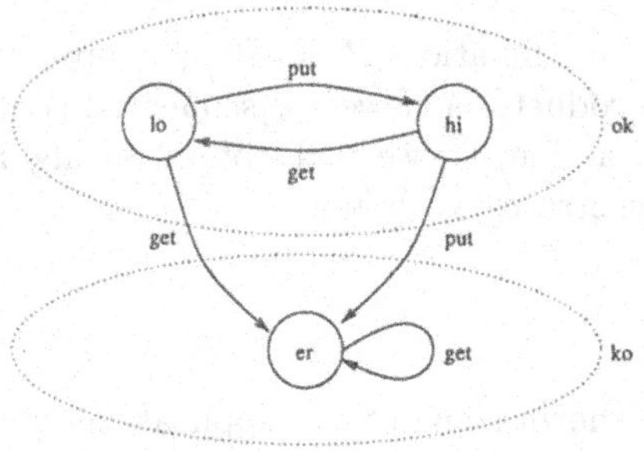

In this picture *lo*, *hi* and *er* denote states, and *put* and *get* are binary relations between states. The set $\{lo, hi\}$ of states is denoted by *ok*, and *ko* is $\{er\}$. The property that any *put* can be followed by a *get* is described by the first-order formula $\forall xy\,(put(x,y) \rightarrow \exists z\,(get(y,z)))$. The same property can also be expressed without use of individual variables x, y, z by the multimodal formula $[put]\langle get\rangle\mathsf{true}$. However, the same virtue is shared by the more expressive relational calculus, which was introduced as a means to formalize mathematics without variables. In the example, the relational equation $put\,;\overline{get\,;\mathbb{T}} = \mathbb{L}$ expresses the same condition as above. As we shall see, relation algebra can serve as natural semantics for modal logics, resulting in some easy completeness and correspondence results. In many cases techniques developed for modal logics can be extended to relation algebra, yielding new insights and opening up new areas of interest.

This chapter is organized as follows: After defining the logical and relational languages used, we give syntactical translations from modal logic into relation algebra and predicate logic. In Sect. 5.2 we characterize the expressive power of modal formulas and relational terms by bisimulations. Then, in Sect. 5.3 we present some modal completeness and incompleteness results and explicate their implications for the relational language.

Finally, Sect. 5.4 on correspondence theory describes a novel way to translate modal axioms into variable-free relation algebraic terms. We conclude with showing how this technique can be applied to more expressive modal languages.

5.1 Logics and standard translations

The language of basic modal logic defined in Sect. 1.6 contains, besides the Boolean connectives, only propositions and one unary modal operator. Propositions are interpreted as unary predicates (subsets) of worlds, and the diamond operator corresponds to a binary accessibility relation between worlds. In general, we will want to express properties of several relations like *put*, *get*, ... Moreover, the restriction to at most binary relations sometimes is artificial, cf. [Blackburn, de Rijke+ 1994]. Therefore we define the modal language over an arbitrary algebraic type τ. Function symbols in τ are called *operators*; unary operators include $\Diamond$ and $\langle R \rangle$, where R is from some index set. Propositions are modal constants, i.e. 0–ary operators. *Modal formulas* (of type τ with variables from V) are terms of type τ with additional Boolean connectives:

- Every $p \in V$ is a modal formula;
- false is a modal formula;
- if φ_1 and φ_2 are modal formulas, then $(\varphi_1 \rightarrow \varphi_2)$ is a modal formula;
- if $\varphi_1, \ldots, \varphi_n$ are modal formulas and $\triangle$ is an n–ary operator from τ, then $\triangle\varphi_1 \ldots \varphi_n$ is a modal formula.

Variables are denoted by lowercase letters $\{p, q, ...\}$; formulas not containing any variables are called *sentences*. Whenever we wish to emphasize the fact that φ contains proposition variables we call φ an *axiom*. Other Boolean connectives $\neg$, $\vee$, $\wedge$, $\leftrightarrow$ are defined as usual; for every operator $\triangle$ its *dual* operator ∇ is given by $\nabla\varphi_1 \ldots \varphi_n \triangleq \neg \triangle \neg\varphi_1 \ldots \neg\varphi_n$. The dual of $\langle R \rangle$ is $[R]$. The *basic* modal language defined in Sect. 1.6 is the set of modal sentences of type $\{P, \Diamond\}$, where P is the set of propositions. In *basic multimodal* formulas the arity of operators is at most one.

To define a semantics for the modal language, there are several choices. The most obvious idea is to extend the notion of Kripke structure as defined in Sect. 1.6 to include n–ary relations: A *standard frame* $\mathcal{F} = (U, \mathcal{I})$ of type τ consists of a nonempty universe U of "worlds", and an interpretation $\mathcal{I}$ assigning to every n–ary operator $\triangle$ from τ an $(n+1)$–ary relation $\triangle^{\mathcal{F}} \triangleq \mathcal{I}(\triangle)$ on U. (For sake of conciseness we will henceforth assume that $\mathcal{I}(\langle R \rangle) = R$.) A *standard model* $\mathcal{M} = (\mathcal{F}, \mathsf{v})$ for formulas of type τ with variables from V is a standard frame $\mathcal{F} = (U, \mathcal{I})$ of type τ together with a *valuation* v assigning to every variable from V a subset of U. The model $\mathcal{M} = (U, \mathcal{I}, \mathsf{v})$ is said to be *based* on the frame

$\mathcal{F} = (U, \mathcal{I})$. Validity of a formula φ in a standard model $\mathcal{M} = (U, \mathcal{I}, \mathsf{v})$ and a world $w \in U$ is defined inductively:

- $\mathcal{M}, w \models p$ if $w \in \mathsf{v}(p)$;
- $\mathcal{M}, w \not\models \mathsf{false}$;
- $\mathcal{M}, w \models (\varphi_1 \rightarrow \varphi_2)$ if $\mathcal{M}, w \models \varphi_1$ implies $\mathcal{M}, w \models \varphi_2$;
- $\mathcal{M}, w \models \Delta\varphi_1 \ldots \varphi_n$ if there are $w_1, \ldots, w_n \in U$ with $(w, w_1, \ldots, w_n) \in \Delta^{\mathcal{F}}$ and $\mathcal{M}, w_i \models \varphi_i$ for all $i = 1, \ldots, n$.

The formula φ is *valid* in the model $\mathcal{M} = (U, \mathcal{I}, \mathsf{v})$ if $\mathcal{M}, w \models \varphi$ for all $w \in U$; it is valid in the frame $\mathcal{F} = (U, \mathcal{I})$ if it is valid in all models $(U, \mathcal{I}, \mathsf{v})$ based on $\mathcal{F}$. Note that any *sentence* is valid in a frame iff it is valid in some model based on that frame.

Alternatively, one can extend v to the set of all formulas by defining

- $\mathsf{v}(\mathsf{false}) \triangleq \emptyset$;
- $\mathsf{v}((\varphi_1 \rightarrow \varphi_2)) \triangleq (U \setminus \mathsf{v}(\varphi_1)) \cup \mathsf{v}(\varphi_2)$;
- $\mathsf{v}(\Delta\varphi_1 \ldots \varphi_n) \triangleq \{w : \exists w_1, ..., w_n\, ((w, w_1, ..., w_n) \in \Delta^{\mathcal{F}} \text{ and } w_i \in \mathsf{v}(\varphi_i))\}$.

Then $\mathcal{M}, w \models \varphi$ iff $w \in \mathsf{v}(\varphi)$, hence φ is valid in $\mathcal{M}$ iff $\mathsf{v}(\varphi) = U$.

To get an intuitive understanding of how relations are coded as modal operators, consider a binary relation R over U. Define for any $w' \in U$ the set

$$\langle R \rangle(w') \triangleq \{w \in U : (w, w') \in R\},$$

and for any $\varphi \subseteq U$ the set

$$\langle R \rangle(\varphi) \triangleq \bigcup\{\langle R \rangle(w') : w' \in \varphi\}.$$

That is, $\langle R \rangle(\varphi)$ is just the Peirce product $R\!:\!\varphi$ introduced in Sect. 1.2. Then we have $(w, w') \in R$ iff $w \in \langle R \rangle(w')$; therefore there is $w' \in \varphi$ with $(w, w') \in R$ iff there is $w' \in \varphi$ with $w \in \langle R \rangle(w')$, i.e. iff $w \in \langle R \rangle\varphi$.

Unary relations (subsets of U) are coded as 0–ary operators (propositions). In our example, $\langle get \rangle ko = \langle get \rangle(\{er\}) = \{er, lo\}$, and $\langle put \rangle ok = \langle put \rangle(\{lo, hi\}) = \langle put \rangle(\{lo\}) \cup \langle put \rangle(\{hi\}) = \emptyset \cup \{lo\}$. Thus $\mathcal{M}, w \models \langle get \rangle ko$ iff $w \in \{er, lo\}$, and $\mathcal{M}, w \models \langle put \rangle ok$ iff $w = lo$. As examples of valid sentences, consider $[put]\langle get \rangle\mathsf{true}$ and $\langle get \rangle\langle get \rangle ko$; as an axiom, $(\langle put \rangle p \rightarrow [put]p)$ is valid since it is valid for every valuation of p with some subset of $\{lo, hi, er\}$.

Standard frames of type τ are precisely relational structures of type τ', where τ' contains an $(n+1)$–ary relation symbol for every n–ary function symbol from τ. But, relational structures of type τ' are also the semantical basis for the first and second-order language of type τ'. The *standard translation* ST of modal formulas of type τ with variables from V into predicate logic formulas of type τ' with proposition variables from V and one free individual variable x is defined as follows:

- $ST(p) \triangleq p(x)$;
- $ST(\mathsf{false}) \triangleq (x \neq x)$;
- $ST((\varphi_1 \rightarrow \varphi_2)) \triangleq (ST(\varphi_1) \rightarrow ST(\varphi_2))$;

- $ST(\Delta\varphi_1 \dots \varphi_n) \triangleq \exists x_1 ... x_n \, (\Delta(x, x_1, ..., x_n) \wedge ST(\varphi_1)[\![x /\!/ x_1]\!] \wedge \dots$
$\dots \wedge ST(\varphi_n)[\![x /\!/ x_n]\!])$

Here $\varphi[\![x /\!/ y]\!]$ denotes the formula derived by *simultaneously* substituting y for all occurrences of x, and x for all occurrences of y in φ, respectively. (This simultaneous substitution which "reuses" variable names was introduced in Sect. 2.8) If the highest arity of any operator in φ is n, then $ST(\varphi)$ contains one free variable x and at most $(n+1)$ bound variables $\{x, x_1, ..., x_n\}$. The number of free individual variables of a first-order formula is sometimes called its *dimension*; hence the standard translation of any modal formula is one-dimensional. $ST(\varphi)$ is a first-order formula iff φ is a modal sentence; if φ contains free proposition variables then so does $ST(\varphi)$. (In this case the axiom φ represents a so-called *monadic* Π^1_1*-property*, that is, one which is expressed by a universal second-order sentence.)

As an example for the standard translation, we calculate

$$\begin{aligned} ST(\langle get\rangle\langle get\rangle ko) &= \exists x_1 \, (get(x, x_1) \wedge ST(\langle get\rangle ko)[\![x /\!/ x_1]\!]) \\ &= \exists x_1 \, (get(x, x_1) \wedge (\exists x_1 \, (get(x, x_1) \wedge (ST(ko)[\![x /\!/ x_1]\!])))[\![x /\!/ x_1]\!]) \\ &= \exists x_1 \, (get(x, x_1) \wedge (\exists x_1 \, (get(x, x_1) \wedge ko(x_1)))[\![x /\!/ x_1]\!]) \\ &= \exists x_1 \, (get(x, x_1) \wedge \exists x \, (get(x_1, x) \wedge ko(x))) \ . \end{aligned}$$

Validity of a modal formula φ in a standard model $(\mathcal{F}, \mathsf{v})$ and world w could have been defined as validity of the predicate logic formula $ST(\varphi)$ in the relational structure $\mathcal{F}$ with second-order valuation v and first-order valuation w where $\mathsf{w}(x) = w$. Of course this new definition for the semantics matches the old one, i.e. $\mathcal{F}, \mathsf{v}, w \models \varphi$ iff $\mathcal{F}, \mathsf{v}, \mathsf{w} \models ST(\varphi)$.

Our standard translation maps modal logic into predicate logic. However, there is another formalism which is equally well suited to serve as semantics for the modal language: the relational calculus.

Since in general we are working with relations of various arities, we have to extend the underlying relation algebra appropriately. There are several approaches to do so.

Any formula of the basic multimodal language can be regarded as a term of a Boolean module (cf. Sect. 1.3), where the Boolean algebra is built from propositions and the relation algebra is built from the unary operators. Of course, Boolean modules provide a much richer algebraic structure than modal formulas, since they include a relation type with operations on operators. In particular, since complementation of relations is present, no general representation result for Boolean modules is available.

As remarked in Sect. 1.3, Boolean modules can be mapped into relation algebras by associating a right ideal element of the relation algebra with each element of the Boolean algebra. Therefore there exists a translation of basic multimodal formulas into relation algebraic terms. Whereas previously formulas were interpreted as sets of worlds, in this new perspective propositions and formulas are viewed as elements of a relation algebra, i.e. as binary relations! This approach to a relational semantics appears, e.g., in [Orlowska 1991], and can be used to give relational formalizations even for more expressive logics such as program and information logics, cf. Chapt. 6. We extend this approach for n-ary operators by extending the dimension of any relation or predicate to the maximum dimension of any relation in the signature.

The *inverse projection* $\pi_i(R)$ of the first component of an $(n+1)$–ary relation R onto the i-th component $(0 \leq i \leq n)$ is defined by

$$\pi_i(R) \triangleq \{(x_0, ...x_{i-1}, y_0, x_{i+1}, ..., x_n) : \exists y_1, ..., y_n\, (R(y_0, y_1, ..., y_n))\}$$

For a binary relation R we have $\pi_0(R) \triangleq \{(x_0, x_1) : \exists y_1\, (R(x_0, y_1))\} = R ; \mathbb{T}$, and $\pi_1(R) \triangleq \{(x_0, x_1) : \exists y_1\, (R(x_1, y_1))\} = (R ; \mathbb{T})^{\smile} = \mathbb{T} ; R^{\smile}$.

Using these operations, the *right ideal translation* RI of modal formulas into relational terms is defined as follows. If φ is a modal formula of type τ with (propositional) variables from V, then $RI(\varphi)$ is a term of type τ'' with variables from V, where τ'' contains Boolean operations, unary projections π_i, and for every modal operator $\triangle$ a relation constant $\triangle$.

- $RI(p) \triangleq \pi_0(p)$ for every proposition variable from V;
- $RI(\mathsf{false}) \triangleq \mathbb{L}$;
- $RI((\varphi_1 \to \varphi_2)) \triangleq (\overline{RI(\varphi_1)} \sqcup RI(\varphi_2))$;
- $RI(\triangle\varphi_1 \ldots \varphi_n) \triangleq \pi_0(\triangle \sqcap \pi_1(RI(\varphi_1)) \sqcap \ldots \sqcap \pi_n(RI(\varphi_n)))$.

For the basic (multi-)modal language this means

- $RI(P) \triangleq \pi_0(P) = P ; \mathbb{T}$ for any proposition P, and
- $RI(\Diamond\varphi_1) \triangleq \pi_0(\Diamond \sqcap \pi_1(RI(\varphi_1)) = (\Diamond \sqcap (RI(\varphi_1);\mathbb{T})^{\smile});\mathbb{T} = \Diamond ; RI(\varphi_1)$.

In the last transformation we used the fact that $RI(\varphi)$ is a right ideal relation, i.e. $RI(\varphi) = RI(\varphi) ; \mathbb{T}$, which is proved by induction on φ. For the above example formula we have $RI(\langle get\rangle\langle get\rangle ko) = get ; get ; ko ; \mathbb{T}$.

A model for terms of type τ'' with projections $\pi_0, ..., \pi_n$ assigns an $(n+1)$–ary relation to every operator and proposition variable. Validity of a term $RI(\varphi)$ in a model $\mathcal{M}$ and worlds $w_0, ..., w_n$ is defined inductively according to the above clause for the projection operation. Now any standard model $\mathcal{M}$ can be *cylindrified* to a model $\mathcal{M}''$ for terms of type τ'' by setting, for an m–ary operator $\triangle$ (where $m \leq n$),

$$\mathcal{I}''(\triangle) \triangleq \{(x_0, ..., x_n) : (x_0, ..., x_m) \in \mathcal{I}(\triangle)\}.$$

E.g., *ko* becomes the binary relation $\{(er, lo), (er, hi), (er, er)\}$. v'' is defined similarly. Then it is an easy exercise to prove that $\mathcal{M}, w \models \varphi$ iff $\mathcal{M}'', w, w_1, ..., w_n \models RI(\varphi)$ for all $w_1, ..., w_n \in U''$. Thus the above relational translation preserves satisfiability. Moreover, since $RI(\varphi) = \pi_0(RI(\varphi))$, we conclude: If $RI(\varphi)$ is valid for all those models $\mathcal{M}''$ which are extensions of some standard model, then $RI(\varphi)$ is valid for all models $\mathcal{M}$ of type τ''. Hence the translation is also validity-preserving: $\models \varphi$ iff $\models RI(\varphi)$.

Cylindification by domain elements is not the only possibility to extend standard frames for modal logics. Another common choice (mainly adopted in dynamic logics) is to associate a constant element $a \sqsubseteq \mathbb{I}$ with any proposition P, i.e. the translation of P could be $P \sqcap \mathbb{I}$. Another possibility is to assign some constant value to every new dimension. Using *relation algebras* as models for modal formulas, however, there is no need for a particular cylindrification.

Consider a relation algebra $\mathcal{A}$ with operators, i.e. a Boolean algebra with operators τ (cf. Sect. 1.3 for the definition of BAO) such that τ contains the relational operators $;$, $\breve{}$, and $\mathbb{I}$. A *relational representation* ρ of the n-ary operator $\triangle$ is any relational expression $\rho(\triangle)$ containing n variables such that for all $a_1, ..., a_n \in \mathcal{A}$ we have $\triangle(a_1, ..., a_n) = \rho(\triangle)(a_1, ..., a_n)$. According to this definition, propositions (0-ary operators) must be represented by their denotation in the algebra. In general, most binary operators will not admit a relational representation.

Call a unary operator $\diamond$ *associative* if $\diamond(p;q) = \diamond(p);q$. Then $\diamond$ is associative if and only if $\diamond(p)$ is represented by $R;p$, where $R \triangleq \diamond(\mathbb{I})$ is an element of $\mathcal{A}$.

Every associative operator is conjugated: If $\diamond(p) = R;p$, then $\diamond(p) \sqcap q = \perp\!\!\!\perp$ iff $p \sqcap \diamond^{\smile}(q) = \perp\!\!\!\perp$ for $\diamond^{\smile}(q) \triangleq R^{\smile};q$.

In the modal encoding of relations as operators we defined for any binary relation R an operator $\langle R\rangle$ by requiring for all $x, y \in U$ that $x \in \langle R\rangle(y)$ iff $(x, y) \in R$. Given a relation algebra $\mathcal{A}$ with operators τ, we call a unary operator $\diamond$ *internal*, if there exists a relation $R \in \mathcal{A}$ such that for all points $x, y \in \mathcal{A}$ we have $x \sqsubseteq \diamond(y)$ iff $x;y^{\smile} \sqsubseteq R$. (Recall that a *point* is any element $y \neq \perp\!\!\!\perp$ with $y = y;\top\!\!\!\top$ and $y;y^{\smile} \sqsubseteq \mathbb{I}$.)

Any associative operator is internal. For, if y is a point, then for all x it holds that $x \sqsubseteq R;y$ iff $x;y^{\smile} \sqsubseteq R$. Proof: Let $x \sqsubseteq R;y$. Then $x;y^{\smile} \sqsubseteq R;y;y^{\smile}$, which implies $x;y^{\smile} \sqsubseteq R$ since y is a point. To prove the other direction, we note that for any point y, it holds that $\top\!\!\!\top;y = \top\!\!\!\top;y;\top\!\!\!\top = \top\!\!\!\top$. Therefore $(R;y) \sqcup (\overline{R};y) = \top\!\!\!\top$, which can be written as $x \sqcap \overline{R;y} \sqsubseteq x \sqcap (\overline{R});y$ by Boolean transformations. Assuming $x;y^{\smile} \sqsubseteq R$, or, equivalently, $x \sqcap (\overline{R});y = \perp\!\!\!\perp$, we have $x \sqcap \overline{R;y} = \perp\!\!\!\perp$, showing that $x \sqsubseteq R;y$. To sum up, $\diamond$ is internal if for all y which are points, $\diamond(y) = R;y$. (A similar proof can also be found in [Schmidt, Ströhlein 1989].)

Any relational representation ρ of all operators in τ induces a translation RT^{ρ} of modal formulas into relation algebraic terms. The above considerations suggest that operators $\langle R\rangle$ coding binary relations are represented by the term $R;p$:

- $RT^{\rho}(P) \triangleq P$ for any proposition or variable P, and
- $RT^{\rho}(\langle R\rangle\varphi) \triangleq R;RT^{\rho}(\varphi)$ for associative operators.

Note that in this translation $RT^{\rho}(\varphi)$ is not necessarily a right ideal element.

5.2 Algebraic characterisation of definability

We say that a formula φ *defines* a set of standard models or frames $\mathcal{F}_{\varphi}$ iff $\mathcal{F}_{\varphi} = \{\mathcal{F} : \mathcal{F} \models \varphi\}$. Several languages were introduced above which can be compared with respect to definability:

MS: basic multimodal sentences of type τ (e.g. $\langle get\rangle\langle get\rangle ko$);
RA: variable-free relation algebraic equations (e.g. $get \sqsubseteq \overline{put}$);
$\mathcal{L}_o$: first-order logic of type τ' (e.g. $\forall xyzu\,(get(x,y) \wedge x{\neq}y{\neq}z{\neq}u \rightarrow ko(y))$);
ML: basic multimodal axioms (e.g. $(\langle put\rangle p \rightarrow [put]p)$);
qRA: relation algebraic equations with relation variables (e.g. $p;p^{\smile} \sqsubseteq put$);
$q\mathcal{L}_o$: monadic second-order logic (e.g. $\exists p\,(\forall x\,(p(x) \wedge \forall y\,(put(x,y) \rightarrow \neg p(y)))))$.

Standard translations like those of Sect. 5.1 establish a syntactic containment between these languages as follows:

$$\begin{array}{ccc} \mathcal{L}_o & \longrightarrow & q\mathcal{L}_o \\ \uparrow & & \uparrow \\ RA & \longrightarrow & qRA \\ \uparrow & & \uparrow \\ MS & \longrightarrow & ML \end{array}$$

As we shall see below, all of the above inclusions are proper.

Since proposition variables do not appear in any formula of *MS*, *RA*, or $\mathcal{L}_o$, with these formalisms there is no difference between definability of frame classes or model classes.

A first characterization of the expressivity of modal sentences can be given as follows: precisely those classes are definable which are defined by first-order sentences elementarily equivalent (i.e. equivalent for all first-order sentences) to the standard translation of a modal sentence. Equivalence to standard translations of modal sentences, however, can also be formulated without the notion of elementary equivalence. For frames $\mathcal{F} = (U, \mathcal{I})$, $\mathcal{F}' = (U', \mathcal{I}')$ and $w \in U$, $w' \in U'$, we say that $(\mathcal{F}, w)$ and $(\mathcal{F}', w')$ are *modally equivalent* ($(\mathcal{F}, w) \equiv_{\mathrm{MS}} (\mathcal{F}', w')$), if $\mathcal{F}, w \models \varphi$ iff $\mathcal{F}', w' \models \varphi$ for all modal sentences φ. A function f between frames $\mathcal{F} = (U, \mathcal{I})$ and $\mathcal{F}' = (U', \mathcal{I}')$ is called a *p-morphism*, if for all $\Delta \in \tau$ and all $w_0 \in U$, $w_0' \in U'$ such that $f(w_0) = w_0'$ we have:

- For all $w_1, ..., w_n \in U$ with $\mathcal{I}(\Delta)(w_0, w_1, ..., w_n)$ there are $w_1', ..., w_n' \in U'$ such that $\mathcal{I}'(\Delta)(w_0', w_1', ..., w_n')$ and $f(w_i) = w_i'$ for $1 \leq i \leq n$;
- For all $w_1', ..., w_n' \in U'$ with $\mathcal{I}'(\Delta)(w_0', w_1', ..., w_n')$ there are $w_1, ..., w_n \in U$ such that $\mathcal{I}(\Delta)(w_0, w_1, ..., w_n)$ and $f(w_i) = w_i'$ for $1 \leq i \leq n$.

A frame $\mathcal{F}'$ is called a p-morphic image of $\mathcal{F}$, if there is a p-morphism from U to U'. We write $\mathcal{F} \hookrightarrow \mathcal{F}'$ (or $(\mathcal{F}, w) \hookrightarrow (\mathcal{F}', w')$) if $\mathcal{F}'$ is a p-morphic image of $\mathcal{F}$ (with $f(w) = w'$). For the basic multimodal language, $\mathcal{F} \hookrightarrow \mathcal{F}'$ iff

- $w \in P^{\mathcal{F}}$ iff $f(w) \in P^{\mathcal{F}'}$ for all propositions $P \in \tau$, and
- $R;f = f;R'$ for all unary operators $\langle R \rangle \in \tau$.

In computer science, *bisimulation* is a concept which often replaces the notion of p-morphisms. Bisimulations are equivalence relations satisfying an analogous condition as above, that is, if w_0 is bisimilar to w_0' and $\Delta^{\mathcal{F}}(w_0, w_1, ..., w_n)$ holds, then there must be $w_1', ..., w_n'$ bisimilar to $w_1, ..., w_n$ such that $\Delta^{\mathcal{F}}(w_0', w_1', ..., w_n')$ holds. If f is a bisimulation on $\mathcal{F}$, then the function mapping any world to its equivalence class is a p-morphism $\mathcal{F} \hookrightarrow \mathcal{F}/_f$. Vice versa, if f is a p-morphism $\mathcal{F} \hookrightarrow \mathcal{F}'$, then the equivalence relation defined by $x \simeq y$ iff $f(x) = f(y)$ is a bisimulation on $\mathcal{F}$ (see also Sect. 3.4).

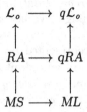

p-morphisms are precisely those homomorphisms which preserve modal equivalence: If $(\mathcal{F}, w) \hookrightarrow (\mathcal{F}', w')$, then $(\mathcal{F}, w) \equiv_{\mathrm{MS}} (\mathcal{F}', w')$. This can be easily shown by induction on the structure of φ. Hence, it is "safe" to substitute a program by a bisimilar one in a structured software development process: All modal specification formulas valid for the original will remain valid for the substituted program. For example, we can "refine" the state *er* in our example program by replacing it with two states er_1 and er_2 such that $get(lo, er_1)$, $get(er_1, er_1)$, $put(hi, er_2)$, $get(er_2, er_1)$, and $ko = \{er_1, er_2\}$. Due to the above preservation theorem this refinement does not change the modal semantics of the program.

For any set Φ of modal formulas, we write $\mathcal{F}, w \models \Phi$ if $\mathcal{F}, w \models \varphi$ for all $\varphi \in \Phi$. Note that $\mathcal{F}, w \models \Phi$ only if $\mathcal{F}, w \models \Phi_0$ for all finite $\Phi_0 \subseteq \Phi$. A frame $\mathcal{F} = (U, \mathcal{I})$ for the basic modal language is *modally saturated* if for all $w \in U$ and all sets Φ of modal formulas the following holds: there is a $w' \in U$ with $\Diamond^{\mathcal{F}}(w, w')$ and $\mathcal{F}, w' \models \Phi$ iff $\mathcal{F}, w \models \Diamond \bigwedge \Phi_0$ for all finite $\Phi_0 \subseteq \Phi$. The frame $\mathcal{F}$ is *image-finite* if for all $w_0 \in U$ the set $\{w_1 : \Diamond^{\mathcal{F}}(w_0, w_1)\}$ is finite. All finite frames are image-finite; all image-finite frames are modally saturated.

For a partial converse of the above theorem, let $\mathcal{F}$, and $\mathcal{F}'$ be modally saturated frames. Then $(\mathcal{F}, w) \equiv_{\mathrm{MS}} (\mathcal{F}', w')$ implies $(\mathcal{F}, w) \hookrightarrow (\mathcal{F}', w')$. Thus modally saturated frames can be characterized "up to p-morphism" by modal sentences. The restriction to modally saturated frames reflects the finiteness of the language; it provides a compactness argument, which could be dropped if modal languages contained infinite conjunctions. Another way to arrive at an exact algebraic characterization of modal sentences is to weaken the notion of p-morphism:

A *partial* p-morphism is a partial function which is a p-morphism on its domain. Let $\mathcal{F} = (U, \mathcal{I})$ and $\mathcal{F}' = (U', \mathcal{I}')$ be frames. Then $(\mathcal{F}', w')$ is a *finitely p-morphic image* of $(\mathcal{F}, w)$, if there is an ω–sequence (F_n) of sets of partial p-morphisms $(\mathcal{F}, w) \hookrightarrow (\mathcal{F}', w')$ such that

- for any $f \in F_{n+1}$, and $w_0, \ldots, w_n \in U$ such that $w_0 \in domf$ and that for some $\triangle \in \tau$ we have $\mathcal{I}(\triangle)(w_0, \ldots, w_n)$, there is an $f' \in F_n$ such that $domf \subseteq domf'$, $ranf \subseteq ranf'$ and $w_1, \ldots, w_n \in domf'$, and
- for any $f \in F_{n+1}$, and $w'_0, \ldots, w'_n \in U$ such that $w'_0 \in ranf$ and that for some $\triangle \in \tau$ we have $\mathcal{I}(\triangle)(w'_0, \ldots, w'_n)$, there is an $f' \in F_n$ such that $domf \subseteq domf'$, $ranf \subseteq ranf'$ and $w'_1, \ldots, w'_n \in ranf'$.

If $(\mathcal{F}, w)$ is a finitely p-morphic image of $(\mathcal{F}', w')$, then it can be shown by induction on the nesting of modal operators, that for any sentence φ it holds that $\mathcal{F}, w \models \varphi$ iff $\mathcal{F}', w' \models \varphi$. Conversely, if $(\mathcal{F}, w)$ and $(\mathcal{F}', w')$ are modally equivalent, then we can construct a sequence of sets of partial p-morphisms with extending domains and ranges, asserting that $(\mathcal{F}', w')$ is a finitely p-morphic image of $(\mathcal{F}, w)$.

This idea can be transformed into an algorithm that checks whether the two finite frames $(\mathcal{F}, w_0)$ and $(\mathcal{F}', w'_0)$ are modally equivalent: Start with the mapping $f(w_0) \triangleq w'_0$, and then recursively try to extend this mapping for all worlds reachable from any world in the domain or range of the already constructed mapping. If for some w and w' we have $f(w) = w'$, but w and w' do not satisfy the same propositions, then backtrack. Upon termination this algorithm delivers a partial p-morphism f from the strongly connected component containing w_0, or a modal sentence distinguishing the two frames. Many computer-aided verification systems incorporate an algorithm for checking bisimulation equivalence which is based on this method.

As another application of p-morphisms we prove that the term $R^{\smile};R \sqsubseteq \mathbb{I}$, or, equivalently, the first-order property of *univalence* ($\forall xyz\,(R(x,y) \wedge R(x,z) \rightarrow y = z)$) cannot be expressed by a modal sentence. We give two frames $\mathcal{F} = (U, \mathcal{I})$ and $\mathcal{F}' = (U', \mathcal{I}')$ such that $\mathcal{F} \hookrightarrow \mathcal{F}'$, and $\mathcal{F}'$ is univalent, but $\mathcal{F}$ is not. Let $U \triangleq \{a_0, a_1, a_2\}$ with $\mathcal{I}(\Diamond) \triangleq \{(a_0, a_1), (a_0, a_2)\}$, and $U' \triangleq \{b_0, b_1\}$ with $\mathcal{I}'(\Diamond) \triangleq \{(b_0, b_1)\}$. The mapping f given by $f(a_0) \triangleq b_0$, $f(a_1) \triangleq b_1$, and $f(a_2) \triangleq b_1$ is the required p-morphism. However, even though the validity of the above term is not preserved under p-morphic *pre-images*, it can be proved that it is preserved under p-morphic *images*; that is, if $\mathcal{F} \hookrightarrow \mathcal{F}'$ and $\mathcal{F}$ is univalent, then $\mathcal{F}'$ is univalent.

Modal axioms, i.e. modal formulas with propositional variables, essentially are Π^1_1-sentences. Although axioms in general are not invariant under p-morphisms, they are preserved: If $(\mathcal{F}, w) \hookrightarrow (\mathcal{F}', w')$, then $\mathcal{F}, w \models \varphi$ implies $\mathcal{F}', w' \models \varphi$. As we will see, univalence can be defined by the axiom $(\Diamond p \rightarrow \Box p)$. An example of a relational property which is not definable by any modal axiom is irreflexivity: $R \sqsubseteq \overline{\mathbb{I}}$. This can be shown using preservation of modal axioms by p-morphisms: The frame $(\{a\}, \{(a, a)\})$ is a reflexive p-morphic image of the frame (ω, S), where ω is the set of natural numbers, S is the successor relation (irreflexive !), and $f(i) \triangleq a$ for all $i \in \omega$.

The concept of bisimulation can be extended to relation algebra. As we have seen, the standard translation of a basic multimodal formula yields a first-order formula with at most two individual variables. Similarly, every relation algebraic term t can be translated into an $\mathcal{L}_o$ formula $ST(t)$ with two free individual variables x_0, x_1 and at most one bound individual variable (cf. Sect. 2.8). As an example, $(Q \sqcap S;R^{\smile});R \sqsubseteq Q$ can be translated into

$$\exists x_2\,(Q(x_0, x_2) \wedge \exists x_1\,(S(x_0, x_1) \wedge R(x_2, x_1)) \wedge R(x_2, x_1) \rightarrow Q(x_0, x_1)).$$

Thus, the relational calculus is contained in the *3-variable fragment* of $\mathcal{L}_o$. Givant and Tarski[Tarski, Givant 1987] showed that also the converse direction is true: For every $\mathcal{L}_o$-sentence with at most three variables there is an equivalent variable-free relation algebraic term. From this theorem, an algebraic characterization of definability with relational terms can be developed by defining a suitable extended version of finite p-morphisms:

Let $\mathcal{F} = (U, \mathcal{I})$ and $\mathcal{F}' = (U', \mathcal{I}')$ be frames. A partial function f from $\mathcal{F}$ to $\mathcal{F}'$ is called a *partial isomorphism*, if

- f is injective, i.e. $w_0 = w_1$ iff $f(w_0) = f(w_1)$, and
- $R(w_0, ..., w_n)$ iff $R'(f(w_0), ..., f(w_1))$ for all $\langle R \rangle \in \tau$.

Compared to partial p-morphisms, partial isomorphisms yield a stronger condition on f, as they additionally require $R'(f(w_0); f(w_1)) \implies R(w_0, w_1)$. Two frames $\mathcal{F}$, $\mathcal{F}'$ are called *bounded finitely isomorphic*, if there is an ω-sequence (F_n) of sets of partial isomorphisms such that:

- for all $f_{n+1} \in F_{n+1}$ and $w_0 \in U$ there is an $f_n \in F_n$ and $w_1 \in \mathit{dom} f_n$ with $\mathit{dom} f_n = \mathit{dom} f_{n+1} \setminus \{w_1\} \cup \{w_0\}$;
- for all $f_{n+1} \in F_{n+1}$ and $w'_0 \in U'$ there is an $f_n \in F_n$ and $w'_1 \in \mathit{ran} f_n$ with $\mathit{ran} f_n = \mathit{ran} f_{n+1} \setminus \{w'_1\} \cup \{w'_0\}$.

$\mathcal{F}$, $\mathcal{F}'$ are *finitely isomorphic with bound* k, if there is such a sequence such that k is the maximal number of elements in $dom f$ for any n and $f \in F_n$. The frames $\mathcal{F}$ and $\mathcal{F}'$ for the basic multimodal language are *relationally isomorphic* if they are finitely isomorphic with bound 3. $(\mathcal{F}, w_0, w_1)$ and $(\mathcal{F}', w_0', w_1')$ are relationally isomorphic if $\mathcal{F}$ and $\mathcal{F}'$ are relationally isomorphic, where each F_n contains an isomorphism mapping w_0 to w_0' and w_1 to w_1'. Using the proof of Givant and Tarski, we can show that two frames satisfy the same relational terms iff they are relationally isomorphic. Therefore, we can develop a similar algorithm as above for testing whether two finite frames can be distinguished by a relational term.

The theorem of Givant and Tarski can also be used to show non-definability of certain properties in relation algebra. For example, consider the first-order frame property $\forall x, y, z, u\,(Q(x,y) \wedge R(x,z) \wedge S(x,u) \rightarrow \exists v\,(T(y,v) \wedge U(z,v) \wedge V(u,v)))$. To express this property we need at least four individual variables, so there is no equivalent term in RA. However, it can be defined in qRA by the axiom $R;\overline{U;\overline{q}} \sqcap S;\overline{V;\overline{p}} \sqcap Q;\overline{T;(p \sqcap q)} = \perp\!\!\!\perp$. These matters will be pursued further in Sect. 5.4.

5.3 Completeness and incompleteness

Whereas in the previous section we focussed on the definitional power of modal formulas, in this section we describe their deductive capabilities.

Allowing an additional proposition for each state, we can describe each finite-state program by a finite set of modal sentences. For our example program this set could be

$$\begin{array}{lll} \Phi \triangleq \{ & (lo \rightarrow \langle put\rangle hi \wedge [put]hi), & (hi \rightarrow \langle get\rangle lo \wedge [get]lo), \\ & (lo \rightarrow \langle get\rangle er \wedge [get]er), & (hi \rightarrow \langle put\rangle er \wedge [put]er), \\ & (er \rightarrow \langle get\rangle er \wedge [get]er), & (er \rightarrow [put]\mathsf{false}) \qquad \} \end{array}$$

(In this modelling we adopt the so-called *branching time* approach. A different modelling with *linear time* will be given below.) As we shall see, additional properties like "R^* is the reflexive transitive closure of $get \cup put$" can be formulated by appropriate modal axioms **X**. Now our program satisfies a specification φ, if φ is valid in any model in which all of the above sentences and properties are valid. Subsequently, we investigate the completeness of this notion. Firstly, we would like to make sure that any specification which is satisfied by a program can be proved, provided the specification is expressible. Secondly, in many cases algorithms for the automated verification of systems can be obtained from the completeness proof for the specification language.

Let any type τ and set of variables V be given. Formally, a *modal logic* $\mathcal{L}$ is any set of modal formulas closed under propositional tautologies (**taut**), modus pones (**MP**), normality (**N**: $\Diamond\mathsf{false} \leftrightarrow \mathsf{false}$), additivity (**add**: $\Diamond(p \vee q) \leftrightarrow (\Diamond p \vee \Diamond q)$), and substitution of formulas for propositional variables. The smallest modal logic is **K**; the largest logic is the one containing false, which consists of all formulas. If **X** is any set of formulas, then $\mathcal{L}(\mathbf{X})$ denotes the smallest logic containing **X**.

This definition allows us to identify a logic $\mathcal{L}(\mathbf{X})$ with its axioms $\mathbf{X}$: For a formula φ, a set Φ of sentences and a set $\mathbf{X}$ of axioms, we write $\Phi \vdash^{\mathbf{X}} \varphi$, if $\varphi \in \mathcal{L}$, where $\mathcal{L}$ is the smallest logic with $\Phi \cup \mathbf{X} \subseteq \mathcal{L}$. If Φ is empty, it is omitted; also $\mathbf{X}$ is omitted whenever no confusion can arise. In contrast to predicate logic, modal logic does not provide the usual deduction theorem: From $\Phi \cup \{\varphi\} \vdash \psi$ we are not allowed to conclude $\Phi \vdash (\varphi \to \psi)$.

To prove $\Phi \vdash^{\mathbf{X}} \varphi$ one has to give a *derivation* of φ from the assumptions Φ, i.e. a sequence of formulas such that the last element of this sequence is φ, and every element of this sequence is either from Φ, or a substitution instance of an axiom from $\mathbf{X}$, or the substitution instance of the consequence of a rule, where all premisses of the rule for this substitution appear already in the derivation.

As an example, let us derive monotonicity in $\mathbf{K}$. First, we note that the rule of replacement of provably equivalent subformulas (**Repl**: $(p \leftrightarrow q) \vdash (\varphi(p) \leftrightarrow \varphi(q))$) is *admissible*, i.e. does not increase the set of derivable formulas. We proceed as follows:

1.	$(\Diamond\neg q \to \Diamond\neg p \vee \Diamond\neg q)$	(**taut**)
2.	$(\Diamond\neg q \to \Diamond(\neg p \vee \neg q))$	(1, **add**)
3.	$(\Diamond\neg q \to \Diamond(\neg p \vee (p \wedge \neg q)))$	(2, **Repl**)
4.	$(\Diamond\neg q \to \Diamond\neg p \vee \Diamond(p \wedge \neg q)))$	(3, **add**)
5.	$(\neg\Diamond\neg p \wedge \Diamond\neg q \to \Diamond(\neg(p \to q)))$	(4, **Repl**)
6.	$(\Box(p \to q) \to (\Box p \to \Box q)$	(5, **taut**)

A formula φ *follows* from a set Φ of sentences in a class G of standard frames or models ($\Phi \Vdash^{G} \varphi$), if φ is valid in every frame or model of G which validates all $\psi \in \Phi$. $\mathbf{X}$ is called *correct* for G, if $\Vdash^{G} \varphi$ whenever $\vdash^{\mathbf{X}} \varphi$. As any deduction can use only finitely many premisses, $\mathbf{X}$ is correct for G iff for all Φ it holds that $\Phi \Vdash^{G} \varphi$ whenever $\Phi \vdash^{\mathbf{X}} \varphi$.

Any set of formulas $\mathbf{X}$ is correct for the set $G(\mathbf{X})$ of frames in which all elements of $\mathbf{X}$ are valid: $\vdash^{\mathbf{X}} \varphi$ implies $\Vdash^{G(\mathbf{X})} \varphi$.

The converse direction of this statement is the completeness problem: $\mathbf{X}$ is called *complete* for G if $\Vdash^{G} \varphi$ implies $\vdash^{\mathbf{X}} \varphi$. $\mathbf{X}$ is *strongly complete* for G if for all Φ, if $\Phi \Vdash^{G} \varphi$ then $\Phi \vdash^{\mathbf{X}} \varphi$. In contrast to the correctness statement, the two notions of completeness do not coincide: there are axiom systems complete for a certain class of models, but not strongly complete. The situation is the same as in more expressive languages like predicate logic or relational calculus: Some (classes of) models can be described by an infinite set of formulas, but not by any finite subset thereof.

The minimal logic $\mathbf{K}$ is strongly complete for the class G of all standard models, i.e. $\Phi \Vdash \varphi$ iff $\Phi \vdash \varphi$. The proof follows the so-called Henkin/Hasenjäger construction and is completely analogous to the proof of Stone's representation theorem for Boolean algebras (cf. Theorem 1.3.1) and its extension to BAOs by Jónsson and Tarski [Jónsson, Tarski 1951].

A subset $u \subseteq \mathcal{A}$ of a Boolean algebra with operators (BAO) is said to have the *finite intersection property*, if for any finite subset $\{p_1, ..., p_n\} \subseteq u$ it holds that $p_1 \sqcap ... \sqcap p_n \neq \perp\!\!\!\perp$. Any such subset can be extended to an ultrafilter of $\mathcal{A}$

by repeatedly adding either p or $\overline{p}$ for each $p \in \mathcal{A}$. (For uncountable $\mathcal{A}$ the axiom of choice is needed in this construction.) In particular, for any atom a of $\mathcal{A}$ there is exactly one ultrafilter u containing a, namely $u \triangleq \{p : a \sqsubseteq p\}$. For BAOs Stone's representation function $\sigma : \mathcal{A} \to \mathcal{P}(U_{\mathcal{A}})$, mapping every $p \in \mathcal{A}$ to $\sigma(p) \triangleq \{u \in U_{\mathcal{A}} : p \in u\}$ is extended such that it assigns an $(n+1)$-ary relation $\sigma(\triangle)$ on $U_{\mathcal{A}}$ to every n-ary operator $\triangle$:

$$\sigma(\triangle) \triangleq \{(u_0, ..., u_n) : p_1 \in u_1 \wedge ... \wedge p_n \in u_n \implies \triangle(p_1, ..., p_n) \in u_0\}$$

Dually, we have $\sigma(\nabla)(u_0, ..., u_n)$ iff $\nabla(p_1, ..., p_n) \in u_0 \implies p_1 \in u_1 \vee ... \vee p_n \in u_n$. Again, $\sigma(\triangle)$ can be regarded as an n-ary operator on $\mathcal{P}(U_{\mathcal{A}})$ in the usual way; $\sigma(\triangle)(x_1, ...x_n) \triangleq \{u_0 : \exists u_1, ..., u_n\,(\triangle(u_0, ..., u_n) \wedge u_1 \in x_1 \wedge ... \wedge u_n \in x_n)\}$. This definition yields a modal homomorphism from $(\mathcal{A}, \sqcup, \neg, \perp\!\!\!\perp, \tau)$ into $(\mathcal{P}(U_{\mathcal{A}}), \cup, -, \emptyset, \sigma(\tau))$:

$$\sigma(\triangle(p_1, ..., p_n)) = \sigma(\triangle)(\sigma(p_1), ..., \sigma(p_n)).$$

Hence every BAO is isomorphic to a subalgebra of a powerset algebra. As a corollary, any term valid in $\mathcal{A}$ is valid in its ultrafilter algebra. Furthermore, if $\mathcal{A}$ is finite, then σ constitutes an isomorphism. This *representation theorem* is fundamental to modal duality theory.

Let us try to illustrate the ultrafilter construction with our example program and $\tau \triangleq \{ok, ko, \langle put\rangle\}$. The BAO $\mathcal{A}$ contains at least the elements $\{\perp\!\!\!\perp, \top\!\!\!\top, ok, ko\}$. Other elements of $\mathcal{A}$ can be constructed as $lo \triangleq \langle put\rangle ok$, $el \triangleq ko \sqcup lo$, and $hi \triangleq \overline{el}$, and $eh \triangleq ko \sqcup hi$. $\mathcal{A}$ validates the following terms: $lo \sqsubseteq \langle put\rangle hi$, $hi \sqsubseteq \langle put\rangle ko$, and $ko \sqsubseteq [put](\perp\!\!\!\perp)$. Therefore $\mathcal{A}$ can be pictured as follows:

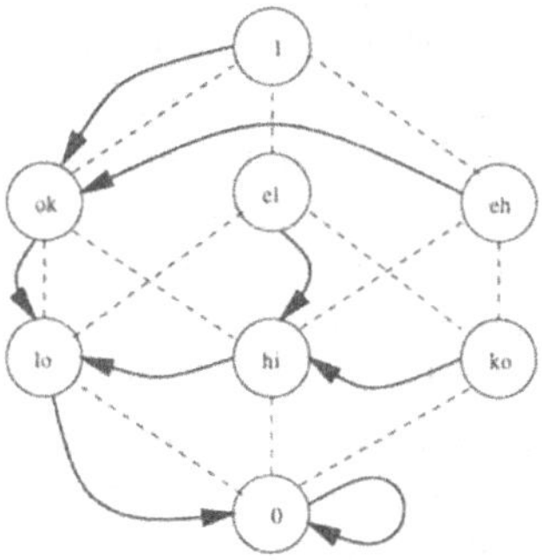

The three ultrafilters of $\mathcal{A}$ are $u_{lo} \triangleq \{lo, ok, el, \top\!\!\!\top\}$, $u_{er} \triangleq \{ko, el, eh, \top\!\!\!\top\}$, and $u_{hi} \triangleq \{hi, ok, eh, \top\!\!\!\top\}$.

The representation function σ is obvious: it maps e.g. $\sigma(lo) = \{u_{lo}\}$ and $\sigma(ok) = \{u_{lo}, u_{hi}\}$. It is easy to see that this map defines a modal isomorphism. Note that for the type $\tau \triangleq \{ok, ko, \langle put \cup get\rangle\}$, no additional elements besides $\{\perp\!\!\!\perp, \top\!\!\!\top, ok, ko\}$ are generated; lo and hi cannot be distinguished in this language.

The logical counterpart of ultrafilters are maximal consistent sets of formulas. A set Ψ of formulas is *consistent with* Φ, if there is no finite subset $\{\psi_1, ..., \psi_n\} \subseteq \Psi$ such that $\Phi \vdash (\psi_1 \wedge ... \wedge \psi_n \leftrightarrow \textsf{false})$. To prove strong completeness, we have to show that every formula consistent with Φ is satisfiable in a model validating

Φ. For, if $\Phi \Vdash \varphi$, then no model validating Φ satisfies $\{\neg\varphi\}$; therefore $\{\neg\varphi\}$ is inconsistent with Φ, hence $\Phi \vdash \varphi$. (Without loss of generality, we can assume here Φ to be consistent with itself, or else $\Phi \vdash \varphi$ holds). Lindenbaum's extension lemma states that any set of formulas consistent with Φ can be extended to a maximal consistent set including Φ (by repeatedly adding ψ or $\neg\psi$, respectively).

The quotient algebra of modal formulas with respect to provable equivalence is called the *Lindenbaum algebra*, named LINDA in [Davey, Priestley 1990]. The analogue of the ultrafilter algebra is the *canonical model* $\mathcal{M} = (U, \mathcal{I}, \mathsf{v})$ for Φ, where U is the set of maximal consistent sets which include Φ, $\mathcal{I}(\Delta) \triangleq \{(w_0, ..., w_n) : p_1 \in w_1 \wedge ... \wedge p_n \in w_n \rightarrow \Delta(p_1, ..., p_n) \in w_0\}$, and $\mathsf{v}(p) \triangleq \{w : p \in w\}$. The fundamental "truth" or "killing" lemma states that for any formula φ and maximal consistent set w it holds that $\varphi \in w$ iff $\mathcal{M}, w \models \varphi$. In the inductive step for this lemma, we have to show that $\Delta\varphi_1...\varphi_n \in w$ iff $\mathcal{M}, w \models \Delta\varphi_1...\varphi_n$. The "if" direction being a direct consequence of the definition, assume that $\Delta\varphi_1...\varphi_n \in w$. We have to find maximal consistent sets $w_1, ..., w_n$ such that $(w, w_1, ..., w_n) \in \mathcal{I}(\Delta)$ and $\varphi_i \in w_i$ for $i \leq n$. Since $\vdash (\Delta\varphi_1...\varphi_n \wedge \nabla\psi_1...\psi_n \rightarrow (\Delta(\varphi \wedge \psi_1)\varphi_2...\varphi_n) \vee ... \vee (\Delta\varphi_1...\varphi_{n-1}(\varphi_n \wedge \psi_n)))$, for every $\nabla\psi_1...\psi_n$ in w there exists some k such that $\Delta\varphi_1...(\varphi_k \wedge \psi_k)...\varphi_n$ is in w. Fix an enumeration (j) of all formulas $\nabla\psi_1...\psi_n$ in w, and define n sequences u_i of consistent sets $u_{i,j}$ such that $\Delta \bigvee u_{1j} ... \bigvee u_{nj}$ is in w for all j. Let $u_{i,0} \triangleq \varphi_i$, and $u_{i,j+1} \triangleq u_{i,j} \cup \psi_i$ if $i = k$, or else $u_{i,j+1} \triangleq u_{i,j}$. (In the basic monomodal language, u_{1j} is just $\{\varphi_1\} \cup \{\psi_j : \Box\psi_j \in w\}$.) Let w_i be any maximal consistent extension of u_{ij}. Then $(w, w_1, ..., w_n) \in \mathcal{I}(\Delta)$, since for formulas $\varphi_1, ...\varphi_n$ the assumptions $\varphi_i \in w_i$ and not $\Delta\varphi_1, ..., \varphi_n \in w_0$ lead to a contradiction. Thus we have achieved our goal of constructing a model for any consistent set.

In fact, we have shown that any set of formulas is strongly complete for its canonical model. To show that a set of formulas is complete for a set G of models, a useful strategy is to show that the canonical model belongs to G, or, that the canonical model can be transformed into a model belonging to G. Such transformations include the unfolding of models into trees, and the collapse of the model with respect to bisimulation equivalence.

The completeness proof can be improved: It is not necessary that maximal consistent sets are maximal in the space of all formulas; it is sufficient to consider maximality with respect to all subformulas of the given consistent set. This idea can be used to transform the above completeness proof into a decision algorithm: For any formula, there are only finitely many different subformulas, and hence only finitely many sets of subformulas. Call such a set w of subformulas *locally maximal consistent*, if

- for any subformula ψ, either $\psi \in w$ or $\neg\psi \in w$, and
- $\mathsf{false} \notin w$, and
- $(\psi_1 \rightarrow \psi_2) \in w$ iff $\psi_1 \in w$ implies $\psi_2 \in w$.

There are two approaches to deciding whether a given formula φ is satisfiable: The first, "local" algorithm, is tableaux-based. We start with the set of all locally maximal consistent sets containing φ and try to systematically extend one of these to a model. Given a locally maximal consistent set w, we construct for any formula $\Delta p_1 \ldots p_n \in w$ as successors all n-tuples of locally maximal consistent sets $(w_1 \ldots w_n)$ which arise in the completeness proof. If there are no such successors, then w is unsatisfiable and we backtrack; otherwise we proceed to extend the successors. Since there are only finitely many locally maximal consistent sets, the process stabilizes. Either all initial nodes are unsatisfiable, or we have constructed a model for the formula.

The second, "global" algorithm for testing satisfiability starts with the set U of all locally maximal consistent sets and the set of all $(n+1)$-tuples for any n-ary operator. It then iteratively deletes "bad arcs" and "bad nodes" until stabilization is reached. Bad arcs are tuples $(w, w_1, \ldots, w_n)$ such that w contains $\nabla \psi_1 \ldots \psi_n$, but for all $i \leq n$ it is not the case that $\psi_i \in w_i$. Bad nodes w contain a formula $\Delta p_1 \ldots p_n$, but there does no longer exist a tuple $(w, w_1, \ldots, w_n)$ with $\psi_i \in w_i$ for all $i \leq n$. The given formula is satisfiable iff upon termination there is a node left in which it is contained.

We proved completeness with respect to sets of models. However, axioms are usually used to define sets of frames. Thus we are looking for completeness statements of the kind "$\Vdash^{G} \varphi$ implies $\vdash^{\mathbf{X}} \varphi$", where G is a class of frames. We know that $\vdash^{\mathbf{X}} \varphi$ iff φ is valid in the set $G(\mathit{Subst}(X))$ of all models satisfying all substitution instances (of propositional variables with formulas) of $\mathbf{X}$. But, this set is much bigger than the set $G(X)$ of all models based on some frame for $\mathbf{X}$, because the latter models have to satisfy all substitution instances of $\mathbf{X}$, where variables are substituted with subsets of worlds. Since in general not all subsets of worlds are described by sentences, we can not infer that validity in all frames for $\mathbf{X}$ implies derivability from $\mathbf{X}$. In fact, there is a finite axiom set $\mathbf{X}$ and formula φ such that the question whether $\Vdash^{G(X)} \varphi$ is Σ^1_1-hard (and thus not recursively enumerable).

Consider our example program as the state transition diagram of a counter machine, which increments and decrements its counter with every *put* and *get* operation, respectively. We show how this machine can be coded by a finite set of formulas, such that every model of these formulas describes the sequence of memory states of a complete run.

Let $\tau \triangleq \{hi, lo, er, put, get, eoq, \langle X \rangle, \langle F \rangle, \langle M \rangle\}$. The operator $\langle X \rangle$ will be used to describe the execution steps of the program in time, the operator $\langle F \rangle$ to denote the transitive closure of $\langle X \rangle$, and the operator $\langle M \rangle$ to access the content of the memory. As we will see in the next section, the following axioms describe that X and M are univalent ($X^{\smile} ; X \sqsubseteq \mathbb{I}$ and $M^{\smile} ; M \sqsubseteq \mathbb{I}$), form a half-grid ($M ; X \sqsubseteq X ; M$), and that F is the transitive closure of X:

$$\langle X \rangle p \to [X] p, \qquad \langle M \rangle p \to [M] p$$
$$\langle M \rangle \langle X \rangle p \to \langle X \rangle \langle M \rangle p$$
$$\langle X \rangle p \vee \langle X \rangle \langle F \rangle p \to \langle F \rangle p, \qquad [F](p \to [X]p) \to ([X]p \to [F]p)$$

Using these relations, we fix the propositions such that (i) the number of M^*-successors in any world labelled *eoq* is the value of the counter, (ii) every world is labelled *hi*, *lo*, or *er*, according to the machine state it denotes, and (iii) every world is labelled *put* or *get*, according to which action is executed next. The relevant sentences are:

put increases the length of the counter by one:
$(put \wedge [M]\mathsf{false} \rightarrow \langle X\rangle\langle M\rangle[M]\mathsf{false})$

get decreases the length of the counter by one:
$(\langle M\rangle(get \wedge [M]\mathsf{false}) \rightarrow \langle X\rangle[M]\mathsf{false})$

Every world is exactly one of $\{put, get\}$ and $\{hi, lo, er\}$:
$(put\ xor\ get) \wedge (hi\ xor\ lo\ xor\ er)$ (*xor* denoting exclusive disjunction)

All worlds reachable by M^* have the same marking:
$(P \rightarrow [M]P)$ for $P \in \{put, get, hi, lo, er\}$

eoq propagates only in one dimension:
$(eoq \rightarrow [X]eoq)$, $[M]\neg eoq$

Transitions:
$(lo \wedge put \rightarrow [X]hi)$, $(hi \wedge get \rightarrow [X]lo)$,
$(lo \wedge get \rightarrow [X]er)$, $(hi \wedge put \rightarrow [X]er)$,
$(er \wedge get \rightarrow [X]er)$, $(er \wedge put \rightarrow [X]\mathsf{false})$

For a conditional transition like "from *er* go to *lo* if counter is zero" we could use the sentence $(er \wedge eoq \wedge [M]\mathsf{false} \rightarrow [X]lo)$. For a multiple counter machine, we can use a similar encoding with several memory access functions $\langle M_i\rangle$. Now there is a computation in which such a machine reaches a certain state (say, *hi*) infinitely often from its initial state iff the sentence $(lo \wedge eoq \wedge [M]\mathsf{false} \wedge [F]\langle F\rangle hi)$ is satisfiable in a model validating all of the above axioms and sentences. Of course, for our example machine, we easily see that the formula is satisfiable; for all single counter machines this *recurrence problem* is decidable. But, for multiple counter machines the problem is Σ^1_1-complete, therefore also the problem whether any sentence follows from a set of axioms is Σ^1_1-hard. Recall that axioms are monadic Π^1_1-properties, so the problem is in Σ^1_1 as well.

However, there is a notion of completeness for frame consequences, which is inspired by the algebraic approach. Even though not every Boolean algebra with operators is isomorphic to its ultrafilter algebra, BAOs can be regarded as models in their own right. This viewpoint gives rise to a semantics more general than standard semantics, which is complete for all modal formulas and relational terms.

5.4 Correspondences: second-order to first-order

Standard translations of modal axioms can be regarded as universal second order formulas. We say that a modal axiom *corresponds to* some monadic Π^1_1 frame property. Sometimes, such a property can be expressed by a first-order sentence or relation algebraic term. Also, in some cases a property which has no equivalent modal *sentence* can be described by a modal *axiom*. Given a modal axiom, does it define a first-order property? And: given a property, is there a modal axiom describing it? Although in general the complexity of these questions is not known, all correspondences which are found in the literature can be derived from simple principles in the relational calculus.

Consider the modal axiom **U**: $\langle R\rangle p \to [R]p$, which we met several times in the previous sections. It defines univalence, i.e. $\forall xyz\,(R(x,y) \wedge R(x,z) \to y = z)$. This correspondence can be established as follows: Assume a frame $\mathcal{F}$ for **U** such that $R(x,y)$ for some x, and a valuation assigning $\{y\}$ to p. Then x satisfies $\Box p$. Hence every z with xRz must satisfy p. But, given the choice of p, we see that z must be equal to y, establishing univalence. Now, assume a univalent frame and a valuation for p which validates the antecedent of **U** in x. Thus, we have a successor y of x in p. Univalence establishes that y is the only successor of x, hence all successors of x are in p, establishing validity of **U**. This proof displays the role of the proposition variables: they can be used as a kind of "register" for a certain individual variable. There are systems that exploit this method to automatically derive a large number of modal correspondences [Ohlbach, Schmidt 1995].

U can also be defined by the relation algebraic sentence $R^{\smile};R \sqsubseteq \mathbb{I}$. This can be shown with the relational translation of modal formulas from Sect. 5.1: **U** is translated into $R;p \sqsubseteq \overline{R;\overline{p}}$, which is by Schröder equivalent to $R^{\smile};R;\overline{p} \sqsubseteq \overline{p}$. Since the relation variable p can be substituted with any relation, this is equivalent to $R^{\smile};R \sqsubseteq \mathbb{I}$.

Assume a relation algebra with operators $(\mathcal{A}, \tau)$, where all $\langle R\rangle$-operators are internal, i.e. represented by the term $R;p$ for some element R of the relation algebra. A relational term $t(p)$ containing a (relation) variable p is valid in $\mathcal{A}$, if $t = \mathbb{T}$ for all valuations assigning a relation from $\mathcal{A}$ to p. Therefore, the following quantifier elimination principle (**qep**) holds:

$$\langle Q\rangle p \to \langle R\rangle p \quad \text{iff} \quad Q;p \sqsubseteq R;p \quad \text{iff} \quad Q \sqsubseteq R\,.$$

The proof is immediate: from right to left it is monotonicity of the relational product; the other direction is achieved by specialization of p to the identity relation $\mathbb{I}$.

Many common modal correspondences can be obtained from (**qep**). For instance, in Sect. 5.3 we defined the property that M and X form a half grid ($M;X \sqsubseteq X;M$). By (**qep**), this is equivalent to $\langle M\rangle\langle X\rangle p \to \langle X\rangle\langle M\rangle p$.

Conjugated operators reflect relational converses via the so-called "tense logic" axiom: $(p \to [Q]\langle R\rangle p \wedge [R]\langle Q\rangle p)$ iff $Q = R^{\smile}$. The proof is an easy relation algebraic deduction: Let $(p \to [Q]\langle R\rangle p)$, or equivalently $p \sqsubseteq \overline{Q;\overline{R;p}}$. Thus we get $Q^{\smile};p \sqsubseteq R;p$, and with (**qep**): $Q^{\smile} \sqsubseteq R$. Symmetrically, the second part gives $R^{\smile} \sqsubseteq Q$, together: $Q^{\smile} = R$.
The following correspondence (**X**) generalizes this idea:

$$Q^{\smile};V \sqsubseteq S;U^{\smile} \quad \text{iff} \quad \langle V\rangle[U]p \to [Q]\langle S\rangle p\,.$$

Relation algebra proves that $V;\overline{U;\overline{p}} \sqsubseteq \overline{Q;\overline{S;p}}$ iff $Q^{\smile};V;p \sqsubseteq S;U^{\smile};p$. By (**qep**) this formula is equivalent to the required axiom. (**X**) can be used as "generic" correspondence scheme by instantiating Q, V, S, and U with appropriate relations; a (nonexhaustive) list of correspondences which can be proved that way together with their traditional names from the literature is given in Table 5.1.

$\mathbb{I} \sqsubseteq R$	$p \to \langle R\rangle p$	**T**
$R;R \sqsubseteq R$	$\langle R\rangle\langle R\rangle p \to \langle R\rangle p$	**4**
$R^{\smile} = R$	$p \to [R]\langle R\rangle p$	**B**
$R^{\smile};R \sqsubseteq \mathbb{I}$	$\langle R\rangle p \to [R]p$	**U**
$R^{\smile};R \sqsubseteq R$	$\langle R\rangle[R]p \to [R]p$	**E**
$R^{\smile};R \sqsubseteq S$	$\langle R\rangle[S]p \to [R]p$	
$R^{\smile};R \sqsubseteq R \sqcup R^{\smile}$	$[R]([R]p \to q) \vee [R]([R]q \to p)$	**Lem**
$R^{\smile};R \sqsubseteq \mathbb{I} \sqcup S \sqcup R$	$\langle R\rangle p \to [R](p \vee \langle R\rangle p \vee \langle S\rangle p)$	
$R^{\smile};R \sqsubseteq R;R^{\smile}$	$\langle R\rangle[R]p \to [R]\langle R\rangle p$	**G**
$R^{\smile};R \sqsubseteq Q;Q^{\smile}$	$\langle R\rangle[Q]p \to [R]\langle Q\rangle p$	
$R^{\smile};R \sqsubseteq \mathbb{I} \sqcup R \sqcup R^{\smile} \sqcup R;R^{\smile}$	$\langle R\rangle(p \wedge [R]p) \to [R](p \vee \langle R\rangle p)$	
$\mathbb{I} \sqsubseteq R;R^{\smile}$	$[R]p \to \langle R\rangle p$	**D**
$\mathbb{I} \sqcup S \sqcup R \sqsubseteq R;R^{\smile}$	$[R]p \vee \langle R\rangle[R]p \vee \langle S\rangle[R]p \to \langle R\rangle p$	
$R^{\smile};(R \sqcap \overline{R^{\smile}}) \sqsubseteq R$	$(\langle R\rangle[R]p \to q) \vee [R]([R]q \to p)$	**F**
$R^{\smile};(R \sqcap \overline{R^{\smile}}) \sqsubseteq R$	$\langle R\rangle[R]p \to (p \to [R]p)$	**R**
$Q;(R \sqcap \overline{S}) \sqsubseteq T$	$\langle Q\rangle(\langle R\rangle p \wedge \neg\langle S\rangle p) \to \langle T\rangle p$	**Y**

Table 5.1 Relation algebraic and modal correspondences

We show how this is done for (**Y**). The variable-free version is obtained by substitution of p with $\mathbb{I}$, and the other direction is proved by the following relational derivation:

1.	$\overline{S;p};p^{\smile} \sqsubseteq \overline{S}$	(ax)
2.	$R;p \sqcap \overline{S;p} \sqsubseteq (R \sqcap \overline{S;p};p^{\smile});(p \sqcap R^{\smile};\overline{S;p})$	
3.	$R,p \sqcap \overline{S;p} \sqsubseteq (R \sqcap \overline{S;p};p^{\smile});p$	(2)
4.	$R;p \sqcap \overline{S;p} \sqsubseteq (R \sqcap \overline{S});p$	(1, 3)
5.	$Q;(R \sqcap \overline{S}) \sqsubseteq T$	(ass)
6.	$Q;(R \sqcap \overline{S});p \sqsubseteq T;p$	(5, **qep**)
7.	$Q;(R;p \sqcap \overline{S;p}) \sqsubseteq T;p$	(4, 6)

Using the same methods, we can derive correspondences even for richer modal languages such as Peirce algebras or dynamic and temporal logics.

To conclude this chapter, we give some correspondences for genuine second order properties expressed by modal and relational derivation rules. In Chapt. 9 it will be shown how such rules can be converted into efficient algorithms which test the respective properties on finite frames.

A path in a frame is a sequence of R-successors of any world for the given accessibility relation R. A frame is called *terminal* if it contains no infinite paths. Finite frames are terminal iff the transitive closure of R is irreflexive ($R^+ \sqcap \mathbb{I} = \perp\!\!\!\perp$). If $R \sqcap \mathbb{I} \neq \perp\!\!\!\perp$, this *trivial loop* gives rise to an infinite path. A frame is called *weakly terminal* if every infinite path ends in a trivial loop. (This notion could be generalized to finite loops of arbitrary fixed length.)

The existence of infinite paths is expressed by the second-order sentence

$$\exists p\,(\exists x\,(p(x) \wedge (\forall y\,(p(y) \to \exists z\,(R(y,z) \wedge p(z))))))\ .$$

Thus the frame $\mathcal{F}$ is terminal if the negation of that sentence holds:

$$\forall p\,(\forall y\,(p(y) \to \exists z\,(R(y,z) \wedge p(z))) \quad \to \quad \forall x\,(\neg p(x)))$$

In qRA this can be formulated as the rule (**LR**):

$$(p \sqsubseteq R;p) \implies p = \mathbb{L}.$$

Its modal version is the so-called Löb-rule:

$$([R]p \to p) \implies p\ .$$

With transitivity it is equivalent to the axiom **W**: $[R]([R]p \to p) \to [R]p$.
Terminality implies irreflexivity ($R \sqcap \mathbb{I} = \mathbb{L}$), which is modally not expressible:

1.	$\mathbb{I} \sqcap R \sqsubseteq \mathbb{I}$	(ax)
2.	$\mathbb{I} \sqcap R \sqsubseteq (\mathbb{I} \sqcap R);(\mathbb{I} \sqcap R)$	(1)
3.	$\mathbb{I} \sqcap R \sqsubseteq R;(\mathbb{I} \sqcap R)$	(2)
4.	$\mathbb{I} \sqcap R = \mathbb{L}$	(**LR**, 3)

All infinite paths in a transitive frame end in a trivial loop, if there is no possibility to alternate the truth value of any proposition p infinitely often:

$$\forall p\,(\forall x\,(p(x) \to \exists y\,(R(x,y) \wedge \neg p(y) \wedge \exists z\,(R(y,z) \wedge p(z)))) \quad \to \quad \forall x\,(\neg p(x))).$$

Therefore, strong terminality can be described by the rule (**Grz**):

$$p \sqsubseteq R;(\overline{p} \sqcap R;p) \implies p = \mathbb{L}.$$

An equivalent modal formula is the Grzegorczyk-axiom:

$$[R]([R](p \to [R]p) \to p) \to [R]p\ .$$

Every transitive standard frame for (**Grz**) is antisymmetrical ($R \sqcap R^{\smile} \sqsubseteq \mathbb{I}$). Antisymmetry can be derived in qRA from **4** ($R;R = R$) and (**Grz**) with point-axiom. Using a nonrepresentable relation algebra, it can be shown that antisymmetry can not be derived from **4** and (**Grz**) alone. So, the *relational* **K4(Grz)** based on qRA is incomplete. However, antisymmetry can not be formulated as a modal formula. In fact, the *modal* **K4(Grz)** is complete.

Terminality is related to discreteness. A path (on a linear frame) is *unbounded*, if every successor of any point on the path has a successor which again is on the path: $p \to [R]\langle R\rangle p$. A frame $\mathcal{F}$ is called *discrete*, if every infinite path is unbounded. So, discreteness is defined by the rule (**Z**):

$$p \sqsubseteq R;p \implies p \sqsubseteq \overline{R;\overline{R;p}}\ .$$

Allowing trivial loops in infinite paths generalizes discreteness to reflexive (transitive) frames as well. A linear frame is weakly discrete iff it satisfies the Dummett rule (**Dum**):

$$p \sqsubseteq R;(\overline{p} \sqcap R;p) \implies p \sqsubseteq \overline{R;\overline{R;p}}\ .$$

With transitivity (**Dum**) is equivalent to the modal axiom:

$$([R]([R](p \to [R]p) \to p) \wedge \langle R\rangle[R]p) \to [R]p\ .$$

Let X be the *next-step* relation of a program (cf. Sect. 5.3), and X^* its reflexive transitive closure. If X is univalent, then X^* satisfies (**Dum**).

The *recursion axiom* $X^* = \mathbb{I} \sqcup X;X^*$ and *induction rule* $X;S \sqsubseteq S \implies X^*;S \sqsubseteq S$ from Sect. 1.2 characterize the reflexive transitive closure of any relation. Its temporal and dynamic logic analogues are the Segerberg axioms **Rec**: $[X^*]p \leftrightarrow (p \wedge [X][X^*]p)$ and **Ind**: $[X^*](p \to [X]p) \to (p \to [X^*]p)$. The recursion axiom forces X^* to be any reflexive transitive relation containing X, and induction determines X^* to be the smallest such relation: Let F be any relation with $\mathbb{I} \sqsubseteq F$, $X \sqsubseteq F$ and $F;F \sqsubseteq F$. Then $X;F \sqsubseteq F;F$, hence $X;F \sqsubseteq F$. Assuming that from $X;S \sqsubseteq S$ we can derive $X^*;S \sqsubseteq S$, we can infer $X^*;F \sqsubseteq F$. Since $\mathbb{I} \sqsubseteq F$, we have $X^*;\mathbb{I} \sqsubseteq X^*;F$, which gives $X^* \sqsubseteq F$.

Even with Kleene star, there are some properties of programs not expressible with modal or relational rules; recall that these are sublanguages of universal second-order logic. For example, (**LR**) expresses terminality, i.e. the property that every execution terminates. To assert the complement, i.e. that some computation loops, we extend modal logic and relational algebra with recursive definitions of operators. By the above, R^* is the smallest relation satisfying $R^* = \mathbb{I} \sqcup X;R^*$; introducing a propositional quantification μ for least fixpoints this can be written as $(\langle R^*\rangle p \leftrightarrow \mu q[p \vee \langle R\rangle q])$ or $(R^* = \mu q[\mathbb{I} \sqcup R;q])$, respectively.

Formally, the μ-calculus can be seen as a sublanguage of monadic second order logic, via the standard translation:

- $ST(\mu q[\varphi]) \triangleq \forall q\,(\forall y\,(ST(\varphi)[x/y] \to q(y)) \to q(x))$

The property that some execution of a program loops can be formulated in the modal μ-calculus as $\neg\mu q[[R]\mathsf{true}]$: The standard translation of this sentence is exactly the second-order characterization of infinite paths given above.

All known program logics which are decidable can be embedded in this language. Moreover, this calculus itself is decidable and was recently shown to be complete by [Walukiewicz 1995]. The relevant recursion axiom and induction rule are $\varphi[q/\mu q[\varphi]] \to \mu q[\varphi]$ and $\varphi[q/p] \to p \implies \mu q[\varphi] \to p$.

The propositional μ-calculus is particularly useful for automatic program verification: Given a μ-calculus *specification* formula, and a model describing (the executions of) the program, we can give simple algorithms for testing whether the specification is satisfied by the program. This *model checking* problem has received much attention, and several elaborate data structures like *binary decision diagrams* for the representation of relational models have been developed. Current research topics in this area are adequate representations of parallel transition relations, arithmetical functions and real-time systems.

5.5 Conclusion

In this chapter we have presented modal logic as a sublanguage of the relational calculus. We gave relational semantics for general modal languages, and compared expressivity, completeness, and correspondence results in both formalisms. In doing so, it turned out that many techniques developed for modal logics – bisimulations, tableaux constructions, derivation systems, undecidability proofs, etc. –

can directly be transferred to relation algebra. On the other hand, the relational approach points out how to increase the limited expressiveness of modal logics: Many properties can be expressed very elegantly using relation algebraic concepts in combination with modal operators.

Modal logics are widely used in computer aided verification of reactive systems. It is in this application that we see a fruitful area for further development of such a combined formalism. Interesting theoretical questions include the search for a largest decidable sublanguage, its expressivity on certain classes of structures (like partial orders), and adequate induction principles for infinite state systems. To be able to apply relational reasoning in an industrial design process, also practical questions have to be solved. Besides the already mentioned search for "good" representations of relational models, heuristics are needed to reduce the average complexity of the analysis of these models. Finally, to support the developer in the construction process, formal tools and design languages are being developed, cf. Chapt. 12. Hopefully, with the help of these tools the reliability of software systems will gradually increase in the future.

Chapter 6

Relational Formalisation of Nonclassical Logics

Ewa Orlowska

The purpose of this chapter is to present and motivate the application of algebras of relations to the formalisation of nonclassical logics. It is shown that a suitably defined relational logic can serve as a general framework for developing nonclassical means of reasoning that are needed in many application areas.

The main advantages of the relational formalisation are uniformity and modularity. The relational logic is a rich formalism that enables us to represent a great variety of nonclassical logical systems. As a consequence, instead of implementing proof systems for each of the logics separately and from scratch, we can create a uniform relational framework in which proof systems for nonclassical logics can be developed in a modular manner. The essential observation leading to the relational formalisation is that the standard algebras of relations – being joins of a Boolean algebra and a monoid – constitute, in a sense, a common core of a great variety of nonclassical logics. The material presented in this chapter is a continuation of work reported in [Orlowska 1991; Orlowska 1992; Orlowska 1993; Orlowska 1994; Orlowska 1995].

The relational formalisation of nonclassical logics is provided by means of a relational semantics of the languages of these logics and relational proof systems. The relational semantics is determined by a class of algebras of relations, possibly with some nonstandard operations and/or constants, and by a meaning function that assigns elements of these algebras to formulas. It follows that under the relational semantics formulas are interpreted as relations. A formula is true in a relational model whenever its meaning is the maximum element of the underlying algebra, and it is valid iff it is true in all of its relational models. The distinguishing feature of the relational semantics dealt with in this chapter (which originated in [Orlowska 1991]) is that the meaning of formulas is provided in terms of right ideal relations, or "vectors", as they are called in Chapt. 3. Another relational semantics is developed for the Lambek calculus in [Mikulás 1992].

Nonclassical logics are two-dimensional in the following sense: they consist of an extensional part and of an intensional part. The extensional part carries declarative information about states of a domain that the logic is intended to model. This information enables us to exhibit the extensions of concepts that the formulas are supposed to represent. The intensional part provides information about relationships between the states of the domain. This information enables us to represent the intensions of concepts. It might also be interpreted as procedural information, i.e. a relationship between states is understood as a transition from one state to another.

Relational formalisation exhibits these two dimensions explicitly. The Boolean reducts of the algebras of relations that determine the relational semantics are the counterpart of the first dimension, and the monoidal parts of these algebras reflect the second dimension. The two logical dimensions are also manifested in the relational proof systems. The decomposition rules for the Boolean operations refer to the first dimension, and the rules for the monoid operations correspond to the second dimension. In this chapter the paradigm of relational representation of the two logical dimensions is illustrated with two families of applied logics. In Sect. 6.2 and 6.3 we present the relational formalisation of a family of logics for reasoning about state transitions. In Sect. 6.4 and 6.5 a family of logics is considered that provide a means of representation of concepts that arise from incomplete information.

6.1 Nonclassical logics

The language of the classical propositional calculus enables us to represent declarative statements that refer to a domain or a universe of discourse, without giving us a possibility of distinguishing between various situations, states of affair or possible worlds that the declared facts might depend upon. One of the reasons for this is that the propositional operations admitted in the classical propositional calculus are *extensional*, which means that in every possible world the truth value of any formula built with any one of these operations is determined by the truth values of its subformulas in that possible world. To increase the expressiveness of logical languages we admit also *intensional* propositional operations. Usually, in nonclassical logics, and in particular in the logics considered in this chapter, some of the propositional operations are intensional. Given a possible world w, the truth value in w of any formula built with intensional operations depends not only on the truth values in w of its subformulas but also on their truth values in some other worlds that might be related to w, for example in the worlds that precede w in a time scale. (Cf. the introduction to modal logic in Chapt. 1.

The semantical presentation of nonclassical logics with intensional operators is usually provided in terms of a *possible worlds semantics* [Kripke 1963; Kripke 1965]: Formulas are understood as subsets of a universe of possible worlds and intensional propositional operations are defined in terms of an accessibility relation or a family of accessibility relations between possible worlds. Next, the inductively defined satisfiability of formulas at worlds provides truth conditions for the formulas. Let a structure $\mathcal{K} = (W, R)$ be given, where W is thought of as a nonempty set of possible worlds and $R \subseteq W \times W$ is a binary relation in W that is referred to as an accessibility relation. In the context of nonclassical logic such a structure is called a *frame*. With any frame $\mathcal{K}$ we associate a family of models $\mathcal{M} = (W, R, m)$, where m is a meaning function that assigns subsets of W to propositional variables. Let VARPROP be a countable set of propositional variables. We consider propositional languages whose formulas are generated by VARPROP by means of various propositional operations. Satisfiability of atomic formulas and formulas built with the classical extensional operations of negation ($\neg$), disjunction ($\vee$), conjunction ($\wedge$), and implication ($\rightarrow$) is defined as follows:

- $\mathcal{M}, w \models p$ iff $w \in m(p)$ for any propositional variable p,
- $\mathcal{M}, w \models \neg A$ iff it is not the case that $\mathcal{M}, w \models A$,
- $\mathcal{M}, w \models A \vee B$ iff $\mathcal{M}, w \models A$ or $\mathcal{M}, w \models B$,
- $\mathcal{M}, w \models A \wedge B$ iff $\mathcal{M}, w \models A$ and $\mathcal{M}, w \models B$,
- $\mathcal{M}, w \models A \rightarrow B$ iff either $\mathcal{M}, w \models A$ does not hold, or $\mathcal{M}, w \models B$.

A formula A is true in $\mathcal{M}$ iff $\mathcal{M}, w \models A$ for all $w \in W$, it is true in the frame $\mathcal{K} = (W, R)$ iff it is true in every model $\mathcal{M} = (W, R, m)$ based on this frame, and A is valid iff it is true in all frames. Given a model $\mathcal{M}$, we extend the meaning function m to the mapping from the set of all the formulas to the subsets of W (for the sake of simplicity we denote this mapping by m as well). For a formula A we define its meaning $m(A)$ in $\mathcal{M}$ to be the set of all the states that satisfy A:

$$m(A) = \{w \in W : \mathcal{M}, w \models A\}.$$

It follows that:

- $m(\neg A) = \overline{m(A)}$ where $\overline{}$ is the complement with respect to the set W.
- $m(A \vee B) = m(A) \sqcup m(B)$,
- $m(A \wedge B) = m(A) \sqcap m(B)$,
- $m(A \rightarrow B) = m(\neg A \vee B)$.

In languages of modal logics we further admit the intensional operations $[R]$ and $\langle R\rangle$, respectively, where R is a relational constant interpreted as an accessibility relation. These operations are semantically defined as follows:

- $\mathcal{M}, w \models [R]A$ iff for all $u \in W$, if $(w, u) \in R$, then $\mathcal{M}, u \models A$,
- $\mathcal{M}, w \models \langle R\rangle A$ iff there is a $u \in W$ such that $(w, u) \in R$ and $\mathcal{M}, u \models A$.

The standard application of these operators is related to representation of propositional attitudes, and in this connection they are interpreted as *necessity* and *possibility* operations, respectively. In applied logics the accessibility relations and the modal operators determined by them receive some other intuitive interpretations.

Various modal systems are obtained by assuming different constraints on the accessibility relations in frames. Let "FRM(*Conditions*)" denote a class of frames (W, R) such that the relation R satisfies the "*Conditions*". Then by L(*Conditions*) we mean the logic of this class of frames, that is the logic whose models are based on frames from FRM(*Conditions*). Validity of a formula A in such a logic is defined as truth of A in all the frames from FRM(*Conditions*). Well-known modal logics are for example K=L(*no restriction*), T=L(*reflexive*), B=L(*reflexive, symmetric*), S4=L(*reflexive, transitive*), S5=L(*reflexive, symmetric, transitive*).

Applied logics are very often *multimodal*, i.e. in their languages a *family* of modal operators is admitted, each of which is determined by an accessibility relation. The respective semantical structures are built with frames in which a family of relations is given, instead of a single relation. Usually, this family of relations is assumed to be closed under some relational operations, and/or to satisfy some constraints. Elements of sets of possible worlds and accessibility relations from such frames receive various intuitive interpretations, depending on the domain of application. In Sect. 6.2 we present a logic of programs, where the elements of the universes of frames and the accessibility relations are interpreted as computation states and programs, respectively. In Sect. 6.4 a family of information logics is given. The elements of the universes of frames of these logics are interpreted as objects in information systems and the accessibility relations reflect types of incompleteness of information that a given logic is intended to model.

In the languages of multimodal logics we admit relational expressions that represent accessibility relations. Let CON be a nonempty, finite or countably infinite set of relational constants, and let the set EREL of relational expressions be generated by CON using all the relational operations admitted in the frames for the logic under consideration. Next, let the set of formulas of the logic consist of formulas generated from propositional variables with the classical propositional operations and the modal operations $[P]$ and $\langle P \rangle$ for all $P \in$ EREL.

The semantics of a multimodal language is determined by models of the form $\mathcal{M} = (W, \{R_P : P \in \text{CON}\}, m)$, where each R_P is a binary relation over W, and m is a meaning function that assigns subsets of W to propositional variables, relations R_P to the respective relational constants P, and extends to all the relational expressions in the homomorphic way. Next, we extend the meaning function m to all other formulas in the same way as for a logic with a single accessibility relation. It follows that each model $\mathcal{M}$ of a multimodal logic determines an algebra of relations generated by set $\{m(P) : P \in \text{EREL}\}$ with the relational operations that are admitted in the language of the given logic. The elements of these algebras in turn determine intensional operations in the language of the logic.

The deep meaning of intensional components of a logical language is related to representation of concepts. According to a well-established logical tradition [Bunge 1967] the two carriers of any concept are its *extension* (or denotation) and *intension* (or connotation). The extension of a concept consists of the objects that are instances of this concept, and the intension of a concept consists of the properties that are meaningful for the objects to which this concept applies. For example, to define the concept "organism" we should list the characteristics of organisms and typical species of organisms. In Sect. 6.4 we show how intensions of concepts can be represented in a calculus of relations.

6.2 Program logics

In various applied logics, modal languages are a medium for describing not only static facts in an application domain but also dynamics in a broad sense, for example changes of an information content of a system caused by performing an action, transitions from one computation state to the other caused by processing a program, etc. In the present section we consider a rich modal logic of programs, where possible worlds in the respective frames are interpreted as computation states, and accessibility relations are interpreted as state transition relations. We assume that the set of programs is closed under a variety of program constructors which cover a wide spectrum of commands from various programming languages.

Let CON be a set of relational constants, interpreted as atomic programs. Compound programs are constructed from the atomic ones by means of the following operations: nondeterministic choice ($\sqcup$), sequential composition ($;$), iteration (*), weakest prespecification ($/$), demonic union ($\|$), demonic composition ($;;$). Let EREL be the set of program expressions generated from CON by means of the above operations. Formulas of the program logic are built from propositional variables with the classical propositional operations and modal operations $\langle P \rangle$ and $[P]$ for every program expression P. It follows that programs play the role of accessibility relations. A *program frame* is a system of the form $\mathcal{K} = (W, \{R_P : P \in CON\})$, where W is a nonempty set of computation states and for each program constant P we have a binary relation R_P in K. In models $\mathcal{M} = (W, \{R_P : P \in CON\}, m)$ based on program frames, the meaning function $m : \mathrm{CON} \to (W \leftrightarrow W)$ extends to all the program expressions in the following way (using the relational operations and notations introduced in Chapt. 1):

- $m(P \sqcup Q) = m(P) \cup m(Q)$ (nondeterministic choice is given by union),
- $m(P ; Q) = m(P) ; m(Q)$ (composition is given by relational composition), etc.

In the same way as for composition, reflexive transitive closure, weakest prespecification, demonic union, and demonic composition are replaced by the corresponding relational operation.

Intuitively, $(w, z) \in m(P)$ means that there exists a computation of program P starting in state w and terminating in state z. Program $P \sqcup Q$ performs P or Q nondeterministically, $P ; Q$ performs first P and then Q, and P^* performs P zero or more times sequentially. R/Q is the weakest prespecification of Q to achieve R, that is the greatest program such that $(R/Q) ; Q \sqsubseteq R$. The interpretation of program specifications as residuations has been suggested in [Hoare, He 1986a]. Demonic program constructors are investigated in [Nguyen 1991], they are motivated by a "demonic" view of nontermination of a program: if it is possible that the program fails to terminate, then it definitely will not terminate.

The satisfiability of formulas of the program logic by states in a model $\mathcal{M} = (W, \{R_P : P \in \mathrm{CON}\}, m)$ is defined in the same way as in Sect. 6.1 for the formulas built with the classical propositional connectives, and for formulas built with intensional connectives $[P]$, $\langle P \rangle$, where for $P \in \mathrm{EREL}$, we have:

- $\mathcal{M}, w \models [P]A$ iff for all $u \in W$ if $(w, u) \in m(P)$, then $\mathcal{M}, u \models A$.
- $\mathcal{M}, w \models \langle P \rangle A$ iff there is a $u \in W$ such that $(w, u) \in m(P)$ and $\mathcal{M}, u \models A$.

Let PRO be the class of program frames defined above, and let L(PRO) be the logic of this class of frames. Formula A is said to be valid in L(PRO) iff A is true in every frame from class PRO. Logic L(PRO) is a combination of dynamic logic [Pratt 1976], dynamic logic with program specifications [Orlowska 1993], and a fragment without demonic iteration of the logic of demonic nondeterministic programs [Demri, Orlowska 1996]. We may adjoin several other program operations and program constants to the language, for example *test* (?), *if _ then _ else _ fi*, *abort*, *skip*, *havoc*. Recall that "*test*" is a function from formulas to program expressions, that is, $A?$ is a program expression for any formula A. The operator *if _ then _ else _ fi* acts on a formula and a pair of programs, that is, if A is a formula, and P, Q are program expressions, then "*if A then P else Q fi*" is a program expression. Given a program model $M = (W, \{R_P : P \in \mathrm{CON}\}, m)$, the semantics of these program expressions is defined as follows:

- $m(A?) = \{(w,w) : w \in W \wedge \mathcal{M}, w \models A\} = I_W \sqcap m(A)$,
- $m(\mathit{if}\ A\ \mathit{then}\ P\ \mathit{else}\ Q\ \mathit{fi}) = m(A?) ; m(P) \sqcup m(\overline{A}?) ; m(Q)$,
- $m(\mathit{abort}) = \emptyset$,
- $m(\mathit{skip}) = I_W$,
- $m(\mathit{havoc}) = V\ (= W \times W)$.

It follows that $A?$ is a command to continue if A is true and to fail otherwise, and *if A then P else Q fi* performs P whenever an input state satisfies A, and performs Q otherwise.

The definition of semantics of the language of L(PRO) makes use of an extension of the standard relational calculus. Each program model $(W, \{R_P : P \in \mathrm{CON}\}, m)$ determines an algebra of relations, namely the algebra generated by $\{m(P) : P \in \mathrm{CON}\}$ with the standard relational operations and operations$/$, *, $\|$, $;;$. The components of the language of L(PRO) that represent dynamics of computation states receive their meaning through interpretation in these algebras. In the following section we develop a relational formalisation of L(PRO), in particular we define a relational semantics of the language of L(PRO) such that both relational expressions that represent programs and formulas receive their meaning through interpretation in some algebras of relations.

6.3 Relational formalisation of program logics

In this section we define a relational semantics and a relational proof system for a logic L(PRO) along the lines suggested in [Orlowska 1991; Orlowska 1994; Orlowska 1995]. Let $\mathcal{K} = (W, \{R_P : P \in \mathrm{CON}\})$ be a program frame. A relational model for L(PRO) based on $\mathcal{K}$ is any system of the form $\mathcal{M} = (W, \{R_P : P \in \mathrm{CON}\}, m)$ such that m is a meaning function that assigns binary relations in W to propositional variables and program constants, and extends to all the program expressions and formulas as follows. $m : \mathrm{VARPROP} \sqcup \mathrm{CON} \to \mathcal{P}(V)$ (where $\mathcal{P}(V) = \mathcal{P}((W \times W)) = (W \leftrightarrow W)$).

- $m(p)$ is a right ideal relation for $p \in \mathrm{VARPROP}$, that is $m(p) = X \times W$ for some $X \subseteq W$,
- $m(P)$ is defined as in Sect. 6.2 for $P \in \mathrm{EREL}$,
- $m(\neg A) = \overline{m(A)}$, $\qquad m(A \vee B) = m(A) \sqcup m(B)$,
  $m(A \to B) = m(\neg A \vee B)$, $\qquad m(A \wedge B) = m(A) \sqcap m(B)$,
- $m(\langle P \rangle A) = m(P) ; m(A)$, $\qquad m([P]A) = \overline{m(P) ; \overline{m(A)}}$.

A formula A is true in the relational model $\mathcal{M}$ iff $m(A) = V$ (the universal relation, $W \times W$). A formula A is relationally valid in L(PRO) iff A is true in every relational model for L(PRO). Now it is not difficult to establish the following facts:

- For every relational model $\mathcal{M} = (W, \{R_P : P \in \text{CON}\}, m)$ for L(PRO), and for every formula A, its meaning $m(A)$ in $\mathcal{M}$ is a right ideal relation in W.
- For every program model $\mathcal{M}$ for L(PRO), there is a relational model $\mathcal{M}'$ such that every formula of L(PRO) is true in $\mathcal{M}$ iff it is true in $\mathcal{M}'$.
- For every relational model $\mathcal{M}$ for L(PRO) there is a program model $\mathcal{M}'$ such that every formula of L(PRO) is true in $\mathcal{M}$ iff it is true in $\mathcal{M}'$.

We can now establish the equivalence of the standard and relational semantics.

Theorem 6.3.1
A formula is valid in L(PRO) iff it is relationally valid in L(PRO). □

It follows that we can view both program expressions and formulas of L(PRO) as expressions that represent binary relations.

The relational proof system for a logic L(PRO) consists of *rules* that apply to *finite sequences* of program expressions and/or formulas labelled with a pair of symbols that might be viewed as individual variables. Let VAR be a countable set of individual variables. We deal with expressions of the form xAy, where A is either a program expression or a formula of the language of L(PRO), or the universal relation V, or the identity relation I, and $x, y \in \text{VAR}$. These expressions are referred to as *relational formulas.* Given a relational model $\mathcal{M} = (W, \{R_P : P \in \text{CON}\}, m)$ for L(PRO), by a *valuation* of individual variables in $\mathcal{M}$ we mean a mapping $v : \text{VAR} \to W$. We say that a relational formula xAy is *satisfied* by v in $\mathcal{M}$ (denoted as $\mathcal{M}, v \models xAy$) whenever $(v(x), v(y)) \in m(A)$. A formula xAy is *true* in $\mathcal{M}$ iff $\mathcal{M}, v \models xAy$ for all v in $\mathcal{M}$. A sequence K of relational formulas is valid iff for every model $\mathcal{M}$ of L(PRO) and for every valuation v in $\mathcal{M}$ there is a formula in K which is satisfied by v in $\mathcal{M}$. It follows that sequences of relational formulas are interpreted as (metalevel) disjunctions of their elements. A relational rule $\frac{K}{H_1 \; \dots \; \dots \; H_n}$ is admissible whenever the sequence K is valid iff all the sequences H_i are valid.

There are two groups of rules: decomposition rules and specific rules. The decomposition rules enable us to decompose some formulas in a sequence into simpler formulas. Decomposition depends on relational and propositional operations occurring in a formula. As a result of decomposition we obtain finitely many new sequences of formulas. The specific rules enable us to modify a sequence to which they are applied; they have the status of structural rules. The role of axioms is played by what is called fundamental sequences. In what follows K and H denote finite, possibly empty, sequences of relational formulas. A variable is said to be *restricted* in a rule whenever it does not appear in any formula of the upper sequence in that rule.

Decomposition rules for relational operations

($\sqcup$) $\dfrac{K, xQ\sqcup Ry, H}{K, xQy, xRy, H}$

($-\sqcup$) $\dfrac{K, x\overline{Q\sqcup R}y, H}{K, x\overline{Q}y, H \quad K, x\overline{R}y, H}$

($\sqcap$) $\dfrac{K, xQ\sqcap Ry, H}{K, xQy, H \quad K, xRy, H}$

($-\sqcap$) $\dfrac{K, x\overline{Q\sqcap R}y, H}{K, x\overline{Q}y, x\overline{R}y, H}$

($;$) $\dfrac{K, x(Q;R)y, H}{K, xQz, H, x(Q;R)y \quad K, zRy, H, x(Q;R)y}$
where z is a variable

($-;$) $\dfrac{K, x\overline{Q;R}y, H}{K, x\overline{Q}z, z\overline{R}y, H}$ where z is a restricted variable

($--$) $\dfrac{K, x\overline{\overline{Q}}y, H}{K, xQy, H}$

(*) $\dfrac{K, xQ^*y, H}{K, xQ^iy, H, xQ^*y}$ where $i \in \mathrm{N}$, $Q^0 = I$, and $Q^{i+1} = Q;Q^i$

Decomposition rules for propositional operations

($\rightarrow$) $\dfrac{K, x(A\rightarrow B)y, H}{K, x\neg Ay, xBy, H}$

($\neg\rightarrow$) $\dfrac{K, x\neg(A\rightarrow B)y, H}{K, xAy, H \quad K, x\neg By, H}$

($[]$) $\dfrac{K, x[Q]By, H}{K, x\overline{Q}z, zBy, H}$
z is a restricted variable

($\neg[]$) $\dfrac{K, x\neg[Q]By, H}{K, xQz, H, x\neg[Q]By \quad K, z\neg By, H, x\neg[Q]By}$
z is a variable

($\langle\rangle$) $\dfrac{K, x\langle Q\rangle By, H}{K, xQz, H, x\langle Q\rangle By \quad K, zBy, H, x\langle Q\rangle By}$
z is a variable

($\neg\langle\rangle$) $\dfrac{K, x\neg\langle Q\rangle By, H}{K, x\overline{Q}z, z\neg By, H}$
z is a restricted variable

The decomposition rules for disjunction and conjunction, that is the rules ($\vee$), ($\neg\vee$), ($\wedge$), ($\neg\wedge$) are obtained from ($\sqcup$), ($-\sqcup$), ($\sqcap$), ($-\sqcap$), respectively, through replacement of $-$, $\sqcup$, $\sqcap$ by $\neg$, $\vee$, $\wedge$ respectively.

Decomposition rules for demonic operations

($||$) $\dfrac{K, xQ||Ry, H}{K, xQz, H, xQ||Ry \quad K, xRt, H, xQ||Ry \quad K, xQy, xRy, H}$ where z, t are variables

($-||$) $\dfrac{K, x\overline{Q||R}y, H}{K, x\overline{Q}z, x\overline{R}t, x\overline{Q}y, H \quad K, x\overline{Q}z, x\overline{R}t, x\overline{R}y, H}$ where z, t are restricted variables

($;;$) $\dfrac{K, xQ;;Ry, H}{K, xQz, H, xQ;;Ry \quad K, zRy, H, xQ;;Ry \quad K, x\overline{Q}t, tRu, H, xQ;;Ry}$
where z, u are variables and t is a restricted variable

($-;;$) $\dfrac{K, x\overline{Q;;R}y, H}{K, x\overline{Q}z, z\overline{R}y, xQt, H, x\overline{Q;;R}y \quad K, x\overline{Q}z, z\overline{R}y, t\overline{R}u, H, x\overline{Q;;R}y}$
where z is a restricted variable and t, u are variables

Specific rules

(I1) $\dfrac{K, xAy, H}{K, xIz, xAy, H \quad K, zAy, xAy, H}$ where z is a variable and $A \in \mathrm{VARPROP} \cup \mathrm{CON} \cup \{I\}$,

(I2) $\dfrac{K, xAy, H}{K, xAz, xAy, H \quad K, zIy, xAy, H}$ where z is a variable and $A \in \mathrm{VARPROP} \cup \mathrm{CON} \cup \{I\}$,

(I3) $\frac{K, xIy, H}{K, yIx, H, xIy}$

(I4) $\frac{K, xIy, H}{K, xIz, H, xIy \quad K, zIy, H, xIy}$ where z is a variable,

(ideal) $\frac{K, xAy, H}{K, xAz, H, xAy}$ where z is a variable and $A \in$ VARPROP.

Fundamental sequences

A sequence of formulas is said to be *fundamental* whenever it contains a subsequence of either of the following forms:

(f1) xAy, $x\overline{A}y$ where A is a program expression, (f3) xVy,
(f2) xAy, $x\neg Ay$ where A is a formula, (f4) xIx.

Theorem 6.3.2
These rules are admissible and the fundamental sequences are valid.

Sketch of the proof: The admissibility of the decomposition rules follows from the definitions of the meanings of the respective relational and propositional operations, and the admissibility of the specific rules that refer to the identity relation follows from its properties. For example, rules (I1) and (I2) are admissible because we have $I;A = A = A;I$ for any relation A. Rules (I3) and (I4) are admissible because of symmetry and transitivity of the identity relation, respectively. Rule (ideal) is admissible because every propositional variable is interpreted as a right ideal relation. □

Relational proofs have the form of trees. Given a relational formula xAy, we successively apply decomposition or specific rules. In this way we form a tree whose root consists of xAy and whose nodes consist of finite sequences of relational formulas. We stop applying rules to formulas in a node after obtaining a fundamental sequence, or when none of the rules is applicable to the formulas in this node. A branch of a proof tree is said to be closed whenever it contains a node with a fundamental sequence of formulas. A tree is closed iff all of its branches are closed.

Theorem 6.3.3 equivalent for a formula A of L(PRO):

(a) A is relationally valid.

(b) For any $x, y \in$ VAR, xAy is true in every relational model for L(PRO).

(c) There is a closed proof tree with root xAy for any $x, y \in$ VAR. □

Proofs of completeness of relational proof systems are based on the method developed in [Rasiowa, Sikorski 1963] and extended to the systems with infinitary rules in [Mirkowska 1977]. The application of these methods to relational proof systems can be found in [Orlowska 1994] and [Demri, Orlowska 1996]. The proof of the above theorem can easily be obtained on the basis of those proofs.

Example 1 We give a relational proof of formula $\langle R;;S\rangle A \rightarrow \langle R\rangle\langle S\rangle A$.

$$x\langle R;;S\rangle A \rightarrow \langle R\rangle\langle S\rangle Ay$$

$$\overline{\qquad\qquad\qquad\qquad} \ (\rightarrow)$$

$$x\neg\langle R;;S\rangle Ay, x\langle R\rangle\langle S\rangle Ay$$

$$\overline{\qquad\qquad\qquad\qquad} \ (\neg\langle\rangle) \text{ with restricted variable } z$$

$$x\overline{R;;S}z, z\neg Ay, x\langle R\rangle\langle S\rangle Ay$$

$$\overline{\qquad\qquad\qquad\qquad} \ (-;;) \text{ with rest. var. } v \text{ and var. } v, z$$

$$x\overline{R}v, v\overline{S}z, t\overline{S}u, z\neg Ay, x\langle R\rangle\langle S\rangle Ay, \ldots \qquad x\overline{R}v, v\overline{S}z, xRt, z\neg Ay, x\langle R\rangle\langle S\rangle Ay, \ldots$$

$$\overline{\qquad\qquad\qquad\qquad} \ (\langle\rangle) \text{ with variable } v \qquad\qquad \textit{fundamental}$$

$$\ldots, xRv, \ldots \qquad\qquad \ldots, v\langle S\rangle Ay, \ldots$$

$$\textit{fundamental} \qquad \overline{\qquad\qquad\qquad\qquad} \ (\langle\rangle) \text{ with variable } z$$

$$\ldots, vSz, \ldots \qquad\qquad \ldots, zAy, \ldots$$

$$\textit{fundamental} \qquad\qquad \textit{fundamental}$$

Observe that the relational proof system defined above is fully modular in the following sense: Whenever we are interested in a reduct of the given language obtained by deleting some relational operations and/or propositional operations and/or relational constants, then the proof system for that reduct is obtained by deleting from the above set of rules and fundamental sequences those rules and fundamental sequences that refer to the deleted operators or constants. A logic L(PRO) can be viewed as the family of its reducts, and the relational proof system given in the present section includes the proof systems for all the logics from that family.

The above observation has a general character. Relational proof theory enables us to build proof systems for nonclassical logics systematically in a modular way. First, deduction rules are defined for the standard relational operations. These rules constitute a core of all the relational proof systems. Next, for any particular logic some new rules are designed and adjoined to the core set of rules. The new decomposition rules correspond to the propositional connectives from the language of the given logic. The new specific rules reflect properties of the accessibility relations that determine the intensional connectives in the logic. Hence, we need not implement each deduction system from scratch; we should only extend the core system with a module corresponding to the specific part of a logic under consideration.

6.4 Information logics

Information logics are multimodal logics that provide a means of representing and reasoning about concepts in the presence of incomplete information. An *information system* is defined as a triple $(\mathcal{OB}, \mathcal{AT}, \{\mathrm{VAL}_a : a \in \mathcal{AT}\})$, such that $\mathcal{OB}$ is a nonempty set of *objects*, $\mathcal{AT}$ is a nonempty set of *attributes*, and each VAL_a is a nonempty set of values of the attribute a. Each attribute is a function $a : \mathcal{OB} \rightarrow \mathcal{P}(\mathrm{VAL}_a)$ that assigns nonempty sets of values (of attributes) to objects. Any set $a(x)$ can be viewed as a *property* of object x. For example, if attribute a is "colour" and $a(x) = \{\text{green}\}$, then x possesses the property "to be green". If a is "age" and x is 25 years old, then $a(x) = \{25\}$. However, we might know the age of x only approximately, say between 20 and 28, and then $a(x) = \{20, \ldots, 28\}$. If $a(x)$ is a singleton set, then we have deterministic information about x, otherwise nondeterministic information. Given an information system, subsets of the set $\mathcal{OB}$ can be interpreted as extensions of concepts and sets $a(x)$ as elements of intensions of concepts. The basic explicit information provided by the system may be enriched by disclosing some implicit relationships between objects. Most often these relationships are expressed in the form of binary relations over the set of objects; if so they are referred to as *information relations*.

Let A be a subset of $\mathcal{AT}$. The following classes of information relations derived from an information system reflect incompleteness of information as manifested by *approximate indistinguishability* relative to the given properties:

Strong indiscernibility: $(x, y) \in \mathrm{ind}(A)$ iff $a(x) = a(y)$ for all $a \in A$,
Weak indiscernibility: $(x, y) \in \mathrm{wind}(A)$ iff $a(x) = a(y)$ for some $a \in A$.

Intuitively, two objects are A-indiscernible whenever all the A-properties that they possess are the same; i.e., up to the discriminative resources of A, these objects are the same. Two objects are weakly A-indiscernible whenever some of their A-properties are the same.

An important application of information relations from the indiscernibility group is related to the representation of *approximations* of sets of objects in information systems. If R is one of the relations $\mathrm{ind}(A)$ or $\mathrm{wind}(A)$, A is a subset of $\mathcal{AT}$, and X is a subset of $\mathcal{OB}$, then the *lower* $R(A)$-approximation $L_{R(A)}(X)$ of X and the upper $R(A)$-approximation $U_{R(A)}(X)$ of X are defined as follows:

$$L_{R(A)}(X) = \{x \in \mathcal{OB} : \text{for all } y \in \mathcal{OB}, \text{ if } (x, y) \in R(A), \text{ then } y \in X\},$$
$$U_{R(A)}(X) = \{x \in \mathcal{OB} : \text{there is } y \in \mathcal{OB} \text{ such that } (x, y) \in R(A) \text{ and } y \in X\}.$$

In the classical theory of *rough sets* [Pawlak 1991], if a relation $R(A)$ is a strong indiscernibility relation, then we obtain the following hierarchy of definability of sets. A subset X of $\mathcal{OB}$ is said to be:

A-definable if $L_{\mathrm{ind}(A)}(X) = X = U_{\mathrm{ind}(A)}(X)$,
roughly A-definable if $L_{\mathrm{ind}(A)}(X) \neq \emptyset$ and $U_{\mathrm{ind}(A)}(X) \neq \mathcal{OB}$,
internally A-indefinable if $L_{\mathrm{ind}(A)}(X) = \emptyset$,
externally A-indefinable if $U_{\mathrm{ind}(A)}(X) = \mathcal{OB}$,
totally A-indefinable if internally A-indefinable and externally A-indefinable.

Another application of information relations is related to modeling uncertain knowledge acquired from information about objects provided in an information system [Orlowska 1983; Orlowska 1989]. Let X be a subset of $\mathcal{OB}$, then we define sets of A-positive ($POS_A(X)$), A-borderline ($BOR_A(X)$), and A-negative ($NEG_A(X)$) instances of X as follows:

$POS_A(X) = L_{\mathrm{ind}(A)}(X)$,
$BOR_A(X) = U_{\mathrm{ind}(A)}(X) - L_{\mathrm{ind}(A)}(X)$,
$NEG_A(X) = \mathcal{OB} - U_{\mathrm{ind}(A)}(X)$.

Knowledge about a set X of objects that can be discovered from information given in an information system can be modelled in the following way:

$K_A(X) = POS_A(X) \sqcup NEG_A(X)$.

Intuitively, A-knowledge about X consists of those objects that are either A-positive instances of X (they are members of X up to properties from A) or

A-negative instances of X (they are not members of X up to properties from A). We say that A-knowledge about X is:

complete if $K_A(X) = \mathcal{OB}$ (otherwise incomplete),
rough if $POS_A(X) \neq \emptyset$, $BOR_A(X) \neq \emptyset$, $NEG_A(X) \neq \emptyset$,
pos-empty if $POS_A(X) = \emptyset$,
neg-empty if $NEG_A(X) = \emptyset$,
empty if pos-empty and neg-empty.

However, in many situations it is more suitable to ask not for indistinguishability but for its opposite. To model degrees of distinguishability in a nonnumerical way we consider a family of diversity-type relations. Incompleteness of information manifested by approximate distinguishability can be modeled with diversity information relations derived from information systems. We define the following classes of these relations:

strong diversity: $(x, y) \in \mathrm{div}(A)$ iff $a(x) \neq a(y)$ for all $a \in A$,
weak diversity: $(x, y) \in \mathrm{wdiv}(A)$ iff $a(x) \neq a(y)$ for some $a \in A$.

Intuitively, objects are A-diverse (weakly diverse) if all (some) of their A-properties are different. Applications of diversity relations are related to algorithms for finding cores of sets of attributes. Let an information system $(\mathcal{OB}, \mathcal{AT}, \{\mathrm{VAL}_a : a \in \mathcal{AT}\})$ be given and let A be a subset of $\mathcal{AT}$. We say that an attribute $a \in A$ is *indispensable* in A iff $\mathrm{ind}(A) \neq \mathrm{ind}(A - \{a\})$, that is there are some objects such that a is the only attribute from A that can distinguish between them. A *reduct* of A is a minimal subset A' of A such that every $a \in A'$ is indispensable in A' and $\mathrm{ind}(A') = \mathrm{ind}(A)$. The core of A is defined as $CORE(A) = \sqcap\{A' \subseteq \mathcal{AT} : A' \text{ is a reduct of } A\}$.

For any pair (x, y) of objects we define the discernibility set $D_{xy} = \{a \in \mathcal{AT} : (x, y) \in \mathrm{div}(\{a\})\}$. It is proved in [Rauszer, Skowron 1992] that

$$CORE(A) = \{a \in A : \text{there are } x, y \in \mathcal{OB} \text{ such that } D_{xy} = \{a\}\}.$$

Logical systems in which information relations play the role of accessibility relations originated in [Orlowska, Pawlak 1984; Orlowska 1984; Orlowska 1985]. Since then a variety of information logics aimed at reasoning with incomplete information have been introduced and investigated [Konikowska 1987; Konikowska 1994; Pomykala 1988; Rasiowa, Skowron 1985; Rasiowa, Marek 1989; Vakarelov 1987; Vakarelov 1989; Vakarelov 1991a; Vakarelov 1991b].

Let $\mathcal{K} = (W, \{R(P) : P \subseteq A\})$ be a relational system such that W is a nonempty set, A is a nonempty finite set, and each $R(P)$ is a binary relation over W. The set A can be thought of as a set of indices, so the relations in $\mathcal{K}$ form a $\mathcal{P}(A)$-indexed family of relations. The need for logics with accessibility relations parameterised with subsets of a set is motivated in [Orlowska 1988]. $\mathcal{K}$ is an *information frame with strong relations of type* A iff for all $P, Q \subseteq A$ the following conditions are satisfied:

$$R(P \sqcup Q) = R(P) \sqcap R(Q) \qquad \text{and} \qquad R(\emptyset) = \top\!\!\!\top$$

K is an information frame with *weak relations of type* A iff for all $P, Q \subseteq A$ the following conditions are satisfied:

$$R(P \sqcup Q) = R(P) \sqcup R(Q) \qquad \text{and} \qquad R(\emptyset) = \perp\!\!\!\perp .$$

We define more specific classes of frames postulating some conditions that the relations are assumed to satisfy:

- Indiscernibility frame (IND): an information frame with strong relations which are equivalence relations.
- Diversity frame (DIV): strong relations whose complements are tolerances and complement of $R(\{a\})$ is transitive for all $a \in A$.
- Weak indiscernibility frame (WIND): relations $R(P)$ are weak tolerances and $R(\{a\})$ is transitive for all $a \in A$.
- Weak diversity frame (WDIV): weak relations whose complements are equivalence relations.

Logics of the classes of frames defined above are modal logics whose languages include relational expressions of the form $R(P)$ that represent relations from the underlying frames. We assume that we have expressions of the form $R(P \sqcup Q)$, $R(\perp\!\!\!\perp)$ among the relational expressions. The formulas of these logics are built from propositional variables with the classical propositional connectives and modal operators $[R(P)]$, $\langle R(P)\rangle$ for all the relational expressions $R(P)$, or operators $[[R(P)]]$, $\langle\!\langle R(P)\rangle\!\rangle$, referred to as sufficiency operators [Gargov, Passy+ 1987]. More exactly, the languages of L(IND) and L(WDIV) include both $[\ ]$, $\langle\ \rangle$ and $[[\ ]]$, $\langle\!\langle\ \rangle\!\rangle$, the language of L(WIND) includes only $[\ ]$, $\langle\ \rangle$, and the language of L(DIV) includes only $[[\ ]]$, $\langle\!\langle\ \rangle\!\rangle$. Semantics of the sufficiency operators is defined as follows.

Let C be a class of information frames. Let $M = (W, \{R(P) : P \subseteq A\}, m)$ be a model for logic L(C) determined by an information frame from class C. Following the presentation of modal logics given in Sect. 6.1, for the sake of simplicity we identify the symbols of relational expressions and the symbols denoting the respective relations, that is we assume that the meaning $m(R(P))$ of a relational expression of the form $R(P)$ equals relation $R(P)$ from M. Then we define:

$$\mathcal{M}, w \models [[R(P)]]F \quad \text{iff} \quad \text{for all } u \in W\text{: } \mathcal{M}, u \models F \text{ implies } (w,u) \in R(P).$$
$$\mathcal{M}, w \models \langle\!\langle R(P)\rangle\!\rangle F \quad \text{iff} \quad \text{there exists } u \in U \text{ such that } (w,u) \notin R(P) \text{ and it is not the case that } \mathcal{M}, u \models F.$$

The sufficiency operators are needed, e.g., to express the fact that relations in an information frame are strong.

6.5 Relational formalisation of information logics

The relational semantics of information logics presented in the previous section is analogous to the relational semantics developed in Sect. 6.3. Let $\mathcal{K} = (W, \{R(P) : P \subseteq A\})$ be an information frame from a class C. A relational model for $L(C)$ based on $\mathcal{K}$ is a system of the form $\mathcal{M} = (W, \{R(P) : P \subseteq A\}, m)$ such that the meaning function m assigns right ideal relations in W to propositional variables and extends to all the formulas in the same way as in Sect. 6.3. Moreover, we define:

$$m(\langle\!\langle R(P)\rangle\!\rangle A) = \overline{\overline{R(P)}; \overline{m(A)}}, \qquad m([[R(P)]]A) = \overline{\overline{R(P)}; m(A)}.$$

A formula A is true in the relational model $\mathcal{M}$ iff $m(A) = V$ $(= W \times W)$, and a formula A is relationally valid in L(C) iff A is true in every relational model for L(C). In what follows we write R instead of $R(P)$ whenever the respective statement holds for any P.

Relational proof systems for information logics include the rules $(\vee)$, $(\neg \vee)$, $(\wedge)$, $(\neg \wedge)$, $(\neg\neg)$, $(\rightarrow)$, $(\neg \rightarrow)$, $([\_])$, $(\neg[\_])$, $(\langle\_\rangle)$, $(\neg\langle\_\rangle)$, $(--)$, (ideal) and some of the following: $([[\_]])$, $(\neg[[\_]])$, $(\langle\langle\_\rangle\rangle)$, $(\neg\langle\langle\_\rangle\rangle)$, and the following additional rules.

Rules for parameterised relations

(s) $\dfrac{K,xR(P \sqcup Q)y,H}{K,xR(P)y,H \quad K,xR(Q)y,H}$ $\qquad$ $(-s)$ $\dfrac{K,x\overline{R(P \sqcup Q)}y,H}{K,x\overline{R(P)}y,x\overline{R(Q)}y,H}$

(w) $\dfrac{K,xR(P \sqcup Q)y,H}{K,xR(P)y,xR(Q)y,H}$ $\qquad$ $(-w)$ $\dfrac{K,x\overline{R(P \sqcup Q)}y,H}{K,x\overline{R(P)}y,H \quad K,x\overline{R(Q)}y,H}$

Rules of symmetry and transitivity

(sym R) $\dfrac{K,xRy,H}{K,yRx,H,xRy}$ $\qquad$ (tran R) $\dfrac{K,xRy,H}{K,xRz,H,xRy \quad K,zRy,H,xRy}$
where z is a variable

Decomposition rules for the sufficiency operators

$(\langle\langle\rangle\rangle)$ $\dfrac{K,x\langle\langle R\rangle\rangle Ay,H}{K,x\overline{R}z,H,x\langle\langle R\rangle\rangle Ay \qquad K,z\neg Ay,H,z\langle\langle R\rangle\rangle Ay}$ where z is a variable

$(\neg\langle\langle\rangle\rangle)$ $\dfrac{K,x\neg\langle\langle R\rangle\rangle Ay,H}{K,xRz,zAy,H}$ where z is a restricted variable

$([[]])$ $\dfrac{K,x[[R]]Ay,H}{K,xRz,z\neg Ay,H}$
where z is a restricted variable

$(\neg[[]])$ $\dfrac{K,x\neg[[R]]Ay,H}{K,x\overline{R}z,H,x\neg[[R]]Ay \qquad K,zAy,H,x\neg[[R]]Ay}$
where z is a variable

The fundamental sequences in relational proof systems for information logics are (f1) and (f2) from Sect. 6.3, and some of the following:

(f5) $xR(\perp\!\!\!\perp)y$, $\qquad$ (f6) $x\overline{R(\perp\!\!\!\perp)}y$,
(f7) xRx, $\qquad$ (f8) $x\overline{R}x$.

Specific rules reflect properties of the underlying relations. For example we have:

Theorem 6.5.1 The rule (sym R) is admissible in $L(C)$ iff in every relational model based on a frame from class C the meaning of R is a symmetric relation.□

The analogous theorem holds for the rule (tran R).

The following sets of rules and fundamental sequences provide relational proof systems for the information logics defined in Sect. 6.4.

- L(IND) *Rules:* $(\vee), (\neg\vee), (\wedge), (\neg\wedge), (\neg\neg), (\rightarrow), (\neg\rightarrow)$, $([\_])$, $(\neg[\_])$, $(\langle\_\rangle)$, $(\neg\langle\_\rangle)$, $([[\_]])$, $(\neg[[\_]])$, $(\langle\langle\_\rangle\rangle)$, $(\neg\langle\langle\_\rangle\rangle)$, (ideal), (s), $(-s)$, (sym R) and (tran R) for all the relations from frames of class *IND*

  Fundamental sequences: (f1), (f2), (f5), (f7).

- L(WIND) *Rules:* $(\vee), (\neg\vee), (\wedge), (\neg\wedge), (\neg\neg), (\rightarrow), (\neg\rightarrow)$, $([\_])$, $(\neg[\_])$, $(\langle\_\rangle)$, $(\neg\langle\_\rangle)$, (ideal), (w), $(-w)$, (sym R) for all the relations from frames of class *WIND*, (tran R) for relations $R(\{a\})$ from these frames, where a is an element of the respective set of indices.
  Fundamental sequences: (f1), (f2), (f6), (f7).
- L(DIV) *Rules:* $(\vee), (\neg\vee), (\wedge), (\neg\wedge), (\neg\neg), (\rightarrow), (\neg\rightarrow)$, $([[\_]])$, $(\neg[[\_]])$, $(\langle\langle\_\rangle\rangle)$, $(\neg\langle\langle\_\rangle\rangle)$, (ideal), (s), $(-s)$, (sym R) for the complements of all the relations from frames of class *DIV*, (tran R) for complements of relations $R(\{a\})$ from these frames, where a is an element of the respective set of indices.
  Fundamental sequences: (f1), (f2), (f5), (f8).
- L(WDIV) *Rules:* $(\vee), (\neg\vee), (\wedge), (\neg\wedge), (\neg\neg), (\rightarrow), (\neg\rightarrow)$, $([\_])$, $(\neg[\_])$, $(\langle\_\rangle)$, $(\neg\langle\_\rangle)$, $([[\_]])$, $(\neg[[\_]])$, $(\langle\langle\_\rangle\rangle)$, $(\neg\langle\langle\_\rangle\rangle)$, (ideal), (w), $(\neg w)$, (sym R), (tran R) for the complements of all the relations from frames of class *WDIV*
  Fundamental sequences: (f1), (f2), (f6), (f8).

Theorems analogous to Theorem 6.3.1 up to Theorem 6.3.3 hold for all the information logics defined above.

Example 2 We show a relational proof of the formula $\langle\langle R\rangle\rangle\neg\langle\langle R\rangle\rangle A \rightarrow \langle\langle R\rangle\rangle A$ in the logic L(WDIV).

$$x\langle\langle R\rangle\rangle\neg\langle\langle R\rangle\rangle A \rightarrow \langle\langle R\rangle\rangle Ay$$

$$\overline{\qquad\qquad\qquad\qquad} \; (\rightarrow)$$

$$x\neg\langle\langle R\rangle\rangle\neg\langle\langle R\rangle\rangle Ay, x\langle\langle R\rangle\rangle Ay$$

$$\overline{\qquad\qquad\qquad\qquad} \; (\neg\langle\langle\rangle\rangle) \text{ with restricted var. } z$$

$$xRz, z\neg\langle\langle R\rangle\rangle Ay, x\langle\langle R\rangle\rangle Ay$$

$$\overline{\qquad\qquad\qquad\qquad} \; (\neg\langle\langle\rangle\rangle) \text{ with restricted var. } t$$

$$xRz, zRt, tAy, x\langle\langle R\rangle\rangle Ay$$

$$\overline{\qquad\qquad\qquad\qquad\qquad\qquad} \; (\langle\langle\rangle\rangle) \text{ with var. } t$$

$$\ldots, x\overline{R}t, \ldots \qquad\qquad \ldots, t\neg Ay, \ldots$$

$$\overline{\qquad\qquad\qquad} \; (\text{tran } \overline{R}) \text{ with var. } z \qquad \textit{fundamental}$$

$$\ldots, x\overline{R}z, \ldots \qquad \ldots, z\overline{R}t, \ldots$$

fundamental *fundamental*

Proof systems developed in this section suggest the relational formalisation of some of the modal logics mentioned in Sect. 6.1.

- K Rules: $(\vee), (\neg\vee), (\wedge), (\neg\wedge), (\neg\neg), (\rightarrow), (\neg\rightarrow)$, (ideal),
  Fundamental sequences: (f1), (f2)
- T Rules: $(\vee), (\neg\vee), (\wedge), (\neg\wedge), (\neg\neg), (\rightarrow), (\neg\rightarrow)$, (ideal),
  Fundamental sequences: (f1), (f2), (f7)
- B Rules: $(\vee), (\neg\vee), (\wedge), (\neg\wedge), (\neg\neg), (\rightarrow), (\neg\rightarrow)$, (ideal), (sym R),
  Fundamental sequences: (f1), (f2), (f7)
- S4 Rules: $(\vee), (\neg\vee), (\wedge), (\neg\wedge), (\neg\neg), (\rightarrow), (\neg\rightarrow)$, (ideal), (tran R),
  Fundamental sequences: (f1), (f2), (f7)

- S5 Rules: $(\vee), (\neg\vee), (\wedge), (\neg\wedge), (\neg\neg), (\rightarrow), (\neg\rightarrow)$, (ideal), (sym R), (tran R), Fundamental sequences: (f1), (f2), (f7)

The relational proof systems presented in this chapter consist of the rules that apply to the (sequences of) relational formulas xAy such that A is either a relational expression or a formula in its original form. Therefore we do not need to transform formulas into relational expressions, as it is required for application of the relational proof systems given in [Orlowska 1991; Orlowska 1995]. The relational formalisation of some other applied logics can be found in [Orlowska 1993; Demri, Orlowska$^+$ 1994; Demri, Orlowska 1996].

6.6 Conclusion

We have presented a method of formalising nonclassical logics within the framework of algebras of relations. This was done in two steps: First, we translated formulas of any nonclassical logic L into terms over a class of algebras of relations. These terms form a relational language for (representation of) L. Next, we defined proof rules for the relational language for L. Relational representation of formulas enabled us to articulate explicitly information about both their syntactic structure and semantic satisfiability condition. Each of the propositional connectives became a "logical" relational operation and in this way an original syntactic form of formulas was preserved. Semantic information about a formula which normally is included in a satisfiability condition consists of the two basic parts: First, we say which states satisfy the subformulas of the given formula, and second, how those states are related to each other by means of an accessibility relation. Those two ingredients of semantic information are of course interrelated and unseparable. In relational representation of formulas, terms representing accessibility relations are included explicitly in the respective relational terms corresponding to the formulas. They become the arguments of the relational operations in a term in the same way as the other of its subterms, obtained from subformulas of the given formula. In this way semantic information is provided explicitly on the same level as syntactic information, and the traditional distinction between syntax and semantics disappears, in a sense. Relational proof systems consist of two kinds of rules: Decomposition rules designed for any relational operation admitted in the underlying relational language for L, and for the complement of that operation; Specific rules that reflect properties of relational constants that are possibly admitted in the relational language for L. We have illustrated the method of relational formalisation with two case studies: relational formalisation of program logics and information logics.

Chapter 7

Linear Logic

Jules Desharnais, Bernard Hodgson, John Mullins[1]

Linear logic, introduced by Jean-Yves Girard [Girard 1987], has aroused considerable interest among logicians and theoretical computer scientists. Among other things, linear logic is said to be a *resource-conscious logic* and a *logic of actions*. According to Girard [Girard 1995], it should be viewed as an extension of classical logic, rather than as an alternative logic.

Most papers on linear logic, e.g. [Bergeron, Hatcher 1995; Girard 1987; Girard 1989; Girard 1995; Lincoln 1992; Troelstra 1992], consider *commutative* linear logic (commutativity refers to the corresponding property of the *par* and *times* operators, to be introduced below), but some of them discuss various species of *noncommutative* logic [Abrusci 1991; Lambek 1993; Lincoln, Mitchell+ 1992; Yetter 1990], including one called *cyclic noncommutative* linear logic. It is not yet clear what is the "right" noncommutative logic; indeed, Girard [Girard 1995] described it as a *Far West*.

For a computer scientist without a strong background in logic, learning linear logic is not an easy task. This is partly due to the fact that the emphasis in the literature is on modelling proofs and that intuitive, classical models are not discussed much[2]. It turns out that *relation algebras are algebraic models of cyclic noncommutative propositional linear logic*. This model constitutes a convenient entrance door to the world of linear logic for the relation algebraist.

The relational model of linear logic has been known for some time. As early as 1989, Pratt has made presentations on it (see [LinLogList]); he has given the connection between linear negation and the relational *complement-of-converse* in [Pratt 1992]. Lambek describes it in his study of bilinear logic [Lambek 1993]. The relational model has also been discussed on the Internet[3].

Our goal in this chapter is to briefly describe linear logic and its relational model. In Sect. 7.1, linear logic is introduced and a sequent calculus is presented. In Sect. 7.2, new relational operators are defined. They correspond to some of

---

[1]All three authors acknowledge the financial support of the NSERC (Natural Sciences and Engineering Research Council) of Canada.

[2]From [Girard 1995]: "The most traditional, and also the less interesting semantics of linear logic associates values to formulas, in the spirit of classical model theory. Therefore it only modelizes provability, and not proofs."

[3]We thank Vaughan Pratt for informing us about the electronic forum on linear logic, and for a compendium of messages about the relational model.

the operators of linear logic, as shown in Sect. 7.3, where the relational model is presented. Finally, in Sect. 7.4, we give additional pointers to the literature.

7.1 A sequent calculus

In 1934, Gentzen introduced the *sequent calculus* for studying the laws of logic. A *sequent* is an expression of the form $\Gamma \vdash \Delta$, where Γ ($= \mathsf{r}_1, \ldots, \mathsf{r}_m$) and Δ ($= \mathsf{t}_1, \ldots, \mathsf{t}_n$) are finite sequences of formulae. The intended meaning of $\Gamma \vdash \Delta$ is

$$\mathsf{r}_1 \text{ and } \ldots \text{ and } \mathsf{r}_m \text{ imply } \mathsf{t}_1 \text{ or } \ldots \text{ or } \mathsf{t}_n$$

but the sense of "and", "imply" and "or" has to be clarified [Girard 1995]. In classical logic, this is partly done by means of three *structural rules*, called *Weakening, Contraction* and *Exchange*:

$$\frac{\Gamma \vdash \Delta}{\Gamma \vdash \mathsf{r}, \Delta} \qquad \frac{\Gamma \vdash \Delta}{\Gamma, \mathsf{r} \vdash \Delta} \qquad \text{(Weakening)}$$

$$\frac{\Gamma \vdash \mathsf{r}, \mathsf{r}, \Delta}{\Gamma \vdash \mathsf{r}, \Delta} \qquad \frac{\Gamma, \mathsf{r}, \mathsf{r} \vdash \Delta}{\Gamma, \mathsf{r} \vdash \Delta} \qquad \text{(Contraction)}$$

$$\frac{\Gamma \vdash \Delta}{\Gamma' \vdash \Delta'} \qquad \text{(Exchange)}$$

where Γ' and Δ' are permutations of Γ and Δ, respectively.

Classical logic may thus be viewed as dealing with static propositions such that each proposition is either true or false. Because of this static nature of propositions in classical logic, one may "duplicate" a proposition at will: $\mathsf{r} \rightarrow (\mathsf{r} \wedge \mathsf{r})$. Also, one may discard propositions, like s in $(\mathsf{r} \wedge \mathsf{s}) \rightarrow \mathsf{r}$. As explained in [Girard 1989; Girard 1995], classical logic deals with stable truth:

if r and $\mathsf{r} \rightarrow \mathsf{s}$, then s, *but* r *still holds*.

This is wrong in real life. For example, if r is to spend \$1 on a pack of cigarettes and s is to get them, you lose \$1 in the process and you cannot do it a second time.

The first modification brought by linear logic is to drop two of the structural rules, Weakening and Contraction; as explained below, this opens the possibility of having two conjunctions (and two disjunctions). In order to be able to deal with stable truths (*situations* [Girard 1989; Girard 1995]), the second modification is to add two connectives, ! and ?, called *exponentials*, which express the iterability of an action (if r means to spend a dollar, then $!\mathsf{r}$ means to spend as many dollars as needed); thus classical logic is embedded in linear logic.

The rule Exchange is kept. However, to obtain cyclic noncommutativity, it has to be altered (see Definition 7.1.1 below).

The operators of linear logic are given in Table 7.1, together with some of the terminology that applies to them.[4] The fourth column gives the neutral element of the corresponding binary operator.

[4]The abbreviations used in the table are the following: *add.* for *additive*, *mult.* for *multiplicative*, *exp.* for *exponential*, *conj.* for *conjunctive* and *disj.* for *disjunctive*.

Operator	Name	Type	Neutral	dual
$\mathbf{0}$		nullary		$\top$
$\top$		nullary		$\mathbf{0}$
$\mathbf{1}$		nullary		$\bot$
$\bot$		nullary		$\mathbf{1}$
$^\bot$	negation, nil	unary		
!	of course	exp., unary		?
?	why not	exp., unary		!
$\oplus$	plus	add., disj., binary	$\mathbf{0}$	&
&	with	add., conj., binary	$\top$	$\oplus$
$⅋$	par	mult., disj., binary	$\bot$	$\otimes$
$\otimes$	times	mult., conj., binary	$\mathbf{1}$	$⅋$
$\multimap$	implication	binary		
$\circ\!-$	retro-implication	binary		

Table 7.1 Operators of linear logic

What follows is a presentation of the sequent calculus of cyclic noncommutative propositional linear logic. This system is drawn from [Girard 1989] (pages 81–83 for commutative linear logic, and pages 85–86 for the modifications to turn it into noncommutative linear logic). See also [Yetter 1990].

The notational conventions are as follows: q and $\mathsf{q}^\bot$ are atomic formulae; r, s are arbitrary formulae; Δ, Γ are finite lists of formulae. Item (1) in Definition 7.1.1 below is the *definition* of the operators $^\bot$, $\multimap$ and $\circ\!-$. The other items present the rules of the sequent calculus. Sequents are right-sided (*i.e.* of the form $\vdash \Delta$); the effect of a general sequent $\Gamma \vdash \Delta$ is obtained via the sequent $\vdash \Gamma^\bot, \Delta$, where $(\mathsf{r}_1, \ldots, \mathsf{r}_n)^\bot \triangleq (\mathsf{r}_n^\bot, \ldots, \mathsf{r}_1^\bot)$.

Definition 7.1.1 The sequent calculus of cyclic noncommutative propositional linear logic is defined as follows:

1. Definition of $^\bot$, $\multimap$ and $\circ\!-$

$$\begin{array}{llll}
\mathbf{1}^\bot \triangleq \bot & \bot^\bot \triangleq \mathbf{1} & (\mathsf{r} \otimes \mathsf{s})^\bot \triangleq \mathsf{s}^\bot ⅋ \mathsf{r}^\bot & (\mathsf{r} ⅋ \mathsf{s})^\bot \triangleq \mathsf{s}^\bot \otimes \mathsf{r}^\bot \\
\top^\bot \triangleq \mathbf{0} & \mathbf{0}^\bot \triangleq \top & (\mathsf{r} \& \mathsf{s})^\bot \triangleq \mathsf{r}^\bot \oplus \mathsf{s}^\bot & (\mathsf{r} \oplus \mathsf{s})^\bot \triangleq \mathsf{r}^\bot \& \mathsf{s}^\bot \\
(\mathsf{q})^\bot \triangleq \mathsf{q}^\bot & (\mathsf{q}^\bot)^\bot \triangleq \mathsf{q} & (!\mathsf{r})^\bot \triangleq ?(\mathsf{r}^\bot) & (?\mathsf{r})^\bot \triangleq !(\mathsf{r}^\bot) \\
 & & \mathsf{r} \multimap \mathsf{s} \triangleq \mathsf{r}^\bot ⅋ \mathsf{s} & \mathsf{r} \circ\!- \mathsf{s} \triangleq \mathsf{r} ⅋ \mathsf{s}^\bot
\end{array}$$

2. Identity group

(a) $\vdash \mathsf{r}, \mathsf{r}^\bot$ (Identity axiom) (b) $\dfrac{\vdash \Gamma, \mathsf{r} \qquad \vdash \mathsf{r}^\bot, \Delta}{\vdash \Gamma, \Delta}$ (Cut rule)

3. Structural group

$$\frac{\vdash \Gamma, \mathsf{r}}{\vdash \mathsf{r}, \Gamma} \quad \text{(Cyclic exchange)}$$

4. Logical group: multiplicatives

(a) $\dfrac{\vdash \Gamma, r \qquad \vdash s, \Delta}{\vdash \Gamma, r \otimes s, \Delta}\ (\otimes)$ (c) $\vdash \mathbf{1}$ (Axiom)

(b) $\dfrac{\vdash \Gamma, r, s}{\vdash \Gamma, r \parr s}\ (\parr)$ (d) $\dfrac{\vdash \Gamma}{\vdash \Gamma, \bot}\ (\bot)$

5. Logical group: additives

(a) $\dfrac{\vdash \Gamma, r \qquad \vdash \Gamma, s}{\vdash \Gamma, r \& s}\ (\&)$ (c) $\dfrac{\vdash \Gamma, s}{\vdash \Gamma, r \oplus s}\ (\oplus^2)$

(b) $\dfrac{\vdash \Gamma, r}{\vdash \Gamma, r \oplus s}\ (\oplus^1)$ (d) $\vdash \Gamma, \top$ (Axiom)

6. Logical group: exponentials

(a) $\dfrac{\vdash ?\Gamma, r}{\vdash ?\Gamma, !r}\ (!)$ (c) $\dfrac{\vdash \Gamma}{\vdash \Gamma, ?r}$ (Weakening)

(b) $\dfrac{\vdash \Gamma, r}{\vdash \Gamma, ?r}$ (Dereliction) (d) $\dfrac{\vdash \Gamma, ?r, ?r}{\vdash \Gamma, ?r}$ (Contraction)

In commutative linear logic, the rule of exchange, coupled with rule (4b) of Definition 7.1.1, makes $\parr$ commutative. Now, according to Girard [Girard 1989], "the natural way of introducing noncommutativity is not to expell exchange, but to restrict it to circular permutations." This is what rule 3 in Definition 7.1.1 does.

Also, in commutative linear logic, one takes $(r \otimes s)^\perp = r^\perp \parr s^\perp$, $(r \parr s)^\perp = r^\perp \otimes s^\perp$ and $r \multimap s = s \multimapinv r = r^\perp \parr s$, which are equivalent to the above definitions when $\parr$ is commutative.

As for the rules about the exponential operators (! and ?), we have kept those of commutative linear logic. Other choices are possible [Lincoln, Mitchell+ 1992], as we will discuss in Sect. 7.3.

One feature worth mentioning is the difference between the rule for $\otimes$ and the rule for &: the latter uses twice the same context (Γ), whereas the former works with disjoint contexts (Γ and Δ) which remain distinct in the conclusion. Thus, the two conjunctions ($\otimes$ and &) are different. It is the very act of dropping the weakening and contraction rules that creates this distinction. Indeed, reintroducing these two rules, we can show that $\otimes$ and & are identified; this is done in the following derivations, where "w" and "c" refer to the use of the weakening and contraction rules, respectively, and "e" is the cyclic exchange rule of Definition 7.1.1. The passage to two-sided sequents is done without mention.

$$\dfrac{\dfrac{\dfrac{\dfrac{\dfrac{\vdash r, r^\perp}{\vdash s^\perp, r^\perp, r}\ (\text{w, e}) \qquad \dfrac{\vdash s, s^\perp}{\vdash s^\perp, r^\perp, s}\ (\text{w, e})}{\vdash s^\perp, r^\perp, r\&s}\ (\&)}{\vdash s^\perp \parr r^\perp, r\&s}\ (\parr, \text{e})}{(s^\perp \parr r^\perp)^\perp \vdash r\&s}}{r \otimes s \vdash r\&s}\ (\text{Definition }^\perp)$$

$$\dfrac{\dfrac{\dfrac{\dfrac{\vdash \mathsf{r}, \mathsf{r}^{\perp}}{\vdash \mathsf{r}^{\perp} \oplus \mathsf{s}^{\perp}, \mathsf{r}}\,(\oplus^1, \mathrm{e}) \qquad \dfrac{\vdash \mathsf{s}, \mathsf{s}^{\perp}}{\vdash \mathsf{s}, \mathsf{r}^{\perp} \oplus \mathsf{s}^{\perp}}\,(\oplus^2)}{\vdash \mathsf{r}^{\perp} \oplus \mathsf{s}^{\perp}, \mathsf{r} \otimes \mathsf{s}, \mathsf{r}^{\perp} \oplus \mathsf{s}^{\perp}}\,(\otimes)}{\vdash \mathsf{r}^{\perp} \oplus \mathsf{s}^{\perp}, \mathsf{r} \otimes \mathsf{s}}\,(\mathrm{e}, \mathrm{c})}{(\mathsf{r}^{\perp} \oplus \mathsf{s}^{\perp})^{\perp} \vdash \mathsf{r} \otimes \mathsf{s}}}{\mathsf{r} \& \mathsf{s} \vdash \mathsf{r} \otimes \mathsf{s}}\,(\text{Definition } {}^{\perp})$$

7.2 Additional relational operators

Our relational model for linear logic will use the operators, notation and terminology introduced in Chapt. 1 and Chapt. 2. We define additional unary relational operators $^{\sim}, !, ?$ as follows:

$$r^{\sim} \triangleq \overline{r}^{\smile}, \qquad !r \triangleq \mathbb{I} \sqcap r, \qquad ?r \triangleq \mathbb{I}^{\sim} \sqcup r. \tag{7.1}$$

We assign to them the same priority as that of the other unary operators (Sect. 1.2). In the proof of their properties and in the next section, we will use the relative addition and residual operators; we briefly recall these notions (see Sect. 2.2 for relative addition, and Sect. 1.2 and 1.3 for residuation). The *relative addition* of r and s, notated $r \dagger s$, is defined by $r \dagger s \triangleq \overline{\overline{r} ; \overline{s}}$. The priority of $\dagger$ is the same as that of $;$. The relative addition operator being the De Morgan dual of the composition operator $;$, many of its properties follow directly from those of the latter. Some of these properties are:

$$r \dagger \overline{\mathbb{I}} = \overline{\mathbb{I}} \dagger r = r, \qquad r \dagger (s \sqcap t) = r \dagger s \sqcap r \dagger t,$$
$$(r \dagger s) \dagger t = r \dagger (s \dagger t), \qquad q \sqsubseteq r \text{ and } s \sqsubseteq t \implies q \dagger s \sqsubseteq r \dagger t.$$
$$r \dagger s \sqcup r \dagger t \sqsubseteq r \dagger (s \sqcup t),$$

As for the *right residual* $\backslash$ and *left residual* $/$, they satisfy the following laws:

$$r ; s \sqsubseteq t \iff r \sqsubseteq t/s \iff s \sqsubseteq r \backslash t, \qquad t/s = \overline{\overline{t} ; s^{\smile}}, \qquad r \backslash t = \overline{r^{\smile} ; \overline{t}} \;.$$

Theorem 7.2.1 Let r and s be relations. Then,

1. $r \sqsubseteq s \iff s^{\sim} \sqsubseteq r^{\sim}$,
2. $r^{\sim\sim} = r$,
3. $\mathbb{I}^{\sim} = \overline{\mathbb{I}}$,
4. $r \dagger \mathbb{I}^{\sim} = \mathbb{I}^{\sim} \dagger r = r$,
5. $r^{\sim} \backslash s = r \dagger s = r / s^{\sim}$,
6. $s^{\sim} ; r \sqsubseteq t \iff r \sqsubseteq s \dagger t \iff r ; t^{\sim} \sqsubseteq s$,
7. $\mathbb{I} \sqsubseteq r \dagger s \iff r^{\sim} \sqsubseteq s \iff s^{\sim} \sqsubseteq r \iff \mathbb{I} \sqsubseteq s \dagger r$.

Proof: Most of these laws follow directly from the definition of relation algebras (Sect. 1.3) and from the properties of the operators involved, given above and in Chapt. 1 and 2. The last three laws are proved as follows.

5. $r^{\sim} \backslash s = \overline{r^{\sim\smile} ; \overline{s}} = \overline{\overline{\overline{r}} ; \overline{s}} = r \dagger s = \overline{\overline{r} ; \overline{s}} = \overline{\overline{r} ; s^{\sim\smile}} = r / s^{\sim}$.

LL	RA	LL	RA	LL	RA	LL	RA
$\mathbf{1}$	$\mathbb{I}$	$\mathsf{r}^{\perp}$	$r^{\sim}$	$\mathsf{r} \otimes \mathsf{s}$	$r;s$	$\mathsf{r} \oplus \mathsf{s}$	$r \sqcup s$
$\perp$	$\mathbb{I}^{\sim}$	$!\mathsf{r}$	$!r$	$\mathsf{r} ⅋ \mathsf{s}$	$r \dagger s$	$\mathsf{r} \& \mathsf{s}$	$r \sqcap s$
$\mathbf{0}$	$\perp\!\!\!\perp$	$?\mathsf{r}$	$?r$	$\mathsf{r} \multimap \mathsf{s}$	$r \backslash s$	$\mathsf{r} \multimapinv \mathsf{s}$	r/s
$\top$	$\top\!\!\!\top$						

Table 7.2 Correspondence linear logic (LL) – relation algebra (RA)

6. Using properties of residuals and law 5,

$$s^{\sim};r \sqsubseteq t \iff r \sqsubseteq s^{\sim}\backslash t \iff r \sqsubseteq s \dagger t \iff r \sqsubseteq s/t^{\sim} \iff r;t^{\sim} \sqsubseteq s.$$

7. This is a special case of law 6. □

7.3 The relational model of linear logic

The correspondence between the language of linear logic and that of relation algebra is the following.

An atomic formula of the form q is interpreted by a relation q and an atomic formula of the form $\mathsf{q}^{\perp}$ is interpreted by the relation $q^{\sim}$, where q is the interpretation of q. The interpretation of a formula is then constructed inductively, using the interpretation of the linear logic operators given in Table 7.2[5]. This semantics is extended to lists of formulae as follows: the semantics of the empty list is $\perp$ (the neutral element for ⅋) and that of the list Γ, Δ is $r \dagger s$, where r interprets Γ and s interprets Δ. Because $\dagger$ interprets ⅋, this amounts to considering a list of formulae as a single formula where all "," are replaced by "⅋". This is perfectly justified, given Girard's rule ⅋ (rule (4b) in Definition 7.1.1). Thus, the interpretation of a list of formulae is a relation. Finally, the interpretation of a sequent is the relation interpreting its list of formulae.

A sequent $\vdash \Gamma$ is defined to be *valid* iff $\mathbb{I} \sqsubseteq r$, where r is the relation interpreting the sequent. The validity of the equations, rules and axioms of linear logic is expressed in the statement of Theorem 7.3.1 (corresponding formulae bear the same label in Definition 7.1.1 and Theorem 7.3.1). Hence, Theorem 7.3.1 shows the soundness of the sequent calculus of cyclic noncommutative linear logic with respect to the relational model.

Theorem 7.3.1 Let q, r, s, t be relations.

1. $\mathbb{I}^{\sim} = \mathbb{I}^{\sim}$ $\quad \mathbb{I}^{\sim\sim} = \mathbb{I}$ $\quad (r;s)^{\sim} = s^{\sim} \dagger r^{\sim}$ $\quad (r \dagger s)^{\sim} = s^{\sim};r^{\sim}$
   $\top\!\!\!\top^{\sim} = \perp\!\!\!\perp$ $\quad \perp\!\!\!\perp^{\sim} = \top\!\!\!\top$ $\quad (r \sqcap s)^{\sim} = r^{\sim} \sqcup s^{\sim}$ $\quad (r \sqcup s)^{\sim} = r^{\sim} \sqcap s^{\sim}$
   $(q)^{\sim} = q^{\sim}$ $\quad (q^{\sim})^{\sim} = q$ $\quad (!r)^{\sim} = ?(r^{\sim})$ $\quad (?r)^{\sim} = !(r^{\sim})$
   $r\backslash s = r^{\sim} \dagger s$ $\quad r/s = r \dagger s^{\sim}$
2. (a) $\mathbb{I} \sqsubseteq r \dagger r^{\sim}$ (b) $\mathbb{I} \sqsubseteq q \dagger r$ and $\mathbb{I} \sqsubseteq r^{\sim} \dagger t \implies \mathbb{I} \sqsubseteq q \dagger t$

[5] Note that the additive operators of linear logic correspond to the *absolute* or *Boolean* operators of relation algebra and the multiplicative operators to the *relative* or *Peircean* operators [Tarski 1941].

3. $\mathbb{I} \sqsubseteq q \dagger r \implies \mathbb{I} \sqsubseteq r \dagger q$
4. (a) $\mathbb{I} \sqsubseteq q \dagger r$ and $\mathbb{I} \sqsubseteq s \dagger t \implies \mathbb{I} \sqsubseteq q \dagger (r ; s) \dagger t$
   (b) $\mathbb{I} \sqsubseteq q \dagger r \dagger s \implies \mathbb{I} \sqsubseteq q \dagger r \dagger s$
   (c) $\mathbb{I} \sqsubseteq \mathbb{I}$
   (d) $\mathbb{I} \sqsubseteq q \implies \mathbb{I} \sqsubseteq q \dagger \mathbb{I}^{\smile}$
5. (a) $\mathbb{I} \sqsubseteq q \dagger r$ and $\mathbb{I} \sqsubseteq q \dagger s \implies \mathbb{I} \sqsubseteq q \dagger (r \sqcap s)$
   (b) $\mathbb{I} \sqsubseteq q \dagger r \implies \mathbb{I} \sqsubseteq q \dagger (r \sqcup s)$
   (c) $\mathbb{I} \sqsubseteq q \dagger s \implies \mathbb{I} \sqsubseteq q \dagger (r \sqcup s)$
   (d) $\mathbb{I} \sqsubseteq q \dagger \mathbb{T}$
6. (a) $\mathbb{I} \sqsubseteq ?q \dagger r \implies \mathbb{I} \sqsubseteq ?q \dagger !r$
   (b) $\mathbb{I} \sqsubseteq q \dagger r \implies \mathbb{I} \sqsubseteq q \dagger ?r$
   (c) $\mathbb{I} \sqsubseteq q \implies \mathbb{I} \sqsubseteq q \dagger ?r$
   (d) $\mathbb{I} \sqsubseteq q \dagger ?r \dagger ?r \implies \mathbb{I} \sqsubseteq q \dagger ?r$

Proof: The laws needed in the following proofs are those of Chapt. 1, Chapt. 2 and Sect. 7.2.

1. These equalities are trivial or have been proved in Theorem 7.2.1.
2. (a) $\mathsf{true} \iff \mathbb{I} ; r \sqsubseteq r \iff \mathbb{I} \sqsubseteq r / r \iff \mathbb{I} \sqsubseteq r \dagger r^{\smile}$.
   (b) $\mathbb{I} \sqsubseteq q \dagger r$ and $\mathbb{I} \sqsubseteq r^{\smile} \dagger t \iff q^{\smile} \sqsubseteq r$ and $r \sqsubseteq t \implies q^{\smile} \sqsubseteq t$ $\iff \mathbb{I} \sqsubseteq q \dagger t$.
3. See Theorem 7.2.1(7).
4. (a) $\mathbb{I} \sqsubseteq q \dagger r$ and $\mathbb{I} \sqsubseteq s \dagger t \iff q^{\smile} \sqsubseteq r$ and $t^{\smile} \sqsubseteq s \implies q^{\smile} ; t^{\smile} \sqsubseteq r ; s$ $\iff q^{\smile} \sqsubseteq (r ; s) \dagger t \iff \mathbb{I} \sqsubseteq q \dagger (r ; s) \dagger t$.
   (b) Trivial.
   (c) Trivial.
   (d) Direct from Theorem 7.2.1(4).
5. (a) $\mathbb{I} \sqsubseteq q \dagger r$ and $\mathbb{I} \sqsubseteq q \dagger s \iff \mathbb{I} \sqsubseteq q \dagger r \sqcap q \dagger s \iff \mathbb{I} \sqsubseteq q \dagger (r \sqcap s)$.
   (b) By monotonicity of $\dagger$.
   (c) By monotonicity of $\dagger$.
   (d) $\mathsf{true} \iff q^{\smile} \sqsubseteq \mathbb{T} \iff \mathbb{I} \sqsubseteq q \dagger \mathbb{T}$.
6. (a) By Equation 7.1 and 7.3.1(1), $(?q)^{\smile} = !(q^{\smile}) = \mathbb{I} \sqcap q^{\smile} \sqsubseteq \mathbb{I}$, hence
   $$\begin{aligned} \mathbb{I} \sqsubseteq ?q \dagger r &\iff (?q)^{\smile} \sqsubseteq r \iff (?q)^{\smile} \sqsubseteq \mathbb{I} \sqcap r \iff (?q)^{\smile} \sqsubseteq !r \\ &\iff \mathbb{I} \sqsubseteq ?q \dagger !r. \end{aligned}$$
   (b) This follows from $r \sqsubseteq ?r$ and the monotonicity of $\dagger$.
   (c) $\mathbb{I} \sqsubseteq q \iff \mathbb{I} \sqsubseteq q \dagger \mathbb{I}^{\smile} \implies \mathbb{I} \sqsubseteq q \dagger ?r$.
   (d) $?r \dagger ?r = ((?r)^{\smile} ; (?r)^{\smile})^{\smile} = (!(r^{\smile}) ; !(r^{\smile}))^{\smile} = ((\mathbb{I} \sqcap r^{\smile}) ; (\mathbb{I} \sqcap r^{\smile}))^{\smile}$ $= (\mathbb{I} \sqcap r^{\smile})^{\smile} = (!(r^{\smile}))^{\smile} = (?r)^{\smile\smile} = ?r$. □

This theorem settles the question of soundness. On the side of completeness, we only mention that linear logic is not complete for relation algebra; for example, it is not possible to derive the distributivity of $\sqcap$ over $\sqcup$ using the rules of linear logic. This is another reason why the relational model does not attract much attention.

At first sight, it is not clear whether property 3 in Theorem 7.3.1 corresponds to the classical exchange rule or to the cyclic exchange rule of Definition 7.1.1. Letting $q \triangleq q_1 \dagger q_2$ in Theorem 7.3.1(3) yields

$$\mathbb{I} \sqsubseteq q_1 \dagger q_2 \dagger r \implies \mathbb{I} \sqsubseteq r \dagger q_1 \dagger q_2,$$

which is a cyclic permutation. Moreover, the following law, describing an exchange, does not hold:

$$\mathbb{I} \sqsubseteq q_1 \dagger q_2 \dagger r \implies \mathbb{I} \sqsubseteq q_1 \dagger r \dagger q_2.$$

This is easily shown by choosing the following relations on $\{0,1\}$:

$$\begin{aligned} q_1 &\triangleq \{(0,1),(1,0),(1,1)\}, \\ q_2 &\triangleq \{(0,0),(0,1),(1,1)\}, \\ r &\triangleq \{(0,0),(1,0),(1,1)\}. \end{aligned}$$

The expression given in Equation 7.1 for $?r$ (and that for its dual $!r$) may seem arbitrary. But, assuming the interpretation of the other operators to be as given above, one finds that $\mathbb{I}^\sim \sqcup r$ is the least solution for $?r$. Indeed, from Theorem 7.3.1(2a,6b) and Theorem 7.2.1(2,7),

$$\mathsf{true} \iff \mathbb{I} \sqsubseteq r^\sim \dagger r \implies \mathbb{I} \sqsubseteq r^\sim \dagger ?r \iff r \sqsubseteq ?r.$$

Also, from Theorems 7.2.1(7) and 7.3.1(6c),

$$\mathsf{true} \iff \mathbb{I} \sqsubseteq \mathbb{I} \implies \mathbb{I} \sqsubseteq \mathbb{I} \dagger ?r \iff \mathbb{I}^\sim \sqsubseteq ?r.$$

We have mentioned in Sect. 7.1 that, in a noncommutative context, the rules for the exponential operators could be different. For example, Lincoln *et al.* [Lincoln, Mitchell+ 1992] add a rule allowing the exchange of an exponentiated formula with an arbitrary formula. Translated in the relational formulation as above, it would read $\mathbb{I} \sqsubseteq r \dagger s \dagger ?t \iff \mathbb{I} \sqsubseteq r \dagger ?t \dagger s$; this does not generally hold for relations.

We now come back to the motivations that we gave in the introduction to this chapter and in Sect. 7.1. It was said that linear logic is a *logic of actions*. Clearly, the relational model reinforces this view: after all, a concrete relation is a set of transitions, and relations have been used to model action systems. Linear logic deals with *stable truths*, or *situations*, by means of the exponentials; but, in the relational model, $!r = \mathbb{I} \sqcap r$, which indeed denotes stability. Moreover, one sees why $?t$ and $!t$ can be "reused": $?t \dagger ?t = ?t$ and $!t ; !t = !t$. This is not the case for an arbitrary relation.

We also said that the intended meaning of a sequent $(\mathsf{r}_1, \ldots, \mathsf{r}_m) \vdash (\mathsf{s}_1, \ldots, \mathsf{s}_n)$ is that r_1 and ... and r_m imply s_1 or ... or s_n, but that the sense of "and", "or" and "imply" had to be clarified. We have seen that in linear logic, "or" is $\parr$ ($\dagger$ in the relational model), whereas "and" is its dual $\otimes$ ($;$ in the relational model). Consider the sequent $\vdash \mathsf{r} \multimap \mathsf{s}$; in the relational model, it is valid when $\mathbb{I} \sqsubseteq r \backslash s$, that is, when $r \sqsubseteq s$. Now, recall that in classical logic, $\mathsf{r} \to \mathsf{r} \wedge \mathsf{r}$ holds (duplication of propositions). The linear logic analog is $\mathsf{r} \multimap \mathsf{r} \otimes \mathsf{r}$; it does not hold, as can be seen by considering its relational interpretation, $r \sqsubseteq r ; r$. Duplication becomes possible by using exponentials: $!\mathsf{r} \multimap !\mathsf{r} \otimes !\mathsf{r}$ (in the relational model: $\mathbb{I} \sqcap r \sqsubseteq (\mathbb{I} \sqcap r) ; (\mathbb{I} \sqcap r)$). In classical logic, we also have $\mathsf{r} \wedge \mathsf{s} \to \mathsf{r}$ (discarding of propositions). The linear logic correspondent $\mathsf{r} \otimes \mathsf{s} \multimap \mathsf{r}$ does not hold, witness the relational interpretation $r;s \sqsubseteq r$. Discarding is possible with exponentials: $\mathsf{r} \otimes !\mathsf{s} \multimap \mathsf{r}$ (in the relational model: $r;(\mathbb{I} \sqcap s) \sqsubseteq r$).

7.4 Conclusion

There is much more to linear logic than this simple presentation can reveal. We cannot close this chapter without at least mentioning what is probably the most important topic of linear logic: *proof nets*. A proof net is a graph representing equivalent proofs of a sequent in such a way that the inessential *sequentializations* disappear (equivalent proofs use the same rules, but in different orders). Many papers treat this subject; for an overview, see [Girard 1995]. Despite its short existence, there is already a vast amount of literature about linear logic. Of course, the original paper by Girard [Girard 1987] is a must. There are also a few tutorials [Lincoln 1992; Troelstra 1992; Troelstra 1993]. Alexiev [Alexiev 1994] describes many applications of linear logic to computer science (functional programming, logic programming, concurrency, Petri nets, state-oriented programming, ...). Complexity issues are investigated in [Lincoln, Mitchell$^+$ 1992]. The connection between linear logic and the relational model is touched upon in [Pratt 1992] and described in [Lambek 1993]. A different relational model is presented by Pratt in [Pratt 1993]; it is based on Chu's construction of *-autonomous categories [Barr 1979]. Quantales are models of linear logic [Yetter 1990] and it is shown in [Brown, Gurr 1993] that every quantale is isomorphic to a relational quantale. Further information can also be obtained on the Internet.

Chapter 8

Relational Semantics of Functional Programs

Rudolf Berghammer, Burghard von Karger

The natural meaning of a program written in a functional language like Lisp or ML is a (possibly partial) function. Functions are relations, and this chapter is a gentle introduction to the idea of regarding functional programs as elements of a relational algebra. Using relations rather than functions we avoid the complexity introduced by artificial bottom elements denoting undefinedness. Relations are also natural candidates for modelling non-determinate or set-valued functions. However, the *main* reason for using relations is that they can be calculated with so well. Using equational reasoning for functional programs, we can construct proofs that are intuitive, memorable and machine-checkable. One basic ingredient of functional programs – composition of functions – is a built-in operator of the relational calculus. Other control constructs as well as concrete data must be translated into the language of relations. We shall attempt to do this in a modular fashion, treating one construct at a time.

In Sect. 8.1 we collect relational descriptions of those data domains, the programs defined in later sections will operate on. In Sect. 8.2 we introduce the idea that a functional program *is* a relation, and look at some ways of combining simple programs into more interesting ones. Section 8.3 contains the formal syntax and relational semantics of a small functional language. In Sect. 8.4 we explore the various kinds of nondeterminism and their significance for termination problems.

8.1 Data

Reasoning about programs must take advantage of specific laws satisfied by its data. Using unrestricted set theory for defining concrete data types and their operations could lead to an undesirable mixture of relation-algebraic and free-style reasoning. Therefore, data domains will be modelled as types and their operations as relations satisfying certain axioms. To illustrate this principle, we shall specify the domain of Boolean constants and the domain of natural numbers. Hopefully, this will enable the reader to axiomatize other basic domains in a similar fashion. Composite domains can be constructed from simpler ones, e.g., as a direct product (see also Sect. 3.6). The exposition draws on material from [Berghammer, Gritzner^{+} 1993; Berghammer, Zierer 1986; Zierer 1991], where also more advanced constructions (including power set and function space) may be found.

Sets and functions

We assume as given a set $\mathcal{T}$ of types and a relational algebra $\mathcal{R}$ with object set $\mathcal{T}$. Recall from Chapt. 1 that $[A \leftrightarrow B]$ denotes the set of all (typed) relations with source A and target B. Instead of $R \in [A \leftrightarrow B]$ we write $R : A \leftrightarrow B$ or, in case R is total and univalent (a function), $R : A \to B$.

Let $A \in \mathcal{T}$. Relation algebra itself does not provide an *is-element* relation $\in$ but we can borrow a coding trick from category theory. Let $\mathbf{1}$ be a fixed singleton set and $B \in \mathcal{T}$. Then an element of B can be modeled as a total function $x : \mathbf{1} \to B$. To remind ourselves of this interpretation we shall abbreviate $x : \mathbf{1} \to B$ as $x \mathrel{\hat{\in}} B$.

Note that the proposition "$\mathbf{1}$ is a singleton" can be expressed in relation algebraic terms, namely by $I_\mathbf{1} = V_\mathbf{1} \neq \emptyset_\mathbf{1}$. We can then use Tarski's rule to prove that V is the only nonzero relation on $\mathbf{1}$: If $R \neq \emptyset$ then $R = I;R;I = V;R;V = V$. Consider a function $f : A \to B$ and an element $x \mathrel{\hat{\in}} A$. Then the result of applying f to x should be an element $y \mathrel{\hat{\in}} B$ such that the marginal diagram commutes. Obviously, we take $y = x;f$. Please remember the following slogan:

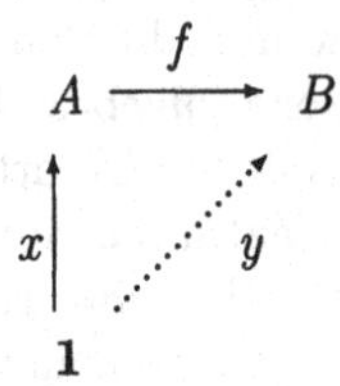

Application is multiplication from the right.

It follows that function application is associative. This property is very convenient for reasoning about functional programs and clearly distinguishes the relational calculus from the lambda calculus.

For any given $x \mathrel{\hat{\in}} A$, the look-up functional (also known as application functional) which maps every function (or relation) $f : A \leftrightarrow B$ to its value $x;f$ at x is a Boolean homomorphism. See also Theorem 3.2.9.

Lemma 8.1.1 If R is univalent then $R;(S \cap T) = R;S \cap R;T$ for all S, T and, if R is also total, $R;\overline{S} = \overline{R;S}$ for all S. □

We shall see later that look-up also distributes over pairing and conditionals. These laws are important because they allow us to eliminate look-ups from functional definitions, thereby paving the way for point-free reasoning.

The treatment of functions varying in several arguments is deferred until after we have defined direct products.

The Boolean constants

In order to calculate with programs that manipulate concrete data, these must be encoded as relations. For example, to provide the Boolean datatype, we assume that there is a distinguished type $\mathbb{B} \in \mathcal{T}$ with distinguished elements $\mathbf{T}, \mathbf{F} \mathrel{\hat{\in}} \mathbb{B}$ satisfying $\mathbf{T} = \overline{\mathbf{F}}$.

Using relation-algebraic reasoning we can show that $\mathbf{T}$ and $\mathbf{F}$ are the only elements of $\mathbb{B}$. The triple $(\mathbb{B}, \mathbf{T}, \mathbf{F})$ and the axioms $\mathbf{T}, \mathbf{F} \mathrel{\hat{\in}} \mathbb{B}$ and $\mathbf{T} = \overline{\mathbf{F}}$ constitute a relational specification of a data structure. This specification has a desirable uniqueness property: If $\mathbb{B}'$ is another type and $\mathbf{T}', \mathbf{F}' \mathrel{\hat{\in}} \mathbb{B}'$ with $\mathbf{T}' = \overline{\mathbf{F}'}$ then

$(\mathbb{B}, \mathbf{T}, \mathbf{F})$ and $(\mathbb{B}, \mathbf{T}', \mathbf{F}')$ are isomorphic. A specification which determines its model in this way (up to isomorphism, see Sect. 3.6) is called *monomorphic.*

The natural numbers

Let $\mathbf{N} \in \mathcal{T}$ be equipped with a constant $0 \mathrel{\hat{\in}} \mathbf{N}$ and a unary successor relation $succ : \mathbf{N} \to \mathbf{N}$ satisfying the following axioms

$succ;succ^{\smile} = I$	($succ$ is total and injective)
$succ^{\smile};succ \subseteq I$	($succ$ is univalent)
$0;succ^{\smile} = \emptyset$	(0 is not a successor)
$\mathbf{N} = \mu(X \mapsto 0 \cup X;succ)$	(induction rule)

where $\mu(X \mapsto f(X))$ denotes the least fixed point of f. Whenever we write this, it is understood that f is order-preserving so that Tarski's fixed point theorem (see Sect. 1.3) ensures the existence of its least fixed point. The above specification is monomorphic [Berghammer, Zierer 1986]. To improve readability we shall write *pred* instead of $succ^{\smile}$ and we let

$$x + 1 \triangleq x;succ \qquad x + n \triangleq x;succ^n \tag{8.1}$$

$$x - 1 \triangleq x;pred \qquad (x = 0) \triangleq x;0^{\smile}. \tag{8.2}$$

Here is a weaker but useful version of the induction rule. It can be proved within the calculus from the above.

Lemma 8.1.2 If $X \subseteq pred;X$ then $X = \emptyset$. □

Direct products

To model a function f of two arguments, say $f : A_1 \times A_2 \to C$, as a binary relation we have to regard an argument pair $a_1 \mathrel{\hat{\in}} A_1$ and $a_2 \mathrel{\hat{\in}} A_2$ as a single element $[a_1, a_2] \mathrel{\hat{\in}} A_1 \times A_2$ of the Cartesian product[1]. (Compare Sect. 3.6)

So assume that the set $\mathcal{T}$ of types is equipped with a binary associative operation $\times$: whenever A_1 and A_2 are types then so is $A_1 \times A_2$. To relate a direct product to its factors, we assume the existence of distinguished functions $\pi_i^{A_1,A_2} : A_1 \times A_2 \to A_i$ $(i = 1, 2)$, called the *projections* from $A_1 \times A_2$ to A_1 and A_2, respectively. Usually, A_1 and A_2 are evident from the context, so that we can abbreviate $\pi_i^{A_1,A_2}$ to π_i. The intended interpretation of π_1 and π_2 is, of course,

$$[x_1, x_2];\pi_i = x_i \quad \text{for all } x_1 \mathrel{\hat{\in}} A_1 \text{ and } x_2 \mathrel{\hat{\in}} A_2 \quad (i = 1, 2)\ . \tag{8.3}$$

To make (8.3) true we shall first recall the projection axioms from Sect. 3.6 and then define the pairing operator appropriately. First of all, we require that the projections are surjective functions:

$$\pi_i^{\smile};\pi_i \ = \ I \qquad I \subseteq \pi_i;\pi_i^{\smile} \qquad (i = 1, 2)\ . \tag{8.4}$$

Secondly, we demand that both components of a pair can be varied independently:

$$\pi_i^{\smile};\pi_j \ = \ V \qquad (\{i, j\} = \{1, 2\})\ . \tag{8.5}$$

[1]Following [Berghammer, Zierer 1986] we use square brackets to denote pairing and reserve parentheses for ordinary grouping.

The last axiom states that two distinct pairs have distinct images under at least one projection:

$$\pi_1;\pi_1^{\smile} \cap \pi_2;\pi_2^{\smile} \subseteq I \,. \tag{8.6}$$

Now we wish to solve (8.3). Given $x_1 \mathrel{\hat{\in}} A_1$ and $x_2 \mathrel{\hat{\in}} A_2$ we have to construct $[x_1 \,,\, x_2] \mathrel{\hat{\in}} A_1 \times A_2$ such that

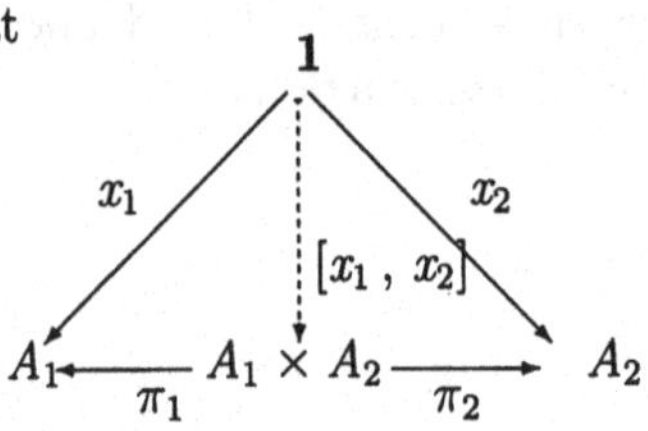

commutes. A glance at this diagram suggests

$$[x_1 \,,\, x_2] \triangleq x_1;\pi_1^{\smile} \cap x_2;\pi_2^{\smile} \,. \tag{8.7}$$

We shall use Definition 8.7 also when x_1 and x_2 are not points but arbitrary relations R and S with the same source:

$$[R \,,\, S] \triangleq R;\pi_1^{\smile} \cap S;\pi_2^{\smile} \,. \tag{8.8}$$

The following lemmas show that the pairing operator has the expected properties; their proofs illustrate the relational calculus in action.

Lemma 8.1.3 Let $R_i : A \leftrightarrow A_i$ for $i = 1,2$ such that R_2 is total. Then $[R_1 \,,\, R_2];\pi_1 = R_1$.

Proof:

$$
\begin{array}{ll}
 & [R_1 \,,\, R_2];\pi_1 \\
= & \quad \{ \text{ Definition of pairing (8.8) } \} \\
 & (R_1;\pi_1^{\smile} \cap R_2;\pi_2^{\smile});\pi_1 \\
= & \quad \{ \ \pi_1 \text{ is univalent, by axiom } (8.4)^2 \ \} \\
 & R_1 \cap R_2;\pi_2^{\smile};\pi_1 \\
= & \quad \{ \ \pi_2^{\smile};\pi_1 = V \text{ by axiom (8.5) and } R_2;V = V \text{ because } R_2 \text{ is total } \} \\
 & R_1 \,.
\end{array}
$$

□

Lemma 8.1.4 If R_1 and R_2 are total, then so is $[R_1 \,,\, R_2]$.

Proof: Recall that a relation R is total iff $R;V = V$. Using the previous lemma we obtain

$$[R_1 \,,\, R_2];V \supseteq [R_1 \,,\, R_2];\pi_1;V = R_1;V = V \,.$$

□

Lemma 8.1.5 If R_1 and R_2 are univalent, then so is $[R_1 \,,\, R_2]$.

Proof: Recall that a relation R is univalent iff $R^{\smile};R \subseteq I$.

---

[2]If S is univalent then $(R;S^{\smile} \cap T);S = R \cap T;S$. The inclusion from left to right holds because composition is order-preserving and $S^{\smile};S \subseteq I$. The other inclusion is an instance of the Dedekind rule.

$$
\begin{array}{ll}
 & [R_1\,,\,R_2]^{\smile};[R_1\,,\,R_2] \\
= & \quad \{\text{ Definition of pairing (8.8), distributivity of converse }\} \\
 & (\pi_1;R_1^{\smile} \cap \pi_2;R_2^{\smile})\;;\;(R_1;\pi_1^{\smile} \cap R_2;\pi_2^{\smile}) \\
\subseteq & \quad \{\text{ Composition is order-preserving }\} \\
 & \pi_1;R_1^{\smile};R_1;\pi_1^{\smile} \;\cap\; \pi_2;R_2^{\smile};R_2;\pi_2^{\smile} \\
\subseteq & \quad \{\ R_1 \text{ and } R_2 \text{ are univalent }\} \\
 & \pi_1;\pi_1^{\smile} \;\cap\; \pi_2;\pi_2^{\smile} \\
\subseteq & \quad \{\text{ Axiom (8.6) }\} \\
 & I\ .
\end{array}
$$

□

By virtue of the last two lemmas, any pair of functions is itelf a function. In particular, if $x_1 \mathrel{\hat\in} A_1$ and $x_2 \mathrel{\hat\in} A_2$ then $[x_1\,,\,x_2] \mathrel{\hat\in} A_1 \times A_2$. Moreover, $[x_1\,,\,x_2]$ is the only $x \mathrel{\hat\in} A_1 \times A_2$ with $x;\pi_i = x_i$ for $i = 1, 2$. For, if x satisfies this condition we have

$$
\begin{array}{ll}
 & [x_1\,,\,x_2] \\
= & \quad \{\text{ Definition of pairing (8.8) }\} \\
 & x_1;\pi_1^{\smile} \cap x_2;\pi_2^{\smile} \\
= & \quad \{\text{ Assumption on } x\ \} \\
 & x;\pi_1;\pi_1^{\smile} \cap x;\pi_2;\pi_2^{\smile} \\
= & \quad \{\text{ Univalent relations distribute leftwards into conjunctions }\} \\
 & x(\pi_1;\pi_1^{\smile} \cap \pi_2;\pi_2^{\smile}) \\
= & \quad \{\text{ Axiom (8.6) }\} \\
 & x\ .
\end{array}
$$

Lemma 8.1.6 Look-up distributes over pairing. In fact, we have

$$Q;[R\,,\,S] \;=\; [Q;R\,,\,Q;S] \quad \text{provided } Q \text{ is univalent.}$$

Proof: By Lemma 8.1.1 □

Direct products of several types can be built up by nesting binary products. To ensure that the projections $\pi_i : A_1 \times A_2 \times A_3 \to A_i$ $(i = 1, 2, 3)$ are well-defined it is necessary to require a certain associativity condition. In the diagram of direct products and projections

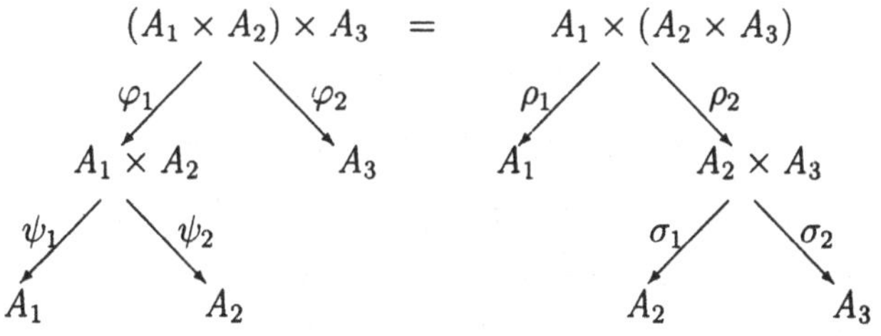

we postulate

$$\varphi_1;\psi_1 = \rho_1 \qquad \varphi_1;\psi_2 = \rho_2;\sigma_1 \qquad \varphi_2 = \rho_2;\sigma_2\ . \tag{8.9}$$

The projections on $A_1 \times A_2 \times A_3$ are now defined by

$$\pi_1 \triangleq \rho_1 \qquad \pi_2 \triangleq \varphi_1;\psi_2 \qquad \pi_3 \triangleq \varphi_2\ .$$

Then we can prove from the axioms governing binary direct products that the projections π_1, π_2, π_3 satisfy

$$\pi_i\breve{}{;}\pi_i = I \qquad \pi_i\breve{}{;}\Big(\bigcap_{j\neq i} \pi_j{;}\pi_j\breve{}\Big) = V \qquad \bigcap_i \pi_i{;}\pi_i\breve{} = I \qquad \text{for } i = 1,2,3\ . \tag{8.10}$$

This generalizes in the obvious way to products of more than three factors.

We now have defined the direct product of n types, where $2 \leq n < \infty$. To complete the picture, we consider direct products of zero or one types. The direct product of one type is this type itself, and the associated projection is the identity. With this definition, the equations (8.10) are valid. This is trivial for the first and for the last equation. The second equation holds because, by convention, the intersection of the empty family of relations equals the universal relation. The direct product of zero types is, by convention, the distinguished type $\mathbf{1}$. There are no projections, so the first two equations in (8.10), which are quantified over i, hold vacuously. The last one states that $V = I$, which is the characteristic equation of the type $\mathbf{1}$.

8.2 Programs as relations

We regard functional programs as denotations for certain relations. As we wish to use arbitrary relational connectives as program constructors, we shall pretend that *all* relations are programs. In this section we introduce conditionals, recursion, parallel composition and nondeterministic choice. To keep calculations simple we delay the usual separation of syntax and semantics until Sect. 8.3.

Applicative definitions

In the previous section we constructed domains for reasoning about functions of two arguments, but we left open the question of how such functions might be defined conveniently. For example, we would like to introduce a function *swap* : $A \times A \to A \times A$ by saying that *swap* must satisfy

$$[x_1\,,\, x_2]{;}swap \;=\; [x_2\,,\, x_1] \quad \text{for all } x_1, x_2 \mathrel{\hat{\in}} A\ . \tag{8.11}$$

Given that $x_i = [x_1\,,\, x_2]{;}\pi_i$ for $i = 1,2$ we can substitute x for $[x_1\,,\, x_2]$ and rewrite (8.11) as:

$$x{;}swap \;=\; [x{;}\pi_2\,,\, x{;}\pi_1] \quad \text{for all } x \mathrel{\hat{\in}} A \times A\ . \tag{8.12}$$

Since look-up distributes over pairing, (8.12) is equivalent to

$$x{;}swap \;=\; x{;}[\pi_2\,,\, \pi_1] \quad \text{for all } x \mathrel{\hat{\in}} A \times A\ . \tag{8.13}$$

Thus the following definition satisfies the specification (8.11):

$$swap \;\triangleq\; [\pi_2\,,\, \pi_1]\ . \tag{8.14}$$

We were able to move from the applicative specification (8.11) to the relational definition (8.14) because look-up distributes over all operators applied to x in the right-hand side of (8.11). The transition is in fact entirely mechanical, so that we are justified in regarding (8.11) as a sugared version of (8.14).

We shall use the *swap* function in a later example; so it is convenient to derive some of its properties now. It is immediate from the definitions (8.14) and (8.8) that *swap* is self-dual:

$$swap^{\smile} = swap \ . \tag{8.15}$$

As a corollary, we obtain:

Lemma 8.2.1 *swap* is a bijective function.

Proof: We know from Lemmas 8.1.4 and 8.1.5 that any pair of functions is itself a function. Moreover, the converse of any function is bijective. Therefore the claim follows from (8.14) and (8.15). □

Now we can prove that the fundamental property (8.11) holds for arbitrary arguments

$$[R\,,\,S];swap \;=\; [S\,,\,R] \quad \text{for all } R, S : B \leftrightarrow A \ . \tag{8.16}$$

$$\begin{aligned}
& [R\,,\,S];swap \\
= \quad & \{ \text{ Definition 8.8, injective relations distribute leftwards into meets } \} \\
& R;\pi_1^{\smile};swap \cap S;\pi_2^{\smile};swap \\
= \quad & \{ \ swap^{\smile} = swap \ \} \\
& R;(swap;\pi_1)^{\smile} \cap S;(swap;\pi_2)^{\smile} \\
= \quad & \{ \ swap = [\pi_2\,,\,\pi_1], \text{ Lemma 8.1.3 } \} \\
& R;\pi_2^{\smile} \cap S;\pi_1^{\smile} \\
= \quad & \{ \text{ Definition 8.8 } \} \\
& [S\,,\,R].
\end{aligned}$$

Conditional expressions

Sometimes one expression R must be used to compute the result of a function when the argument satisfies a certain condition b, and another action S when it does not. Most programming languages provide some conditional construct to deal with this situation.

Let us represent the condition as a column vector $b : A \leftrightarrow \mathbf{1}$, and assume $R, S : A \leftrightarrow B$.

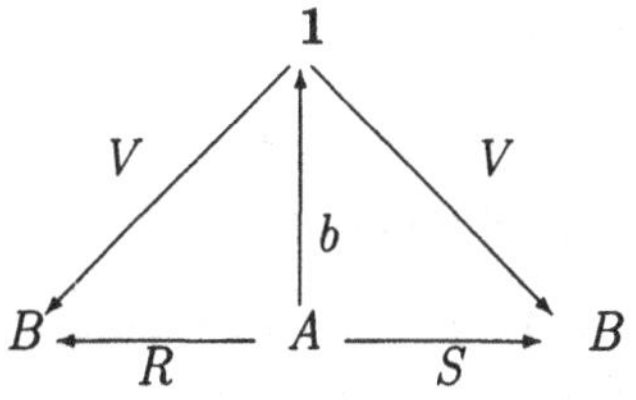

Restricting R to the domain of b (the states where b evaluates to true) computes to $R \cap b;V$. Similarly, restricting S to the states where b is false gives $S \cap \overline{b};V$. Therefore, we define the conditional (R if b else S) by

$$R \triangleleft b \triangleright S \ \triangleq \ (R \cap b;V) \cup (S \cap \overline{b};V) \ . \tag{8.17}$$

It is now straightforward to check that conditionals satisfy the expected algebraic laws, such as listed, for example, in [Hoare⁺ 1987]. By Lemma 8.1.1 we have:

Lemma 8.2.2 Look-up distributes over conditionals

$$x;(R \lhd b \rhd S) \;=\; x;R \lhd x;b \rhd x;S \quad \text{provided } x \text{ is total and univalent.}$$ □

Recursion

Suppose we wish to enrich the domain of natural numbers defined on page 117 by an addition operator $sum : \mathbf{N} \times \mathbf{N} \to \mathbf{N}$. Then we might write the program

$$[x_1\,,\,x_2];sum \;\triangleq\; x_2 \lhd x_1 = 0 \rhd (x_1 \lhd x_2 = 0 \rhd [x_1 - 1\,,\,x_2 - 1];sum + 2)\;. \tag{8.18}$$

Recall that the unary operations " $-\,1$" and " $+\,2$" are just abbreviations for " $;succ^{\smile}$ " and " $;succ;succ$ ". Eliminating look-ups from (8.18) yields a closed expression for sum:

$$sum \;\triangleq\; \pi_2 \lhd \pi_1 = 0 \rhd (\pi_1 \lhd \pi_2 = 0 \rhd [\pi_1 - 1\,,\,\pi_2 - 1];sum + 2)\;. \tag{8.19}$$

This is a fixed point equation, rather than an ordinary definition. To distinguish a solution, we note that it defines an order-preserving function τ that maps any $g : \mathbf{N} \times \mathbf{N} \leftrightarrow \mathbf{N}$ to

$$\tau(g) \;\triangleq\; \pi_2 \lhd \pi_1 = 0 \rhd (\pi_1 \lhd \pi_2 = 0 \rhd [\pi_1 - 1\,,\,\pi_2 - 1];g + 2) \;:\; \mathbf{N} \times \mathbf{N} \leftrightarrow \mathbf{N}\,. \tag{8.20}$$

So we shall take (8.18) as a sugared version of the definition

$$sum \triangleq \mu\tau\;, \tag{8.21}$$

with τ defined by (8.20). We stress that τ is not regarded as an element of the underlying relation algebra, but as a function (in the everyday sense of the word) on the set $[\mathbf{N} \times \mathbf{N} \leftrightarrow \mathbf{N}]$.

An example proof

To show relational semantics in action we shall prove that the addition operator defined in the previous section is commutative:

$$[x_1\,,\,x_2];sum \;=\; [x_2\,,\,x_1];sum \quad \text{for all } x_1, x_2 \mathrel{\hat{\in}} \mathbf{N}\;. \tag{8.22}$$

By symmetry, we need to show only one inclusion. Eliminating look-ups, we simplify our proof obligation to

$$sum \;\subseteq\; swap;sum\;. \tag{8.23}$$

Since $sum = \mu\tau$ is the *least* fixed point of τ, it is sufficient to prove that

$$\tau(swap;\mu\tau) \;=\; swap;\mu\tau\;. \tag{8.24}$$

We calculate both sides of (8.24) separately. For the left-hand side,

$$\begin{aligned}
& \tau(swap;\mu\tau) \\
= \quad & \{ \text{ Unfold the definition of } \tau \ \} \\
& \pi_2 \lhd \pi_1 = 0 \rhd (\pi_1 \lhd \pi_2 = 0 \rhd [\pi_1 - 1\,,\,\pi_2 - 1];swap;\mu\tau + 2) \\
= \quad & \{ \ [x\,,\,y];swap = [y\,,\,x] \text{ by (8.16) } \} \\
& \pi_2 \lhd \pi_1 = 0 \rhd (\pi_1 \lhd \pi_2 = 0 \rhd [\pi_2 - 1\,,\,\pi_1 - 1];\mu\tau + 2)\;.
\end{aligned}$$

For the right-hand side,

$$
\begin{array}{ll}
 & swap;\mu\tau \\
= & \quad \{ \text{ Substitute } \tau(\mu\tau) \text{ for } \mu\tau \text{ and then unfold the definition of } \tau \ \} \\
 & swap;(\pi_2 \lhd \pi_1 = 0 \rhd (\pi_1 \lhd \pi_2 = 0 \rhd [\pi_1 - 1\,,\ \pi_2 - 1];\mu\tau + 2)) \\
= & \quad \{ \text{ Distributivity of look-up } \} \\
 & swap;\pi_2 \\
 & \lhd\, swap;\pi_1 = 0 \rhd \\
 & (swap;\pi_1 \lhd swap;\pi_2 = 0 \rhd [swap;\pi_1 - 1\,,\ swap;\pi_2 - 1];\mu\tau + 2) \\
= & \quad \{ \ swap;\pi_1 = \pi_2 \text{ and } swap;\pi_2 = \pi_1 \ \} \\
 & \pi_1 \lhd \pi_2 = 0 \rhd (\pi_2 \lhd \pi_1 = 0 \rhd [\pi_2 - 1\,,\ \pi_1 - 1];\mu\tau + 2)\ .
\end{array}
$$

The terms we have obtained are equal except for the order in which π_1 and π_2 are tested for zero. In general, the order of two tests may be interchanged, provided that their associated results coincide in the case where both tests are successful. This is stated formally in the following lemma, the proof of which we leave as an exercise.

Lemma 8.2.3 Let $P_i : A \leftrightarrow B$ and $b : A \leftrightarrow \mathbf{1}$. Then

$$P_1 \lhd b_1 \rhd (P_2 \lhd b_2 \rhd Q) \;=\; P_2 \lhd b_2 \rhd (P_1 \lhd b_1 \rhd Q)$$

provided that $P_1 \lhd (b_1 \cap b_2) \rhd \emptyset \;=\; P_2 \lhd (b_1 \cap b_2) \rhd \emptyset$. □

We proceed to verify the premise of Lemma 8.2.3. Let $i \in \{1,2\}$. Then

$$
\begin{array}{ll}
 & \pi_i \lhd (\pi_1 = 0 \ \cap\ \pi_2 = 0) \rhd \emptyset \\
= & \quad \{ \text{ Definitions of conditional (8.17) and of ``}= 0\text{'' (8.2) } \} \\
 & \pi_i \ \cap\ (\pi_1;0^{\smile} \ \cap\ \pi_2;0^{\smile});V \\
= & \quad \{ \text{ Definition of pairing (8.8) } \} \\
 & \pi_i \ \cap\ [0\,,\,0]^{\smile};V \\
= & \quad \{ \ [0\,,\,0] \text{ is univalent (Lemma 8.1.5)}\ ^3 \ \} \\
 & [0\,,\,0]^{\smile} ; ([0\,,\,0];\pi_i \cap V) \\
= & \quad \{ \text{ Lemma 8.1.3 } \} \\
 & [0\,,\,0]^{\smile};0\ .
\end{array}
$$

Thus $(\pi_1 \lhd (\pi_1 = 0 \cap \pi_2 = 0) \rhd \emptyset) \;=\; (\pi_2 \lhd (\pi_1 = 0 \cap \pi_2 = 0) \rhd \emptyset)$, as required.

Parallel composition

Functional languages enjoy built-in parallelism in the sense that several arguments of a function call may be executed concurrently. A closely related kind of parallelism is the *direct product* $R_1 \parallel R_2$ of two functions which maps every pair (a_1, a_2) to $(R_1(a_1), R_2(a_2))$. In the relational calculus R_1 and R_2 may be arbitrary relations, and their parallel composition can be defined in terms of the appropriate projections (see diagram):

$$R_1 \parallel R_2 \ \triangleq\ \pi_1;R_1;\varphi_1^{\smile} \ \cap\ \pi_2;R_2;\varphi_2^{\smile}\ . \tag{8.25}$$

When R_1 and R_2 are total, both squares in the diagram commute.

[3] If S is univalent then $R \cap S^{\smile};T = S^{\smile};(S;R \cap T)$ for all R and T, cf. footnote on p. 118.

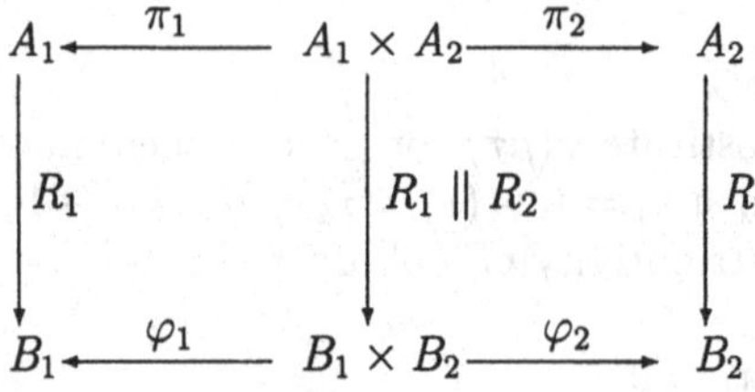

The relational parallel operator models simultaneous evaluation without interaction, which makes it less powerful than concurrency with shared variables or message passing. Its beauty lies in its simplicity; an operator that can be defined in terms of the existing ones introduces no additional complexity into the theory. In conjunction with sequencing, parallel composition is a powerful structuring aid. For an example see the definition of *prod* in the next subsection.

Nondeterminism

There is nothing to say that a relational program must always define a (single-valued) function. For example, we can define a "program" *less* that takes a single input $x \mathrel{\hat{\in}} \mathbf{N}$ and returns an arbitrary $y \mathrel{\hat{\in}} \mathbf{N}$ that is lower than or equal to x:

$$x;less \;=\; x \cup (x-1);less \quad \text{for all } x \mathrel{\hat{\in}} \mathbf{N} . \tag{8.26}$$

This recursive definition explains *less* in terms of the successor function (recall that $x - 1 = x;succ^{\smile}$). Rewriting (8.26) in non-applicative style yields

$$less \;=\; \mu(g \mapsto I \cup succ^{\smile};g) . \tag{8.27}$$

One attraction of the relational calculus is economy of expression. For example, x is lower than or equal to $y + z$ if and only if we can write $x = y' + z'$ with $y' \leq y$ and $z' \leq z$. In relations, this is written simply as

$$sum;less \;=\; [less\,,\, less];sum \;.$$

It is a challenging exercise to prove this law within the relational calculus.

Even if the final goal is a deterministic program, nondeterminism may still be useful in program design. Consider the task of programming multiplication in terms of addition. The distributivity law

$$a * (b + c) \;=\; a * b + a * c \tag{8.28}$$

shows that multiplying a by a large number can be achieved by multiplying a with two smaller numbers and adding the results. Translating (8.28) into relational notation and eliminating look-ups yields the recursive equation

$$prod \;=\; (I \parallel sum^{\smile}) \;;\; [[\pi_1, \pi_{21}], [\pi_1, \pi_{22}]] \;;\; (prod \parallel prod) \;;\; sum \;, \tag{8.29}$$

with appropriately defined projection functions π_1, π_{21} and π_{22}. The first statement transforms an initial state (x, y) to some state $(x, (u, v))$ with $u + v = y$. The second statement rearranges the state to $((x, u), (x, v))$. Then the products xu and xv are computed in parallel and added.

Actually, the right-hand side of (8.29) is a purely functional expression, but we could not resist the temptation to read it as a sequence of four statements. By emphasizing the sequential structure of functional programs, the relational calculus blurs the distinction between the functional and the imperative style, potentially combining the advantages of both.

We can obtain a program from (8.29) by specifying the termination case and refining $sum^{\smile}$ into some constructive statement. The latter task could be accomplished by splitting y into $y-1$ and 1, or else by dividing y into two roughly equal parts, which is more efficient but requires case distinction. In modern software engineering such choices are delayed as long as possible, and nondeterminism allows us to do so without abandoning mathematical rigour.

8.3 A simple applicative language

Our choice of notations was governed by the desire for a smooth algebraic calculus, but designers of real programming languages have other concerns. To enable mechanical translation, syntax must be given in more formal style, and often syntactic sugar is added to aid visual structuring and parsability.

Interpreters and compilers operate on syntactical objects. To reason about their correctness, we need a function that associates with each syntactic program P a relation $\mathcal{S}(P)$, called the *semantics* of P. Equations between relations become transformation rules for programs or their syntactic substructures

$$P \equiv Q \quad \text{iff} \quad \mathcal{S}(P) = \mathcal{S}(Q)\ .$$

The transformational approach to program transformation has been studied extensively in the literature. One advantage of a separate syntax is that it makes it possible to state transformation rules with syntactical side conditions, like "x is not free in P". Formal syntax also provides a basis for relating *different* semantics of the same language – for example a denotational semantics is sometimes justified by proving its consistency with an operational one. Another typical application is the comparison between different semantics for nondeterminism. Finally, formal syntax is a prerequisite for mechanical proof checking or discovery.

As an example, we shall formally define the syntax and relation-algebraic semantics of a simple applicative language.

Syntax

We assume as given a set *Relid* the elements of which we use as identifiers for relations. The set of *relational programs* is defined by the following grammar:

1. If $t_1, \ldots, t_k$ are relational programs, then so is the tuple $(t_1, \ldots, t_k)$. This includes the empty tuple $()$.
2. Let $f \in$ *Relid*. If t is a relational program, then so is the application $f(t)$. In the case where t is a tuple, say $t = (t_1, \ldots, t_k)$, we write $f(t_1, \ldots, t_k)$ instead of the clumsy $f((t_1, \ldots, t_k))$. In particular, we omit the parentheses in $f()$.
3. If s and t are relational programs, then so is the choice $s \,[\!]\, t$.

4. If b, s and t are relational programs, then so is the conditional

$$\texttt{if } b \texttt{ then } s \texttt{ else } t \ .$$

5. If t is a relational program, $x_1, \ldots, x_n, f \in \mathit{Relid}$ and $A_1, \ldots, A_n, B \in \mathcal{T}$, then the following (possibly recursive) declaration is a relational program:

$$\texttt{relation } f(x_1 : A_1, \ldots, x_n : A_n) : B \ ; \ t .$$

For example, here is a program for the relation *less* defined on page 124):

$$\texttt{relation } \mathit{less}(x : \mathbb{N}) : \mathbb{N} \ ; \ x \,[]\, \mathit{less}(\mathit{pred}(x)) \ .$$

Semantics

An environment is a mapping that assigns a relation to every $f \in \mathit{Relid}$. Not every term is well-typed with respect to every environment (and some terms cannot be well-typed at all), so one has to give a set of rules to decide well-formedness. This is standard and we assume it has been done. For any program t that is well-typed with respect to an environment ρ we define the semantics $\mathcal{A}(t, \rho)$. To make the defining rules below more readable, we have omitted the second parameter of $\mathcal{A}$, except where it is textually different from ρ.

1. The relational calculus has a tupling operation (see Definition 8.8), which we can use to explain the semantics of a tuple $(t_1, \ldots, t_k)$. For $i = 1, \ldots, k$ let $R_i = \mathcal{A}(t_i) : A \leftrightarrow B_i$. With $\pi_1, \ldots, \pi_k$ denoting the projections of $B_1 \times \cdots \times B_k$ we define

$$\mathcal{A}(t_1, \ldots, t_k) = \textstyle\bigcap_i R_i ; \pi_i\breve{} \ .$$

2. To evaluate $f(t)$ in the environment ρ, evaluate t and then apply $\rho(f)$ to the result:

$$\mathcal{A}(f(t)) = \mathcal{A}(t) ; \rho(f) \ .$$

3. $s \,[]\, t$ may behave like s or like t, so choice is modelled by disjunction:

$$\mathcal{A}(s \,[]\, t) = \mathcal{A}(s) \cup \mathcal{A}(t) \ .$$

4. Relational calculus has a conditional, so we use it:

$$\mathcal{A}(\texttt{if } b \texttt{ then } s \texttt{ else } t) = \mathcal{A}(s) \lhd \mathcal{A}(b) \rhd \mathcal{A}(t) \ .$$

5. Let $s = (\texttt{relation } f(x_1 : A_1, \ldots, x_k : A_k) : B \ ; \ t)$. The parameter declaration binds every x_i to the ith projection π_i from $A_1 \times \cdots \times A_k$, so let

$$\rho' \triangleq \rho[\pi_1/x_1, \ldots \pi_k/x_k] \ .$$

Then $\mathcal{A}(s, \rho)$ is the least solution of the equation

$$R = \mathcal{A}(t, \rho'[R/f]) \ ,$$

where R ranges over $[A_1 \times \cdots \times A_k \leftrightarrow B]$. Its existence follows from Tarski's fixed point theorem and the fact that $\mathcal{A}(t, \rho)$ is monotonic in ρ (which follows by induction on the structure of t).

8.4 Nondeterminism revisited

In this section we discuss various forms of nondeterminism, their uses, and how they can be modelled in the calculus of relations.

Angelic nondeterminism

Reconsider the choice $s \mathbin{[\!]} t$ when s or t may loop. Since we modelled choice as union, the result is defined when either s or t terminates. It is as if the choice between s and t were made by an angel who can predict either outcome and does its best to save us from waiting forever. To implement angelic nondeterminism, evaluate both branches concurrently until either terminates. Of course, in a program with many branching points, the angelic evaluation strategy is terribly inefficient.

Now assume that a program has been proved correct with respect to the angelic semantics, but no angel is available at execution time. Even if choices are entrusted to some unknown agent, we shall never get an erroneous result. So, angelic nondeterminism may be seen as an abstraction of the "real" behaviour of programs, obtained by disregarding the possibility of infinite loops. In some cases, this information may be all we need. For example, a certification agency should be satisfied when the compiler that has been used to generate some safety-critical code has been shown "angelically" correct.

Angelic nondeterminism is worthwhile because it is far simpler than its alternatives, but it can only be used for establishing claims of *partial* correctness:

If t terminates then the result will be correct.

The semantic function defined in the previous section is of the angelic type (which explains why we called it $\mathcal{A}$).

Failures

In order to reason about termination, we need to extend our semantic theory in such a way that we can express the possibility of its absence. There are various ways to do this, including the following:

- Augment each data type with a special element denoting failure. With the help of *flat lattices* this approach can be encoded within the relational calculus. This approach is elaborated in [Gritzner, Berghammer 1996]. A variation on this theme is the demonic calculus presented in Chapt. 11.
- Introduce a second semantic function $\mathcal{F}(s)$ to describe the set of arguments which might cause s to loop. This idea was used in the semantics of CSP [Brookes, Roscoe 1984], and first applied to relational semantics in [Berghammer, Zierer 1986].
- Assign to each program a higher order relation, usually called a *predicate transformer*. This method is advocated by Dijkstra [Dijkstra, Scholten 1990], and Maddux [Maddux 1992] translated it into the calculus of relations.

Among these we choose the second alternative because it requires the least increase in complexity. Also, it allows us to relate angelic, demonic and erratic nondeterminism in a simple and clear way. We assume a *failure environment* φ that assigns to every $f \in Relid$ a column vector $\varphi(f)$ representing the inputs which may lead to desaster. We define for every relational program t its failure vector $\mathcal{F}(t, \rho, \varphi)$, which depends on the environment ρ and the failure environment φ. As before, we shall suppress the environment arguments, unless they are modified. We require that ρ and φ map any given variable to relations of the same type.

1. A tuple may fail as soon as one component does:

$$\mathcal{F}(t_1, \ldots, t_k) = \bigcup_i \mathcal{F}(t_i);\pi_i^{\smile} .$$

2. An application $f(t)$ fails when t cannot be evaluated, or else the value of t is in the failure set $\varphi(f)$ of f:

$$\mathcal{F}(f(t)) = \mathcal{F}(t);V \cup \mathcal{A}(t);\varphi(f) .$$

3. A choice fails when either alternative fails:

$$\mathcal{F}(s \,[\!]\, t) = \mathcal{F}(s) \cup \mathcal{F}(t) .$$

4. A conditional fails if its condition cannot be evaluated, or if the selected branch fails:

$$\mathcal{F}(\texttt{if } b \texttt{ then } r \texttt{ else } s) = \mathcal{F}(b);V \cup (\mathcal{F}(s) \lhd \mathcal{A}(b) \rhd \mathcal{F}(t)) .$$

5. Let $s \triangleq (\texttt{relation } f(x_1 : A_1, \ldots, x_k : A_k) : B \; ; \; t)$. We define ρ' as before. Since look-ups cannot fail, the x_i are bound to $\emptyset$ in the failure environment

$$\varphi' = \varphi[\emptyset/x_1, \ldots, \emptyset/x_k] .$$

Then $\mathcal{F}(s, \rho, \varphi)$ is the *greatest* solution of the equation

$$R = \mathcal{F}(t, \rho'[\mathcal{A}(s)/f], \varphi'[R/f]) .$$

The reader may wonder why we use the smallest fixed point for defining the set of possible results, but the largest one for the set of possible failures. It is beyond the scope of this book to give a full operational justification, but to get the idea consider an unguarded recursion

$$\texttt{relation } f(x : \mathbb{N}) : \mathbb{N} \; ; \; x .$$

Both the semantic functional and the failure functional of t are the identity, so every relation is a fixed point. But calling f will never produce a result (smallest fixed point) and always fail (greatest fixed point).

Erratic and demonic nondeterminism

The semantics of a program is now given as a pair of relations

$$\mathcal{E}(t) \;\triangleq\; (\mathcal{F}(t)\;,\; \mathcal{A}(t)) .$$

$\mathcal{E}(t)$ is known as the *erratic semantics* of t. Traditionally, erratic semantics is defined using least fixed points with respect to a special order (e.g. in [Hennessy, Ashcroft 1976; Plotkin 1976], called the Egli-Milner order. Our definition has the advantage of simplicity and modularity, and it can be shown that it is equivalent to one using the Egli-Milner order. A somewhat similar approach (but for imperative languages) can be found in [Doornbos 1994].

Now assume that all program specifications are of the form

t will terminate *and* produce a correct result.

A program satisfying such a requirement is said to be *totally correct*. Since semantics should be as simple as possible, it is counterproductive to distinguish between two programs when there is no specification that is satisfied by only one of them. With respect to total correctness, a result is only relevant when termination can be guaranteed, and a program that might fail is just as bad as one that is bound to fail. Two programs t_1 and t_2 are equivalent in this sense iff

$$\mathcal{F}(t_1) = \mathcal{F}(t_2)$$

and

$$\mathcal{A}(t_1) \cup \mathcal{F}(t_1) = \mathcal{A}(t_2) \cup \mathcal{F}(t_2)\ .$$

So a semantics for total correctness is defined by

$$\mathcal{D}(t)\ \triangleq\ (\mathcal{F}(t)\ ,\ \mathcal{A}(t) \cup \mathcal{F}(t))\ .$$

$\mathcal{D}(t)$ is known as the *demonic semantics* of t.

Termination

The failure semantics can also be used to prove termination separately from partial correctness. Let us reconsider the recursively defined addition function. Expressed in the formal syntax defined on page 125, it reads

$$\begin{array}{ll}\texttt{relation}\ \mathit{sum}(x_1 : \mathbf{N}, x_2 : \mathbf{N}) : \mathbf{N}\ ; & \\ \quad\texttt{if} & \mathit{is\text{-}zero}(x_1)\ \texttt{then}\ x_2 \\ \quad\texttt{else if} & \mathit{is\text{-}zero}(x_2)\ \texttt{then}\ x_1 \\ \quad\texttt{else} & \mathit{inc}(\mathit{inc}(\mathit{sum}(\mathit{dec}(x_1), \mathit{dec}(x_2))))\ .\end{array}$$

The semantics and the failures of the basic functions *is-zero*, *dec* and *inc* are predefined in terms of the operations on natural numbers defined on page 117:

$$\begin{array}{ll} \rho(\mathit{is\text{-}zero}) \triangleq 0^{\smile} & \varphi(\mathit{is\text{-}zero}) \triangleq \emptyset \\ \rho(\mathit{inc}) \triangleq \mathit{succ} & \varphi(\mathit{inc}) \triangleq \emptyset \\ \rho(\mathit{dec}) \triangleq \mathit{pred} & \varphi(\mathit{dec}) \triangleq \overline{0}^{\smile} ; V\ . \end{array}$$

Then expanding all the definitions yields

$$\mathcal{A}(\mathit{sum}, \rho) = \mu\tau$$

where τ is the functional defined by (8.20) on page 122. The failures of t are given by

$$\mathcal{F}(\mathit{sum}, \rho, \varphi) = \nu\sigma$$

where $\nu\sigma$ denotes the greatest fixed point of the functional σ defined by

$$\sigma(X) = (pred \parallel pred);X .$$

Assume that $X = \sigma(X)$. Then

$$\pi_1\breve{};X = \pi_1\breve{};\sigma(X) \subseteq \pi_1\breve{};\pi_1;pred;\pi_1\breve{};X = pred;(\pi_1\breve{};X) .$$

By Lemma 8.1.2 this implies

$$\pi_1\breve{};X = \emptyset$$

and we conclude that

$$X = (\pi_1;\pi_1\breve{} \cap \pi_2;\pi_2\breve{});X \subseteq \pi_1;\pi_1\breve{};X = \emptyset .$$

Therefore, $\emptyset$ is the only fixed point of σ, and it follows that *sum* never fails.

8.5 Conclusion

It has been known for some time that the relations are well-suited for modeling sequential computer programs. We hope that this article has convinced some readers that the relational calculus is equally suitable for reasoning about functional programs. For obvious reasons, we have restricted ourselves to a very basic language. Using Zierer's advanced domain constructions [Zierer 1991], we can integrate more advanced features, such as functional parameters into our language.

Our approach is related to the work of Bird and de Moor who use allegories (a categorical version of the relational calculus). Their forthcoming book contains many relational derivations of functional programs.

Current efforts concentrate on using the relational calculus for modeling temporal and reactive aspects of programs and concurrent processes and on deriving parallel processes from temporal specifications [Hoare, von Karger 1995; Berghammer, von Karger 1996].

Chapter 9

Algorithms from Relational Specifications

Rudolf Berghammer, Burghard von Karger

The purpose of a specification is to state a problem as clearly as possible. In many cases, the most direct and intuitive way to specify a problem is by writing down a logical predicate that describes its possible solutions. Here, we employ the calculus of relations for developing efficient algorithms from problem specifications.

The first task is to obtain a *relational specification* from the original logical problem description. There are a number of correspondences between logical and relation-algebraic operations, which in many practical examples yield the desired relational specification easily. In more complex situations one can, for instance, use relational algebra extended with direct products and their associated projections.

Relational specifications are very compact, and if the carrier sets are not too large, they can be executed directly using a relation manipulation system. In this way, the desired algorithm can be prototyped rapidly. In some cases the resulting algorithm is good enough, and then the development is complete. But often the relational specification is not acceptable as an implementation since its evaluation takes too much time and/or space. Then we apply the well-developed apparatus of relational algebra to derive a more efficient algorithm. We illustrate the relational approach to algorithm development with numerous graph- and order-theoretic examples, including bounds and extremal elements, tests for cycle-freeness and reachability, and algorithms to compute transitive closures and kernels for specific classes of directed graphs.

9.1 Preliminaries

To make this chapter self-contained, we first collect some preliminaries.

Matrix notation. A relation R stored on a computer is a Boolean matrix. The y-th element of the x-th row of a matrix R is traditionally denoted R_{xy}. Since the entries in a Boolean matrix range over 0 (false) and 1 (true) the proposition R_{xy} simply states that R relates x to y, or $(x, y) \in R$. In the special case where R is a vector, $R = R;V$, we omit the second subscript. Sets will be represented as relations with a distinguished one-point target $\mathbf{1}$. Every relation $R \in [X \leftrightarrow \mathbf{1}]$ is a vector. Thus R_x is true just when x is a member of the set R.

Residuals. In Boolean matrix notation the definition of residuals reads:

$$(S/R)_{yx} \iff \forall z\ (R_{xz} \to S_{yz}) \qquad (R\backslash S)_{xy} \iff \forall z\ (R_{zx} \to S_{zy}). \tag{9.1}$$

These equivalences are invaluable for removing universal quantifications. There are two special cases that deserve explicit mention, namely

$$(S/V)_y \iff \forall z\ S_{yz} \qquad (\overline{R}\backslash\emptyset)_x \iff \forall z\ R_{zx}. \tag{9.2}$$

We recall the universal property of left residuals

$$Q;S \subseteq T \iff Q \subseteq T/S, \tag{9.3}$$

and its corollary, the *cancellation rule*

$$(Q/S);S \subseteq Q. \tag{9.4}$$

Fixed points. Let $(V, \sqsubseteq)$ and $(W, \sqsubseteq)$ be complete lattices. Recall from Chapt. 1 that every order-preserving function f on a complete lattice has least and greatest fixed points μf and νf (Tarski's fixed point theorem). Often, we can compute μf and νf iteratively using the formulas in the following theorem.

Theorem 9.1.1 Assume f is an order-preserving function on V. If f preserves limits of ascending chains (is *continuous*) then $\mu f = \bigsqcup_{i\geq 0} f^i(\perp\!\!\!\perp)$ and if f preserves limits of descending chains (is *co-continuous*) then $\nu f = \bigsqcap_{i\geq 0} f^i(\top\!\!\!\top)$ □

Another useful way to compute a fixed point is to express it in terms of another – hopefully more tractable – fixed point.

Lemma 9.1.2 (*μ-Fusion Rule*) If $f : V \to V$ and $g : W \to W$ are order-preserving, and $h : V \to W$ is strict $(h(\perp\!\!\!\perp) = \perp\!\!\!\perp)$ and continuous then $h \circ f = g \circ h$ implies $h(\mu f) = \mu g$. □

There are a number of laws concerning least fixed points of composite functions, of which the most useful and well-known is the rolling rule.

Lemma 9.1.3 (*Rolling Rule*) If $f : V \to W$ and $g : W \to V$ are order-preserving then $\mu(f \circ g) = f(\mu(g \circ f))$. □

Graph theory. A *directed graph* is a pair $g = (X, R)$ where X is a nonempty set and $R \in [X \leftrightarrow X]$. The elements of X and R are the *vertices* and *arcs* of the graph. A (finite) *path* of length n from vertex x to vertex y in g is a sequence $p = (p_0, p_1, \ldots, p_n)$ such that $x = p_0$, $y = p_n$, and $R_{p_i,p_{i+1}}$ for all i with $0 \leq i \leq n-1$. In this case, y is called *reachable* from x via p. If $n \neq 0$, then p is said to be *non-empty*, and if $p_0 = p_n$, then p is called *closed*. A *cycle* is a non-empty and closed path. An *infinite path* in $g = (X, R)$ is an infinite sequence $p = (p_0, p_1, \ldots)$ such that $R_{p_i,p_{i+1}}$ for all $i \geq 0$. The following lemma explains the connection between reachability and transitive closure.

Lemma 9.1.4 Let $g = (X, R)$ be a directed graph and $x, y \in X$. Then y is reachable from x if and only if R^*_{xy} and y is reachable from x via a nonempty path if and only if R^+_{xy}. □

9.2 Relational problem specification

In the first stage of every development we rewrite a logical formula into a relational specification and the purpose of this section is to illustrate the underlying technique through a collection of examples involving graphs and partial orders. The translation eliminates quantification over elements of the carrier sets and often produces very concise "code" that lends itself to further formal manipulation. We roughly divide our specifications into three classes: Some are efficient algorithms as they stand, others can be executed as prototypes but are too slow for daily use, and some can only be enacted after further refinements.

Efficiently executable specifications

Sometimes the transition from a predicate logic problem description to a relational expression immediately yields an efficient algorithm.

Example 9.2.1 (*Cycle-freeness*) A directed graph $g = (X, R)$ is *cycle-free* if every path $p = (p_0, \ldots, p_n)$ with $n > 0$ satisfies $p_0 \neq p_n$. Or: For all vertices x and y, if y is reachable from x via a non-empty path then $x \neq y$. Using Lemma 9.1.4 we can rewrite this predicate to

$$\forall x, y\ (R^+_{xy} \to \overline{I}_{xy}).$$

Removing universal quantification, we arrive at a relational form of cyclefreeness:

$$\textit{cycle-free} : [X \leftrightarrow X] \to \mathbb{B} \qquad \textit{cycle-free}(R) \iff R^+ \subseteq \overline{I}.$$

If Warshall's well-known algorithm is used (see Sec. 9.3), the cost of evaluating *cycle-free*(R) is of the same order as for computing the transitive closure of R, namely cubic time. □

Example 9.2.2 (*Difunctionality*) Difunctional relations were introduced in the paper [Riguet 1950] and have found many applications in computer science, see for instance [Schmidt, Ströhlein 1993]. A relation R is *difunctional* if there are partial functions (univalent relations) f and g such that $R = f;g^{\smile}$ or, equivalently, if the Boolean matrix associated with R can be transformed into *block diagonal form* by rearranging rows and columns. The latter property can be restated as follows: Whenever two rows x and y have a "1" in the same place, they must be equal. This is expressed by

$$\forall x, y\ (\exists m\ (R_{ym} \wedge R_{xm}) \to \forall n\ (R_{yn} \leftrightarrow R_{xn})).$$

Using elementary predicate calculus and symmetry, the equivalence sign in the above formula may be replaced with an implication:

$$\forall x, n\ (\exists m, y\ (R_{xm} \wedge R^{\smile}{}_{my} \wedge R_{yn}) \to R_{xn}).$$

According to the definition of relational composition, the antecedent becomes $(R;R^{\smile};R)_{xn}$. We have now arrived at the following relational specification:

$$\textit{difunctional} : [X \leftrightarrow X] \to \mathbb{B} \qquad \textit{difunctional}(R) \iff R;R^{\smile};R \subseteq R.$$

This is an efficient difunctionality test, which has the same complexity as an algorithm for computing the composition of two relations. □

Example 9.2.3 (*Bounds and extremal elements*) J. Riguet discovered that many concepts involving preorders can be expressed relationally; confer [Riguet 1948]. His definitions nowadays are widely used in computer science (see [Zierer 1991; Bird, de Moor 1992; Berghammer, Gritzner+ 1993] for example) and we present them in the sequel as relational specifications. Assume we are given a fixed preorder $E \in [X \leftrightarrow X]$. Then $y \in X$ is a *lower bound* of $s \subseteq X$ if

$$\forall z\ (z \in s \rightarrow E_{yz}).$$

Writing s_z instead of $z \in s$, we can use (9.1) to eliminate the quantification and we see that the set of all lower bounds of s is represented by the vector $E/s^{\smile}$. The formal relational specification is

$$lwbd : [X \leftrightarrow X] \times [X \leftrightarrow \mathbf{1}] \rightarrow [X \leftrightarrow \mathbf{1}] \qquad lwbd(E, s) = E/s^{\smile}.$$

This is an efficient algorithm and its computational complexity depends on the cost of computing the residual. Now, that we have an algorithm for computing the lower bounds of a set, we can exploit it to generate related sets. The set of all *least elements* of s (there may be more than one because E is only a preorder) is obtained by computing the intersection of s with its set of lower bounds:

$$least : [X \leftrightarrow X] \times [X \leftrightarrow \mathbf{1}] \rightarrow [X \leftrightarrow \mathbf{1}] \qquad least(E, s) = s \cap lwbd(E, s).$$

Transposing E, we compute greatest elements by

$$greatest : [X \leftrightarrow X] \times [X \leftrightarrow \mathbf{1}] \rightarrow [X \leftrightarrow \mathbf{1}] \qquad greatest(E, s) = least(E^{\smile}, s).$$

The function *inf* computing the vector of all greatest lower bounds of s is

$$inf : [X \leftrightarrow X] \times [X \leftrightarrow \mathbf{1}] \rightarrow [X \leftrightarrow \mathbf{1}] \qquad inf(E, s) = greatest(E, lwbd(E, s)).$$

For E antisymmetric, $z \in X$ is a *minimal element* of $s \subseteq X$ if

$$z \in s \wedge \forall y\ (E_{yz} \wedge y \neq z \rightarrow y \notin s).$$

Writing $\overline{s}_y$ for $s \notin y$ and $\overline{I}_{yz}$ for $y \neq z$, the second conjunct becomes $\forall y\ ((E \cap \overline{I})_{yz} \rightarrow \overline{s}_y)$ and (9.1) transforms it into $((E \cap \overline{I}) \backslash \overline{s})_z$, giving

$$min : [X \leftrightarrow X] \times [X \leftrightarrow \mathbf{1}] \rightarrow [X \leftrightarrow \mathbf{1}] \qquad min(E, s) = s \cap (E \cap \overline{I}) \backslash \overline{s}.$$

If s is not a vector but an arbitrary relation $R \in [X \leftrightarrow Y]$, the functions derived above compute bounds and extremal elements column-wise. For instance, if E and R are Boolean matrices, every column of $least(E, R) \in [X \leftrightarrow Y]$ is a point or the empty vector from $[X \leftrightarrow \mathbf{1}]$. In the first case the point represents the least element of the corresponding column of R; in the latter case such a least element does not exist. For example, we can take R to be the membership relation in *inf*; we then obtain an algorithm for checking if a given finite partial order (X, E) is a lattice. □

Prototyping relational specifications

Using ideas from [Berghammer, Gritzner+ 1993] we will now present relational specifications of some higher-order objects (sets of sets). Unlike the previous examples, these are not efficient enough to be used in a production environment.

They are rapid prototypes, and as such they serve a double purpose. Firstly, they may be executed directly in the RELVIEW workbench for concrete relations (see [Berghammer, Schmidt 1991]), allowing tests and experiments (with small inputs). Secondly, they provide a formal starting point for the derivation of more efficient algorithms. Examples of this will be presented in Sects. 9.3 and 9.4.

In order to write higher-order specifications, we incorporate the set-theoretic is-element and is-subset-of relations into the calculus. To avoid confusion, we introduce ε and $\preceq$ on the object-level as synonyms for the meta-level symbols $\in$ and $\subseteq$. Hence, ε and $\preceq$ may be used in relation-algebraic expressions and formulae like $\preceq \; = \varepsilon\backslash\varepsilon$ or $\preceq^2 \subseteq \preceq$. Furthermore, the powerset of a set X is denoted by 2^X instead of $\mathcal{P}(X)$.

Example 9.2.4 (*Kernels of directed graphs*) A vertex set $a \subseteq X$ of a directed graph $g = (X, R)$ is *absorbant* if every vertex outside a has at least one successor inside a:

$$\forall x \; (x \notin a \rightarrow \exists y \; (y \in a \wedge R_{xy})).$$

A vertex set $s \subseteq X$ is *stable* if no two elements of s are neighbours:

$$\forall x \; (x \in s \rightarrow \forall y \; (y \in s \rightarrow \overline{R}_{xy})).$$

If k is stable and absorbant, k is said to be a *kernel*. Kernels correspond to winning strategies in certain combinatorial games ("move into the kernel each time it is your turn"); for an overview see [Schmidt, Ströhlein 1993]. Now, let's construct an algorithm that takes a directed graph $g = (X, R)$ as its input and returns the set of its kernels (represented as a vector over 2^X). First we observe that $\forall y \; (y \in s \rightarrow \overline{R}_{xy})$ is equivalent to $(\overline{R;\varepsilon})_{xs}$. Thus, s is stable if and only if $\forall x \; (\overline{\varepsilon \cap R;\varepsilon})_{xs}$. Due to (9.2), this is equivalent to $((\varepsilon \cap R;\varepsilon)\backslash\emptyset)_s$. Thus, a function *stable* returning the vector of all stable vertex sets of its argument is defined by

$$\mathit{stable} : [X \leftrightarrow X] \rightarrow [2^X \leftrightarrow \mathbf{1}] \qquad \mathit{stable}(R) = (\varepsilon \cap R;\varepsilon)\backslash\emptyset.$$

An analogous argument shows how to calculate the set of all absorbant vertex sets:

$$\mathit{absorb} : [X \leftrightarrow X] \rightarrow [2^X \leftrightarrow \mathbf{1}] \qquad \mathit{absorb}(R) = (V\backslash(\varepsilon \cup R;\varepsilon))^{\smile}.$$

These relational specifications are algorithmic (though not very efficient), and they can be combined into an algorithm computing all the kernels of the given graph:

$$\mathit{kernel} : [X \leftrightarrow X] \rightarrow [2^X \leftrightarrow \mathbf{1}] \qquad \mathit{kernel}(R) = \mathit{absorb}(R) \cap \mathit{stable}(R).$$

That this computation is not efficient is not the fault of relation algebra, as the kernel problem is in fact NP-complete. For specific classes of graphs, more efficient algorithms to compute a single kernel of a directed graph will be derived in Sect. 9.4. □

Example 9.2.5 (*Vertex bases*) A *vertex base* of a directed graph $g = (X, R)$ is a minimal set b of vertices from which every vertex in X can be reached via a path. We note that every vertex can be reached from b if and only if

$$\forall x \, \exists y \; (y \in b \wedge R^*_{yx}).$$

Since $b \in y$ can be written as ε_{by}, this is equivalent to $\forall x\ (\varepsilon^{\smile};R^*)_{bx}$. Using the residual properties (9.2), we can further refine this expression to $((\varepsilon^{\smile};R^*)/V^{\smile})_b$, where V is the top element of $[X \leftrightarrow \mathbf{1}]$. Among all b with this property we require one that is minimal wrt. the set inclusion relation $\preceq$. Fortunately, we already have a relational specification for computing minimal elements of a set (see Example 9.2.3). So we obtain the relational specification:

$$base : [X \leftrightarrow X] \to [2^X \leftrightarrow \mathbf{1}] \qquad base(R) = min(\preceq, (\varepsilon^{\smile};R^*)/V^{\smile})$$

of a function that computes the vector of vertex bases. An even simpler specification can be built with the aid of the *kernel* function of Example 9.2.4. Observe that $b \in 2^X$ is a vertex base of g if and only if every vertex in $\overline{b}$ can be reached from a vertex in b and no vertex in b can be reached from some other vertex in b. In other words: A set b is a vertex base of $g = (X, R)$ just when b is a kernel of $g' = (X, (R^{\smile})^+ \cap \overline{I})$. Therefore we have that $base(R) = kernel((R^{\smile})^+ \cap \overline{I})$. □

Non-algorithmic relational specifications

So far, we have considered only relational specifications without quantification, because these can be regarded as algorithmic (provided the underlying sets are finite). To increase expressive power we shall now drop this restriction, and we shall also consider implicit specifications, where the desired relation is described as the solution of some equation or formula using also quantification over relations.

Even though they cannot be executed directly, these specifications are valuable for calculational reasoning about graphs and order relations. But their real interest is that they can serve as starting points for the formal development of efficient algorithms. We shall illustrate this by two examples. In Example 9.2.6 we give a relational definition of progressive finiteness, and in Sect. 9.3 we will see how it can be refined into an algorithm for testing cycle-freeness of a finite graph. Furthermore, in Example 9.2.7 we encounter an implicit specification of the kernel function, and in Sect. 9.4 we shall derive from it efficient algorithms for computing kernels of particular types of graphs.

Example 9.2.6 (*Progressive Finiteness*) A relation $R \in [X \leftrightarrow X]$ is called *progressively finite* if there is no infinite path $p = (p_0, p_1, p_2, \ldots)$ through the directed graph $g = (X, R)$. In the literature, progressively finite relations are also known as noetherian, or definite, see Definition 3.5.3. They were used by D. Park [Hitchcock, Park 1973] for proving program termination. Since then, progressively finite relations have been used in relation-algebraic works on semantics and rewriting systems [Schmidt, Ströhlein 1993], and for "making induction calculational" [Backhouse, Doornbos 1994]. A little reflection shows that R is progressively finite if and only if

$$\neg\exists s\ (s \neq \emptyset \wedge \forall x\ (x \in s \to \exists y\ (y \in s \wedge R_{xy}))).$$

Note that we have used quantification over subsets s of X. In a first step, we rewrite the above predicate to

$$\forall s\ (\forall x\ (x \in s \to \exists y\ (y \in s \wedge R_{xy})) \to s = \emptyset).$$

If we represent sets by vectors, then $x \in s$ and $y \in s$ become s_x and s_y. In combination with the component-wise definition of composition and inclusion of relations we obtain the test

$$progr\_finite : [X \leftrightarrow X] \to \mathbb{B} \qquad progr\_finite(R) \iff \forall s\ (s \subseteq R;s \to s = \emptyset)$$

for progressive finiteness. This specification involves quantification over relations, and therefore cannot be executed directly. Strictly speaking, the quantification should range over vectors only, but it is easy to show that this is irrelevant. □

Example 9.2.7 (*Kernels as fixed points*) In Example 9.2.4 we specified the vector of all kernels of a given directed graph $g = (X, R)$. Now, suppose we wish to compute just one kernel (if one exists). For this task it is appropriate to represent vertex sets as vectors $v \in [X \leftrightarrow \mathbf{1}]$ (rather than points $p \in [2^X \leftrightarrow \mathbf{1}]$). Then we may replace the predicate $x \in v$ by v_x, so a vector a is absorbant if and only if

$$\forall x\ (\overline{a}_x \to \exists y\ (a_y \land R_{xy}))$$

which is equivalent to $\overline{a} \subseteq R;a$. Similarly, s is stable if and only if

$$\forall x\ (s_x \to \neg\exists y\ (R_{xy} \land s_y))$$

or $s \subseteq \overline{R;s}$. As a consequence, a vector $k \in [X \leftrightarrow \mathbf{1}]$ is a kernel of g if and only if $k = \tau(k)$, where the *kernel function* τ is defined by

$$\tau : [X \leftrightarrow \mathbf{1}] \to [X \leftrightarrow \mathbf{1}] \qquad \tau(v) = \overline{R;v}.$$

We would like to define a function that computes a single kernel of a relation implicitly by

$$kernel : [X \leftrightarrow X] \to [X \leftrightarrow \mathbf{1}] \qquad kernel(R) = \tau(kernel(R)).$$

Unfortunately, the kernel function τ is order-*reversing* in the sense that $v \subseteq w$ implies $\tau(w) \subseteq \tau(v)$, so the above task is not solved by Tarski's fixed point theorem. Indeed, not every directed graph has a kernel. □

9.3 Efficient algorithms for path problems

Many practical problems that can be formulated in terms of graph theory are concerned with paths. For example, we might wish to determine if there is a path between two given vertices, or to compute the set of all vertices reachable from a given one, or to check if there are any cyclic paths.

Algorithms for transitive closures

Assume $g = (X, R)$ is a directed graph and we are asked to produce a program for testing whether there is a path between two given vertices. If many such questions will be asked about the same graph, and response time is critical, then it is a good idea to compute the transitive closure R^+ of R once and for all, so that subsequent queries can be answered by simple look-up (see Lemma 9.1.4).

Recall that $R^+ = \bigcup_{i\geq 1} R^i$. A simple induction in combination with the Fixed point theorem 9.1.1 shows that

$$R^+ = \bigcup_{i\geq 1} R^i = \bigcup_{i\geq 1} \varrho_1^i(\emptyset) = \mu\varrho_1,$$

where $\mu\varrho_1$, is the least fixed point of the continuous function

$$\varrho_1 : [X \leftrightarrow X] \to [X \leftrightarrow X] \qquad \varrho_1(Q) = R \cup R;Q.$$

Assume that X has $n < \infty$ elements. Then the iteration $\emptyset \subseteq \varrho_1(\emptyset) \subseteq \varrho_1^2(\emptyset) \subseteq \dots$ computes $\mu\varrho_1$ in $O(n)$ steps. We use Pascal-like notation to present the algorithm we have developed in its final form:

```
function closure (R);
  Q := ∅;    S := R;
  while S ≠ Q do   Q := S; S := R ∪ R;Q end;
  return S
  end.
```

To run this program, we can represent the relation-valued variables Q and S by Boolean matrices and use standard functions on matrices for equality tests, multiplication, and union. The run time complexity is then $O(n^4)$.

Let's look for more efficient algorithms. As noted in Sect. 1.2, the transitive closure of R is the smallest transitive relation above R. Therefore

$$R^+ = \bigcap\{Q : R \subseteq Q \wedge Q \text{ is transitive}\} = \bigcap\{Q : R \cup Q;Q \subseteq Q\}.$$

By Tarski's fixed point theorem, the last term is precisely the least fixed point of the function ϱ_2 defined by

$$\varrho_2 : [X \leftrightarrow X] \to [X \leftrightarrow X] \qquad \varrho_2(Q) = R \cup Q;Q.$$

Now, the least fixed points of the two functionals ϱ_1 and ϱ_2 coincide:

Theorem 9.3.1 $\mu\varrho_1 = \mu\varrho_2$.

We have outlined above a free-style proof of Theorem 9.3.1, and it is a nice exercise to redo the proof within relational calculus. The relational style is much more economic, as it uses no general closure arguments, continuity, infinite intersections and the like. The only rules concerning fixed points are $f(\mu f) = \mu f$ (computation) and $\mu f \subseteq x$ provided $f(x) \subseteq x$ (induction). Therefore, relational proofs are easy to check by machine, and we try to stay within the calculus wherever we can.

The relational proof of Theorem 9.3.1 requires a lemma interesting in itself.

Lemma 9.3.2 $\mu\varrho_1$ is transitive.
Proof: By the definition of ϱ_1 and the cancellation rule (9.4) we have

$$\varrho_1(\mu\varrho_1/\mu\varrho_1);\mu\varrho_1 = R \cup R;(\mu\varrho_1/\mu\varrho_1);\mu\varrho_1 \subseteq R \cup R;\mu\varrho_1 = \varrho_1(\mu\varrho_1) = \mu\varrho_1,$$

whence $\varrho_1(\mu\varrho_1/\mu\varrho_1) \subseteq \mu\varrho_1/\mu\varrho_1$ by the universal property (9.3). Now, the fixed point induction rule yields the inclusion $\mu\varrho_1 \subseteq \mu\varrho_1/\mu\varrho_1$. Applying (9.3) again yields $\mu\varrho_1;\mu\varrho_1 \subseteq \mu\varrho_1$, as desired. □

Proof: (of Theorem 9.3.1). From $R \subseteq R \cup \mu\varrho_2;\mu\varrho_2 = \varrho_2(\mu\varrho_2) = \mu\varrho_2$ we get

$$\varrho_1(\mu\varrho_2) = R \cup R;\mu\varrho_2 \subseteq R \cup \mu\varrho_2;\mu\varrho_2 = \varrho_2(\mu\varrho_2) = \mu\varrho_2$$

which in turn implies $\mu\varrho_1 \subseteq \mu\varrho_2$, by the induction rule. For the converse, we use transitivity of $\mu\varrho_1$ and the inclusion $R \subseteq \mu\varrho_1$ (which follows from the computation rule) and obtain

$$\varrho_2(\mu\varrho_1) = R \cup \mu\varrho_1;\mu\varrho_1 \subseteq R \cup \mu\varrho_1 = \mu\varrho_1.$$

Now, the induction rule yields $\mu\varrho_2 \subseteq \mu\varrho_1$. □

As an immediate consequence, the transitive closure is computed as the limit of the iteration $\emptyset \subseteq \varrho_2(\emptyset) \subseteq \varrho_2^2(\emptyset) \subseteq \ldots$ for the least fixed point $\mu\varrho_2$, which is finite provided X is finite, and we have the second algorithm

```
function closure (R);
  Q := ∅;    S := R;
  while S ≠ Q do   Q := S; S := R ∪ Q;Q end;
  return S
  end.
```

The complexity of this algorithm is only $O(n^3 \log n)$, where n is the cardinality of the set X of vertices.

An even more efficient algorithm for computing transitive closures was proposed by S. Warshall. It employs a clever problem generalization (see below) similar forms of which occur in many transformational developments. With relational algebra we can capture this idea in a concise calculation. Denote the vertices of the finite directed graph $g = (X, R)$ by $x_1, \ldots, x_n$. For $0 \leq i \leq n$ let Q_i be the element of $[X \leftrightarrow X]$ which relates x to y if and only if there exists a path from x to y all "inner" vertices of which are members of $\{x_1, \ldots, x_i\}$. If we interpret the vertices as points $x_i \in [X \leftrightarrow \mathbf{1}]$ in the relational sense, then a purely relation-algebraic definition of Q_i is

$$Q_i = (R;I_i)^*;R,$$

where $I_i = \bigcup_{j=1}^{i} x_j;x_j^{\smile}$ is the partial identity associated with $\{x_1, \ldots, x_i\}$. The following theorem shows how the Q_i can be used for iteratively computing the transitive closure of R. Its third equation states that the problem of computing Q_i is a generalization of the original problem of computing R^+.

Theorem 9.3.3 The relations $Q_0, \ldots, Q_n$ defined above satisfy

(i) $Q_0 = R$ (ii) $Q_i = Q_{i-1} \cup Q_{i-1};x_i;x_i^{\smile};Q_{i-1}$ (iii) $Q_n = R^+$.

Proof: Equation (i) follows from $I_0 = \emptyset$ and $\emptyset^* = I$. In the proof of (ii) we will use the star decomposition rule,

$$(S \cup T)^* = (S^*;T)^*;S^*. \tag{9.5}$$

We also need that, for any vector v,

$$(S;v;v^{\smile})^* = I \cup S;v;v^{\smile}. \tag{9.6}$$

To prove (9.6) note that $(S;v;v^{\smile})^* = I \cup (S;v;v^{\smile})^+$ and $(S;v;v^{\smile})^2 \subseteq S;v;v^{\smile}$. Now, we can calculate the recursive equation (ii) by

$$
\begin{aligned}
Q_i &= (R;I_i)^*;R \\
&= (R;I_{i-1} \cup R;x_i;x_i^{\smile})^*;R && \text{definition of } I_i \\
&= ((R;I_{i-1})^*;R;x_i;x_i^{\smile})^*;(R;I_{i-1})^*;R && \text{star decomposition (9.5)} \\
&= (I \cup (R;I_{i-1})^*;R;x_i;x_i^{\smile});(R;I_{i-1})^*;R && \text{by (9.6)} \\
&= (R;I_{i-1})^*;R \cup (R;I_{i-1})^*;R;x_i;x_i^{\smile};(R;I_{i-1})^*;R \\
&= Q_{i-1} \cup Q_{i-1};x_i;x_i^{\smile};Q_{i-1}.
\end{aligned}
$$

Finally, equation (iii) follows from $I_n = I$ and $R^+ = R^*;R$. □

By this theorem, the iteration $R = Q_0 \subseteq Q_1 \subseteq \ldots \subseteq Q_n = R^+$ yields the transitive closure of R. Using an initialization and a `for`-loop, it can be implemented as

```
Q := R;
for i := 1 to n do
   Q := Q ∪ Q;x_i;x_i˘;Q end.
```

The relation $Q \cup Q;x_i;x_i^{\smile};Q$ contains additionally to the pairs of Q all pairs (y,z) such that y is a predecessor and z is a successor of x_i. Given Q as matrix and the points x_i as the columns of the identity matrix of the same dimension, the "updated" matrix $Q \cup Q;x_i;x_i^{\smile};Q$ can be computed in quadratic time.

Now, we suppose that the `for`-loop of the above algorithm is performed on a Boolean array representation for relations such that single entries Q_{jk} – representing whether Q holds for the pair (x_j, x_k) or not – can be overwritten. Then we are able to implement the assignment $Q := Q \cup Q;x_i;x_i^{\smile};Q$ joining every predecessor to each successor of x_i by two nested loops:

```
function closure (R);
  Q := R;
  for i := 1 to n do
    for j := 1 to n do
      for k := 1 to n do
         Q_jk := Q_jk ∨ (Q_ji ∧ Q_ik) end end end;
  return Q
  end.
```

This is precisely Warshall's algorithm for computing the transitive closure of a relation. In the literature, one usually finds a version in which – by a conditional command – the inner loop is only performed if Q_{ji} holds and the assignment is replaced by $Q_{jk} := Q_{jk} \vee Q_{ik}$.

Reachability algorithms

Let $g = (X, R)$ be a directed graph and $s \subseteq X$. Then $y \in X$ is reachable from s if R^*_{xy} for some $x \in s$. If we represent sets by vectors then $x \in s$ becomes s_x and the existential quantification can be expressed in terms of relational multiplication, viz. by $((R^{\smile})^{*};s)_y$. Removing the index y we arrive at the relational specification

$$reach : [X \leftrightarrow X] \times [X \leftrightarrow \mathbf{1}] \to [X \leftrightarrow \mathbf{1}] \qquad reach(R,s) = (R^{\smile})^{*};s$$

of a function *reach* that computes the set of vertices reachable from s. This specification is executable but suffers from two inefficiencies. First, for computing a vector the big intermediate object "matrix" is used. Secondly, its asymptotic time complexity is dominated by the costs for computing the reflexive-transitive closure R^*.

We can save memory space by avoiding the computation of R^*. There may be a huge speed up as well, particularly when the closure R^* is sparsely populated. Define a function

$$\gamma_1 : [X \leftrightarrow X] \to [X \leftrightarrow X] \qquad \gamma_1(Q) = I \cup R^{\smile};Q \ .$$

Clearly, we have the equation

$$reach(R,s) = (R^{\smile})^{*};s = \mu\gamma_1;s \ .$$

We can eliminate the intermediate data structure by merging the fixed point computation on relations and the multiplication with s into a single iteration. This technique is known as *deforestation*, though in our case it's not trees but matrices that we wish to dispose of. We require the relational version of a well-known lemma from transformational programming. It is a special case of the μ-fusion rule 9.1.2.

Theorem 9.3.4 If γ and σ are two order-preserving functions on relations fulfilling $\gamma(Q);S = \sigma(Q;S)$ for every Q then $\mu\gamma;S = \mu\sigma$. □

Now, let us calculate an appropriate function σ. For every relation Q we must have that the equation

$$\sigma(Q;s) = \gamma_1(Q);s = (I \cup R^{\smile};Q);s = s \cup R^{\smile};(Q;s)$$

holds, so we define the function

$$\sigma : [X \leftrightarrow \mathbf{1}] \to [X \leftrightarrow \mathbf{1}] \qquad \sigma(v) = s \cup R^{\smile};v.$$

This gives us the equation

$$reach(R,s) = \mu\gamma_1;s = \mu\sigma.$$

When the set X is finite, the vector $\mu\sigma$ can be computed as the limit of the iteration $\emptyset \subseteq \sigma(\emptyset) \subseteq \sigma^2(\emptyset) \subseteq \ldots$ which is finite, too. Thus we have derived the following algorithm for computing the set of all vertices of the graph $g = (X,R)$ that can be reached from s:

```
function reach (R, s);
  v := ∅;   w := s;
  while v ≠ w do   v := w; w := s ∪ R˘;v end;
  return w
  end.
```

Using Boolean matrices and vectors for implementation, this reachability algorithm has the same worst case run time as the original algorithmic specification, but we have saved the memory space for storing R^*.

What can we do about time-wise efficiency? The techniques we used in the previous section for speeding up the computation of transitive closures don't seem to work here, so we shall try a different attack. A lot of effort can be saved by keeping track of the vertices already visited. Therefore, consider the function which maps a pair of vertex sets u and v to the set of all vertices contained in u or reachable from v via a path without vertices from u:

$$f(u,v) = u \cup \mathit{reach}(R \cap V;\overline{u}^{\smile}, v).$$

Here $R \cap V;\overline{u}^{\smile}$ contains only those edges of $g = (X, R)$ that end outside u. Since f generalizes the original function *reach* via the equation

$$\mathit{reach}(R, s) = f(\emptyset, s),$$

any implementation of the function f will solve the reachability problem. We already saw that $\mathit{reach}(R, s) = \mu\sigma$ for any R and s. Now, we can use this fact to compute the second disjunct of f as

$$\mathit{reach}(R \cap V;\overline{u}^{\smile}, v) = \mu\sigma(u, v)$$

where the function $\sigma(u, v)$ is defined by

$$\sigma(u,v) : [X \leftrightarrow \mathbf{1}] \to [X \leftrightarrow \mathbf{1}] \qquad \sigma(u,v)(w) = v \cup (R \cap V;\overline{u}^{\smile})^{\smile};w.$$

This definition generalizes our earlier definition of σ. Since $\overline{u}$ is a vector, we can simplify as follows:

$$\sigma(u,v)(w) = v \cup (R \cap V;\overline{u}^{\smile})^{\smile};w = v \cup (\overline{u};V \cap R^{\smile});w = v \cup (\overline{u} \cap R^{\smile};w). \quad (9.7)$$

To implement the function

$$f(u,v) = u \cup \mu\sigma(u,v) \quad (9.8)$$

more efficiently, we aim at fusing the outer operation (joining u) into the fixed point computation. This will yield a recursion which computes $f(u, v)$ directly. We terminate the recursion when $v = \emptyset$. Since $\sigma(u, \emptyset)(\emptyset) = \overline{u} \cap R^{\smile};\emptyset = \emptyset$ implies $\mu\sigma(u, \emptyset) = \emptyset$ we have

$$f(u, \emptyset) = u \cup \mu\sigma(u, \emptyset) = u.$$

As long as $v \neq \emptyset$, we recur, using the following identity:

$$\begin{aligned}
f(u,v) &= u \cup \mu\sigma(u,v) && \text{by (9.8)}\\
&= u \cup \sigma(u,v)(\mu\sigma(u,v)) && \mu\sigma(u,v) \text{ is a fixed point}\\
&= u \cup v \cup (\overline{u} \cap R^{\smile};\mu\sigma(u,v)) && \text{definition of } \sigma \text{ (9.7)}\\
&= u \cup v \cup (\overline{u} \cap \overline{v} \cap R^{\smile};\mu\sigma(u,v)) && v \cup X = v \cup (\overline{v} \cap X)\\
&= u \cup v \cup \mu\sigma(u \cup v, \overline{u \cup v} \cap R^{\smile};v) && \text{by } \mu\text{-Fusion 9.1.2, see below}\\
&= f(u \cup v, \overline{u \cup v} \cap R^{\smile};v) && \text{by (9.8).}
\end{aligned}$$

Now, it remains to verify the μ-fusion step. We used the "transfer" function h defined by

$$h(Q) = \overline{u} \cap \overline{v} \cap R^{\smile};Q.$$

Clearly h is strict and continuous. So the fusion step in the preceding calculation is justified by the μ-fusion rule 9.1.2, provided we prove that

$$h \circ \sigma(u,v) = \sigma(u \cup v, \overline{u \cup v} \cap R^{\smile};v) \circ h.$$

This is achieved by the following calculation:

$$\begin{aligned} h(\sigma(u,v)(w)) &= \overline{u} \cap \overline{v} \cap R^{\smile};(v \cup (\overline{u} \cap R^{\smile};w)) \\ &= \overline{u \cup v} \cap R^{\smile};(v \cup (\overline{u} \cap \overline{v} \cap R^{\smile};w)) \\ &= \overline{u \cup v} \cap R^{\smile};(v \cup h(w)) \\ &= (\overline{u \cup v} \cap R^{\smile};v) \cup (\overline{u \cup v} \cap R^{\smile};h(w)) \\ &= \sigma(u \cup v, \overline{u \cup v} \cap R^{\smile};v)(h(w)). \end{aligned}$$

This calculation completes the proof that for all vectors u and v,

$$f(u,v) = f(u \cup v, \overline{u \cup v} \cap R^{\smile};v).$$

If the vertex set X is finite, then we definitely reach the termination case eventually, because the first argument of f strictly increases with every call. Since the recursion is tail-recursive, we can easily rewrite it as a while loop. Here is the reachability algorithm in its final form:

```
function reach (R, s);
  u := ∅;    v := s;
  while v ≠ ∅ do   u := u ∪ v;  v := ū ∩ R˘;v end;
  return u
  end.
```

In each iteration the set u of vertices visited so far is enlarged by its proper successors v, so the number of iterations it takes to discover some vertex x is just the distance between x and s. This is typical of a breadth-first search. If we implement R by successor lists, the time consumed by this algorithm is quadratic in the number of vertices.

Progressive finiteness and testing cycle-freeness

Recall that in Example 9.2.6 we characterized progressively finite relations R by the property

$$s \subseteq R;s \implies s = \emptyset$$

to hold for every vector $s \in [X \leftrightarrow \mathbf{1}]$. In the following, we aim at an algorithmic test. The antecedent $s \subseteq R;s$ and Tarski's fixed point theorem draw our attention to the function δ_1 defined by

$$\delta_1 : [X \leftrightarrow \mathbf{1}] \to [X \leftrightarrow \mathbf{1}] \qquad \delta_1(v) = R;v.$$

Since $\nu\delta_1 = \delta_1(\nu\delta_1) = R;\nu\delta_1$, progressive finiteness of R implies $\nu\delta_1 = \emptyset$. The reverse implication also holds, because $\nu\delta_1$ is the greatest vector v with $v \subseteq R;v$. Thus

$$\mathit{progr\_finite}(R) \iff \nu\delta_1 = \emptyset.$$

Based on this equivalence, an algorithm for testing progressive finiteness has to compute the chain $V \supseteq \delta_1(V) \supseteq \delta_1^2(V) \supseteq \ldots$ until it becomes stationary and afterwards to test whether $\emptyset$ is the last chain element, i.e., the greatest fixed point $\nu\delta_1$. Using graph-theoretic notation, the kth chain element represents the set of vertices from which a path of length k emerges.

If no vertex of the directed graph $g = (X, R)$ lies on an infinite path, then g is cycle-free. We can state and prove this claim formally within the calculus.

Theorem 9.3.5 Let $g = (X, R)$ be a finite directed graph. If R is progressively finite then g is cycle-free. If X is finite, the converse is also true.

Proof: Assume R is progressively finite. We claim that R^+ is also progressively finite. For assume $x = R^+;x$. Then $x = R;(I \cup R^+);x = R;x$, whence $x = \emptyset$. Thus $x = \emptyset$ is the only solution of $x = R^+;x$ and R^+ is indeed progressively finite. Since we have

$$(R^+ \cap I);V = (R^+ \cap I);(R^+ \cap I);V \subseteq R^+;(R^+ \cap I);V,$$

it follows that $(R^+ \cap I);V = \emptyset$. Now, Tarski's rule yields $R^+ \subseteq \overline{I}$, so $g = (X, R)$ is cycle-free. Conversely, if X is finite and $R^+ \subseteq \overline{I}$ then some power of R vanishes, hence so does $\nu\delta_1 = \bigcap_{i \geq 0} R^i;V$. □

Together with the greatest fixed point characterization of progressive finiteness, this theorem yields an algorithm for testing cycle-freeness of a finite directed graph $g = (X, R)$:

```
function cycle-free (R);
  v := V;   w := R;V;
  while v ≠ w do   v := w;  w := R;v end;
  return w = ∅
  end.
```

If $n = |X|$, then this algorithm runs in time $O(n^3)$ which is the same time complexity as the algorithmic specification $R^+ \subseteq \overline{I}$ given in Example 9.2.1. However, we have saved the memory space for storing the transitive closure.

Instead with the entire vertex set represented by V, the exhaustion process can also be started with the empty set. This approach is taken in [Schmidt, Ströhlein 1993] and is based on a least fixed point characterization of progressive finiteness. A relation R is progressively finite if and only if $\mu\delta_2 = V$, where the function δ_2 on vectors is defined by $\delta_2(v) = \overline{R;\overline{v}}$. Compared with δ_1, the disadvantage of the function δ_2 is that it is non-continuous, making sometimes proofs and calculations more difficult.

9.4 Efficient algorithms for kernels

In Example 9.2.7 we characterized kernels of graphs as fixed points of the function $\tau(v) = \overline{R;v}$. For specific classes of graphs, the rolling rule 9.1.3 allows us to transform this specification into an efficient algorithm. We present two cases in which this approach works.

Progressively finite graphs

The kernel function $\tau(v) = \overline{R;v}$ of the directed graph $g = (X, R)$ is order-reversing. Not every order-reversing function on a complete lattice has a fixed point, but the following theorem helps to find one when one exists. We use the following notation: The square f^2 of a function is defined by $f^2(x) = f(f(x))$.

Theorem 9.4.1 Let V and W be two complete lattices and assume $f : V \to W$ and $g : W \to V$ are order-reversing. Then $f \circ g$ and $g \circ f$ are order-preserving, and we have:

$$\text{(i)}\ \ f(\nu(g \circ f)) = \mu(f \circ g) \qquad \text{(ii)}\ \ f(\mu(g \circ f)) = \nu(f \circ g).$$

In the special case where $V = W$ and $f = g$ we obtain

$$\text{(iii)}\ \ f(\mu f^2) = \nu f^2 \qquad \text{(iv)}\ \ f(\nu f^2) = \mu f^2.$$

In particular, if f^2 has a unique fixed point, then this is also the only fixed point of f.

Proof: (Due to R. Backhouse.) Let W^{op} denote the lattice obtained from W by reversing the order. Then $f : V \to W^{\mathrm{op}}$ and $g : W^{\mathrm{op}} \to V$ are order-preserving and (i) is precisely the rolling rule 9.1.3 (note that least fixed points in W correspond to greatest fixed points in W^{op}). The second claim can be proved by a dual argument and the rest are obvious corollaries. □

The following theorem serves as a tool for establishing below a criterion for kernels to exist.

Theorem 9.4.2 If R is progressively finite then the function $\tau(v) = \overline{R;v}$ has a unique fixed point.

Proof: Starting with the computation rule $\nu\tau^2 \subseteq \tau^2(\nu\tau^2)$ the definition of τ and contraposition yield $R;\overline{R;\nu\tau^2} \subseteq \overline{\nu\tau^2}$ which in turn is equivalent to $R^{\smile};\nu\tau^2 \subseteq R;\nu\tau^2$ due to the Schröder equivalence. Together with the Dedekind rule we get

$$R;\nu\tau^2 \cap \nu\tau^2 \subseteq R;(\nu\tau^2 \cap R^{\smile};\nu\tau^2) \subseteq R;(\nu\tau^2 \cap R;\nu\tau^2).$$

Since R is progressively finite, this shows $R;\nu\tau^2 \cap \nu\tau^2 = \emptyset$ and the definition of τ and Theorem 9.4.1 (iv) imply $\nu\tau^2 \subseteq \tau(\nu\tau^2) = \mu\tau^2$. Hence, τ^2 has a unique fixed point and this is also the only fixed point of τ because of Theorem 9.4.1. □

The following theorem follows immediately from Theorem 9.4.2 and the fact that every order-preserving function between finite sets is continuous.

Theorem 9.4.3 Assume that R is a progressively finite relation. Then the directed graph $g = (X, R)$ has a unique kernel. Moreover, if X is finite then the kernel of g is the limit of the sequence $\emptyset \subseteq \tau^2(\emptyset) \subseteq \tau^4(\emptyset) \subseteq \ldots$, where $\tau(v) = \overline{R;v}$. □

Assume that the vertex set X is finite. We have noted before that in this case R is progressively finite just when $g = (X, R)$ is cycle-free. Thus, a finite acyclic directed graph has a unique kernel, which is computed by the algorithm

$$
\begin{array}{l}
\texttt{function}\ \mathit{kernel}\ (R);\\
\quad k := \emptyset;\quad v := \overline{R;V};\\
\quad \texttt{while}\ k \neq v\ \texttt{do}\quad k := v;\ v := \overline{R;\overline{R;k}}\ \texttt{end};\\
\quad \texttt{return}\ k\\
\quad \texttt{end.}
\end{array}
$$

To estimate the complexity of this program note that the vector k increases in each turnaround of the while loop, so that there can be at most $|X|$ iterations.

Bipartite graphs

A bipartite directed graph is a quadruple $g = (X, Y, R, S)$ where X and Y are disjoint vertex sets, $R \in [X \leftrightarrow Y]$ is the set of arcs from X to Y and $S \in [Y \leftrightarrow X]$ is the set of arcs from Y to X. Let us represent the given bipartite directed graph $g = (X, Y, R, S)$ as an "ordinary" directed graph $g = (V, B)$, where $V = X + Y$, the disjoint union of X and Y, and the matrix $B \in [V \leftrightarrow V]$ has the special form

$$B = \begin{pmatrix} \emptyset & R \\ S & \emptyset \end{pmatrix}.$$

In the previous section we found a kernel of g by looking at the square of the kernel function $\tau(v) = \overline{Bv}$. In the present case, we have

$$\tau^2 \begin{pmatrix} v_X \\ v_Y \end{pmatrix} = \overline{\begin{pmatrix} \emptyset & R \\ S & \emptyset \end{pmatrix} ; \overline{\begin{pmatrix} \emptyset & R \\ S & \emptyset \end{pmatrix} ; \begin{pmatrix} v_X \\ v_Y \end{pmatrix}}} = \begin{pmatrix} \overline{R;\overline{S;v_X}} \\ \overline{S;\overline{R;v_Y}} \end{pmatrix}$$

which suggests considering the pair of order-reversing functions on vectors:

$$
\begin{array}{ll}
\alpha : [Y \leftrightarrow \mathbf{1}] \to [X \leftrightarrow \mathbf{1}] \qquad & \alpha(v) = \overline{R;v} \\
\beta : [X \leftrightarrow \mathbf{1}] \to [Y \leftrightarrow \mathbf{1}] \qquad & \beta(w) = \overline{S;w}.
\end{array}
$$

From this we immediately obtain two kernels (fixed points of τ), namely,

$$k_1 = \begin{pmatrix} \nu(\alpha \circ \beta) \\ \mu(\beta \circ \alpha) \end{pmatrix} \qquad k_2 = \begin{pmatrix} \mu(\alpha \circ \beta) \\ \nu(\beta \circ \alpha) \end{pmatrix}.$$

To prove that, say, k_1 is indeed a fixed point of τ, we appeal to Theorem 9.4.1, and calculate

$$\begin{pmatrix} \emptyset & R \\ S & \emptyset \end{pmatrix} ; \begin{pmatrix} \nu(\alpha \circ \beta) \\ \mu(\beta \circ \alpha) \end{pmatrix} = \begin{pmatrix} R;\mu(\beta \circ \alpha) \\ S;\nu(\alpha \circ \beta) \end{pmatrix} = \begin{pmatrix} \overline{\alpha(\mu(\beta \circ \alpha))} \\ \overline{\beta(\nu(\alpha \circ \beta))} \end{pmatrix} = \begin{pmatrix} \overline{\nu(\alpha \circ \beta)} \\ \overline{\mu(\beta \circ \alpha)} \end{pmatrix}.$$

The above argument relies on a set-theoretical interpretation of the direct sum $V = X + Y$ and uses component-wise reasoning. Following [Berghammer, Gritzner$^+$ 1993], we will now show that this kernel construction can be achieved within pure relational algebra. To achieve this, we need the relational analogue of disjoint union. A direct sum $X + Y$ of X and Y is given by two injections

$$\iota \in [X \leftrightarrow (X + Y)] \qquad \kappa \in [Y \leftrightarrow (X + Y)]$$

satisfying the laws

$$\iota;\iota^{\smile} = I \qquad \kappa;\kappa^{\smile} = I \qquad \iota^{\smile};\iota \cup \kappa^{\smile};\kappa = I \qquad \iota;\kappa^{\smile} = \emptyset \qquad \kappa;\iota^{\smile} = \emptyset.$$

These five equations characterize the direct sum up to isomorphism, see [Zierer 1991] for example. Note that the fourth and fifth equations are equivalent. With the aid of the injections ι and κ, the relation B and the vectors k_1, and k_2 (which we previously defined in terms of components) can be expressed in abstract relational algebra. For the relation B we get the relation-algebraic description

$$B = \iota^{\smile};R;\kappa \cup \kappa^{\smile};S;\iota$$

and without components the two vectors become

$$k_1 = \iota^{\smile};\nu(\alpha \circ \beta) \cup \kappa^{\smile};\mu(\beta \circ \alpha) \qquad k_2 = \iota^{\smile};\mu(\alpha \circ \beta) \cup \kappa^{\smile};\nu(\beta \circ \alpha).$$

We have to show that $B;k_1 = \overline{k_1}$ holds, and in doing so we shall only use relational calculus, the above characteristic properties of ι and κ, and the preceding defining equations.

Using the Dedekind rule in combination with the fourth axiom of the direct sum we get

$$\iota^{\smile};V \cap \kappa^{\smile};V \subseteq \iota^{\smile};(V \cap \iota;\kappa^{\smile};V) = \emptyset.$$

On the other hand we obtain from the third axiom of the direct sum that

$$\iota^{\smile};V \cup \kappa^{\smile};V = (\iota^{\smile};V \cup \kappa^{\smile};V);V \supseteq (\iota^{\smile};\iota \cup \kappa^{\smile};\kappa);V = V$$

holds, which shows that the vectors $\iota^{\smile};V$ and $\kappa^{\smile};V$ are each other's complements. It follows that

$$\kappa^{\smile};V = \overline{\iota^{\smile};V} \subseteq \overline{\iota^{\smile};\nu(\alpha \circ \beta)} \qquad \iota^{\smile};V = \overline{\kappa^{\smile};V} \subseteq \overline{\kappa^{\smile};\mu(\beta \circ \alpha)}. \tag{9.9}$$

From $\iota;\iota^{\smile} = I$ and $\kappa;\kappa^{\smile} = I$ we get that ι and κ are univalent relations. It is easy to show that every univalent relation Q satisfies the equation $Q;\overline{R} = \overline{Q;R} \cap Q;V$, whence

$$\iota^{\smile};\overline{\nu(\alpha \circ \beta)} = \overline{\iota^{\smile};\nu(\alpha \circ \beta)} \cap \iota^{\smile};V \qquad \kappa^{\smile};\overline{\mu(\beta \circ \alpha)} = \overline{\kappa^{\smile};\mu(\beta \circ \alpha)} \cap \kappa^{\smile};V. \tag{9.10}$$

Combining this with the preceding formulas we obtain

$$\begin{aligned}
&\iota^{\smile};\overline{\nu(\alpha \circ \beta)} \cup \kappa^{\smile};\overline{\mu(\beta \circ \alpha)} \\
&\quad = (\overline{\iota^{\smile};\nu(\alpha \circ \beta)} \cap \iota^{\smile};V) \cup (\overline{\kappa^{\smile};\mu(\beta \circ \alpha)} \cap \kappa^{\smile};V) && \text{by (9.10)} \\
&\quad = (\overline{\iota^{\smile};\nu(\alpha \circ \beta)} \cup \overline{\kappa^{\smile};\mu(\beta \circ \alpha)}) \cap (\overline{\iota^{\smile};\nu(\alpha \circ \beta)} \cup \kappa^{\smile};V) \\
&\qquad \cap (\iota^{\smile};V \cup \overline{\kappa^{\smile};\mu(\beta \circ \alpha)}) \cap (\iota^{\smile};V \cup \kappa^{\smile};V) \\
&\quad = (\overline{\iota^{\smile};\nu(\alpha \circ \beta)} \cup \overline{\kappa^{\smile};\mu(\beta \circ \alpha)}) \cap \overline{\iota^{\smile};\nu(\alpha \circ \beta)} \cap \overline{\kappa^{\smile};\mu(\beta \circ \alpha)} && \text{by (9.9)} \\
&\quad = \overline{\iota^{\smile};\nu(\alpha \circ \beta)} \cap \overline{\kappa^{\smile};\mu(\beta \circ \alpha)},
\end{aligned}$$

and this enables us to prove the original claim, since

$$\begin{aligned}
B;k_1 &= (\iota^{\smile};R;\kappa \cup \kappa^{\smile};S;\iota);(\iota^{\smile};\nu(\alpha \circ \beta) \cup \kappa^{\smile};\mu(\beta \circ \alpha)) \\
&= \iota^{\smile};R;\mu(\beta \circ \alpha) \cup \kappa^{\smile};S;\nu(\alpha \circ \beta) && \text{axioms direct sum} \\
&= \iota^{\smile};\overline{\alpha(\mu(\beta \circ \alpha))} \cup \kappa^{\smile};\overline{\beta(\nu(\alpha \circ \beta))} \\
&= \iota^{\smile};\overline{\nu(\alpha \circ \beta)} \cup \kappa^{\smile};\overline{\mu(\beta \circ \alpha)} && \text{Theorem 9.4.1} \\
&= \overline{\iota^{\smile};\nu(\alpha \circ \beta)} \cap \overline{\kappa^{\smile};\mu(\beta \circ \alpha)} && \text{see above} \\
&= \overline{\iota^{\smile};\nu(\alpha \circ \beta) \cup \kappa^{\smile};\mu(\beta \circ \alpha)} \\
&= \overline{k_1}.
\end{aligned}$$

Thus, we have proved a generalization of M. Richardson's well-known theorem, which states that every finite bipartite directed graph has at least one kernel. Richardson's original proof, given in [Richardson 1953], is non-constructive, quite complicated, and is based on the fact that a finite and cyclefree graph has precisely one vertex base, viz. the set of all initial points. The more simple inductive proof of [Schmidt, Ströhlein 1993] uses that a bipartite and strongly connected graph possesses a kernel.

Unlike Richardson, we did not need to assume finiteness. The full statement of our theorem reads as follows:

Theorem 9.4.4 Every bipartite directed graph $g = (X, Y, R, S)$ has two distinguished (but not necessarily distinct) kernels k_1 and k_2, given by

$$k_1 = \iota^{\smile};\nu(\alpha \circ \beta) \cup \kappa^{\smile};\mu(\beta \circ \alpha) \qquad k_2 = \iota^{\smile};\mu(\alpha \circ \beta) \cup \kappa^{\smile};\nu(\beta \circ \alpha),$$

where ι and κ denote the canonical injections from X and Y into $X + Y$ and α and β are given by $\alpha(v) = \overline{R;v}$ and $\beta(w) = \overline{S;w}$. □

An early (and also different) version of this constructive existence theorem for kernels in bipartite graphs goes back to T. Ströhlein, who investigated in [Ströhlein 1970] combinatorial two-player games like chess, go, and hex with relation-algebraic means.

If the vertex set $X+Y$ is finite, the two functions $\alpha\circ\beta$ and $\beta\circ\alpha$ are continuous, and their least and upper fixed points can be obtained by iteration. To be more precisely, the sequence $V \supseteq (\alpha \circ \beta)(V) \supseteq (\alpha \circ \beta)^2(V) \supseteq \ldots$ converges in finitely many steps to $\nu(\alpha \circ \beta)$, and similarly $\emptyset \subseteq (\beta \circ \alpha)(\emptyset) \subseteq (\beta \circ \alpha)^2(\emptyset) \subseteq \ldots$ stabilizes at $\mu(\beta \circ \alpha)$. Here is the final algorithm, which is of polynomial complexity:

```
function kernel (R, S);
  k := V;   v := ‾(R;‾(S;V)‾)‾;
  while k ≠ v do   k := v;  v := ‾(R;‾(S;k)‾)‾ end;
  k := ∅;   w := ‾(S;V)‾;
  while k ≠ w do   k := w;  w := ‾(S;‾(R;k)‾)‾ end;
  return ι˘;v ∪ κ˘;w
  end.
```

Our representation of a bipartite directed graph uses two disjoint sets X and Y and two relations $R \in [X \leftrightarrow Y]$ and $S \in [Y \leftrightarrow X]$. In contrast, [Ströhlein 1970] and [Schmidt, Ströhlein 1993] take an "ordinary" directed graph $g = (V, B)$ and use two complementary vectors v, w to define bipartiteness, and proceed to prove an accordingly modified version of Theorem 9.4.4. Interestingly, their representation yields a different algorithm, which first computes the bounds $\mu\tau^2$ and $\nu\tau^2$ of the set of all kernels using the original kernel function τ and then obtains the specific kernels k_1 and k_2 as

$$k_1 = (v \cap \nu\tau^2) \cup (w \cap \mu\tau^2) \qquad k_2 = (v \cap \mu\tau^2) \cup (w \cap \nu\tau^2).$$

This procedure is less efficient than ours, because during the iterations it works with the entire relation B and a vector representing the entire vertex set, whereas our algorithm uses only the smaller parts R and S and two "shorter" vectors for the disjoint parts of the vertex set.

9.5 Conclusion

We have proposed relational algebra as a practical means for formal problem specification, prototyping, and algorithm development. Relations are best suited for reasoning about discrete programming problems. Many common programming tasks can be specified very tersely in terms of relations, and so relation algebra is a smooth and formal environment for transforming specifications into executable programs. The success of relational reasoning is due to the simple and "linear" nature of relational expressions, which allows formal and often concise manipulations. Quantifiers are packed into relational operations and a law involving these often compresses a series of logical inferences into a single step.

Since the entire development process works only with formal objects, one of the basic requirements for machine support is fulfilled. A typical mechanical task for a computer is validating a specification by building a rapid prototype to see whether the intention is met in special cases. The RELVIEW system (which we have already mentioned in Sect. 9.2) is such a tool.

The relational approach is wonderful when it works, but it is limited by its lack of expressiveness. Its expressive power can be increased enormously by including direct products and projections, but experience has shown that these tend to produce complicated encodings and lose much of the elegant simplicity of the pure calculus. Other authors are experimenting with n-ary relations and even formal languages. Our own work concentrates on *sequential calculus* [Hoare, von Karger 1995] which has slightly weaker axioms than the relational calculus, but a much wider range of computationally relevant models.

Chapter 10

Programs and Datatypes

Henk Doornbos, Netty van Gasteren, Roland Backhouse

We are programmers, in the sense that it is our concern to improve the process of program construction. Therefore we want to answer questions like: What is programming, why is it so difficult and error-prone, and how can we learn what is needed to make the process more manageable? In the following we shall address these issues in a relational framework. Section 10.1 gives an introductory overview explaining the background to our approach. Section 10.2 shows how we deal with (recursive and non-recursive) datatypes in the relational framework. Section 10.3 discusses programs in this context, concentrating on a class of programs characterized by relational equations of a specific but quite general shape. Program termination is the subject of Section 10.4. Finally, Section 10.5 briefly touches on the design and execution of (terminating) programs. For a more extensive treatment see [Doornbos 1996].

10.1 Programs, datatypes, and specifications

For us, programming is the construction of programs from formal specifications. So we do not address the difficult but interesting and important question of how to arrive at such specifications and how to make sure that they capture exactly what the client has in mind.

To be able to discuss programming as we see it, we therefore need an idea of what a specification is. Basically, most formal specification methods such as Z and VDM have in common that in some formalism an input-output relation is specified. (Of course, other aspects can be involved, such as the required efficiency of a solution, but in the present paper we do not address such aspects.) That a specification is a relation rather than a function is important: a functional specification might be too specific. For example, in optimisation problems, an optimum is not necessarily unique. So, for the moment we take the point of view that a specification is a relation, this giving us the possibility to abstract from the formalism in which it is expressed.

We also need to model programs. This topic belongs to the field known as semantics. We do not study semantics in the first place to prove compilers correct or to verify that programs do what they should do, let alone to show that models of programs do exist. Our interest is in how semantics can help in program construction. That this is indeed possible is evidenced by the work that has been done in the field called axiomatic semantics.

Traditionally, in semantics programs are modelled and then one concentrates on the mapping between programs and their models. Another issue that is considered important in semantics is whether models exist or not. As said before, we do not address these issues; instead, we concentrate on the properties of the model. This is motivated by the following. Ideally the model should contain only those aspects of programs that are relevant in the process of programming. It should abstract from the features of programming languages that are irrelevant for this process. Furthermore, the model should have properties that support programming. (An example is the wp-wlp semantics for sequential programs, which is used extensively for *designing* programs as well.) Of course that does not mean that the mapping between concrete and abstract programs is not important. Requirement number one is that it be simple.

What then should such a model of programs be like? Programs and specifications have much in common and indeed there are many formalisms in which programs are viewed as special cases of specification. However, the two are not really the same. One aspect of a program is its input-output behaviour, the responsibility of the computer being to construct this. However, any program also has properties like efficiency and termination, which make no sense for all specifications. It is of course possible that a specification mentions the required efficiency and demands that the program terminates. However, it is too restrictive a view to only study specifications that are terminating and have an efficiency: this would imply that all specifications are executable.

Here we will only study programs that terminate. It will turn out that all three aspects of programs mentioned above (input-output behaviour, efficiency and termination) can be assigned to equations in relations, expressed in the standard syntax. Therefore we advocate the view that programs are equations in relations. Here we concentrate on a limited but wide class of such equations, which model a similar class of sequential and functional programs.

A third concept to be discussed besides programs and specifications are datatypes. In programming they play an important role. First, it has long been observed that data structure influences program structure. Second, efficient implementation of programs often requires the design of ingenious datatypes. Traditionally a datatype is a set with a number of operations on it. We concentrate on the class of datatypes known as the initial datatypes, because primarily these are known to influence program structure. A set can be modelled as a relation and, of course, so can the operations on it. Thus, datatypes fit into the relational framework as well.

Now that we have identified three concepts that play an important role in programming, the question remains, whether this is enough to study at least part of the programming process. For us an important way of going from specifications to programs is by calculation, because calculation offers the possibility of isolating the straightforward parts of the programming process. Programming can be viewed as calculating an equation (the program) that is satisfied by a relation (the specification). Therefore, the relation calculus can be a useful tool in programming. This gives us one system to study programs, specifications, and datatypes.

We want to make use of work done so far, such as Guarded-Command programming [Dijkstra, Feijen 1984; Gries 1981], and functional pogramming in the squiggol-style [Bird 1989; Meertens 1986] and the fold-unfold-style [Burstall, Darlington 1977]. Therefore, concepts like Hoare triples, weakest preconditions, and functor-like operations are introduced in the calculus.

10.2 Datatypes

In computing science there are two different understandings of what a datatype is. The first is: a datatype is a set together with a number of operations on it. This view is taken in the study of abstract datatypes. The standard example of such a datatype is of course the stack. Here the set, usually called the carrier of the datatype, is formed by all stacks. The operations of this datatype are push, pop, etc. Other examples are for instance priority queues and the booleans with operations like "and" and "or".

A second interpretation of a datatype is: it is just a set. This is the one used in typed programming languages like Pascal or Haskell: a variable is of type integer means that it is an element of the set of integers. Similarly, a variable is of type "$A{\rightarrow}B$" says that it is an element of the set of functions from set A to set B. For the moment the second view is adopted here: a datatype is just a set. It turns out, however, that if a new datatype is constructed, usually a number of operations on the datatype emerge from the construction as well.

In programming and mathematics alike, new datatypes are formed from old by datatype constructors, either with or without the use of recursion. Well-known examples of constructions without recursion are the product or pairing operation and the disjoint union or sum of two sets: given two sets A and B, their product $A{\times}B$ is the set of pairs (x,y) such that x and y are elements of A and B, respectively.

The disjoint union $A{+}B$ of the sets A and B consists of elements of the form $tag0(x)$ and $tag1(y)$, where, again, x is an element of A and y is taken from B. The tagging operations are such that it is decidable, whether an element z from $A{+}B$ has the form $tag0(x)$ or $tag1(y)$. Furthermore, it must be possible to strip the tag from its argument, in other words the tagging operations have an inverse. Therefore it is possible to distinguish between the cases "z represents an element of A" and "z represents an element of B". So $A{+}B$ is indeed the *disjoint* union of A and B.

A typical example of a datatype constructor that is defined by recursion is the list constructor. Given a set A, the set $list(A)$ consists of what we call the list-structures over A, i.e. all finite lists over A. The *list* operation is defined by recursion, as is seen in the usual definition in functional programming languages: a list is either a constant (the empty list), or an element from A "consed" with a list (where the "cons" operation is the well-known one from the programming language Lisp). So lists are defined in terms of lists, i.e. by a recursive definition.

In general: every datatype constructor F is a possibly recursive function from datatypes to datatypes, mapping a set A to the set of what are called the F-structures over A.

Associated with each datatype constructor F we have a relation constructor Fr such that, informally speaking, $Fr(R)$ is the relation "apply R to each element". As a typical example consider lists; given the functional relation f, the relation $Listr(f)$ is the well-known function $map(f)$: $Listr(f)([x,y,z]) = [f(x),f(y),f(z)]$. Of course it is not necessary to restrict $Listr$ to functional relations; we have, for instance, $[x,y,z]\langle Listr(R)\rangle[p,q,r] \equiv x\langle R\rangle p \wedge y\langle R\rangle q \wedge z\langle R\rangle r$ [1] for arbitrary relation R.

In general, relation constructor Fr associated with datatype constructor F maps a relation R (on a set A) to a relation $Fr(R)$ on F-structures (over A) that is to be thought of as "R applied element-wise".

Given these datatypes and datatype constructors, how then do we model them? The first choice we have to make is how to model sets. Because we want a system in which datatypes and specifications can be treated on the same footing, and we have already decided to model specifications as relations, we model datatypes or sets as relations too. Here we have two obvious choices: a set is a vector (Definition 3.2.1) or a set is a partial identity relation. From a theoretical point of view the choice is irrelevant: there is a one-to-one correspondence between the two. Here we choose to model sets as partial identity relations, also known as *monotypes*, partly because vectors come in two kinds and the choice between the two is rather irrelevant, and also because monotypes appear to give rise to simpler formulae and manipulations.

So a relation A is a *monotype* iff $A \subseteq \mathbb{I}$. Monotype A represents the set a if $x{\in}a \equiv x\langle A\rangle x$.

We wish to stress that (mono)types are elements of the universe of relations rather than forming a concept outside the system. We have, for instance, for monotypes A and B,

$$R \in A \leftrightarrow B \quad \equiv \quad A;R = R = R;B \ .$$

In order to introduce datatype constructors and the corresponding relation constructors, we need two functions: a function that maps monotypes to monotypes —the type constructor— and a function from relations to relations. Because monotypes are a special case of relations, however, it turns out that these two functions can be combined into one: we only need to consider functions from relations to relations. Of course, not every function of this type will do, because not every such function is the combination of a datatype constructor with its associated relation constructor. Functions that are will be called *relators*. As for the properties that relators should have, we know that $F(R)$ should be thought of as R applied element-wise. This means that relators are monotonic with respect to the standard order on relations. It also means that they distribute over relation composition and commute with the converse operator. (It is instructive to verify these properties for $Listr$.)

[1] Square brackets denote lists; $x\langle S\rangle y$ denotes that relation S relates x to y.

Of course we still want a relator to map sets to sets; this is the case if we require that $F(\mathbb{I}) \subseteq \mathbb{I}$: given the monotonicity of F this implies (even equivales) that F maps monotypes to monotypes. (It is necessary to check that in this respect it is harmless to identify sets with partial identity relations, in other words that if a monotype A is viewed as a set, monotype $F(A)$ is indeed the set of F-structures over A, whereas if A is viewed as a relation, relation $F(A)$ is "relation A applied element-wise". Again, checking this for *List* is instructive.)

Definition 10.2.1 (*Relators*) A function F from relations to relations is a *relator* if it enjoys the following properties.

(a) F is monotonic

(b) F distributes over ";", i.e. over the composition operator

(c) F distributes over $\smile$, i.e. over the converse operator

(d) $F(\mathbb{I}) \subseteq \mathbb{I}$ □

Examples of relators are the identity relator Id ($\mathsf{Id}(R) = R$) and the constant relators, i.e. the relators F such that $F(R) = A$ for a fixed monotype A. Note that for a binary relator, that is for a function from pairs of relations to relations, the composition and converse operators are to be applied component-wise. Thus, for instance, the distribution of binary relator $\oplus$ over composition takes the shape

$$R;U \oplus S;V \quad = \quad (R\oplus S);(U\oplus V) \ .$$

Datatypes defined without recursion

The basic type constructors we consider are the constant relators, product, and disjoint sum. We shall not introduce them axiomatically, but merely present their relational interpretations and their main properties. (From these properties it can be proved that the constructors are, indeed, relators. For a more extensive treatment see [Backhouse, de Bruin$^+$ 1991b; Backhouse, de Bruin$^+$ 1992].)

The unit (mono)type, denoted **1** , corresponds with a set with one (anonymous) element. Next, for each monotype A, the constant function mapping arbitrary relation R to A, is a relator. (By convention, variables A, B, C, etc. will henceforth denote monotypes; variables R, S, T, etc. will denote arbitrary relations.) It is furthermore assumed that the universe is closed both under summing and under product. (Note that given the unit type and the sum, we may conclude that the universe is infinite.)

For monotypes A and B, disjoint sum $A+B$ and product $A\times B$ have their familiar interpretations. The left and right projection functions are denoted $\ll$ and $\gg$, respectively, and the left and right tagging (or, injection) functions are denoted $tag0$ and $tag1$, respectively . Thus $\ll \in A\times B \leftrightarrow A$ and $tag0 \in A \leftrightarrow A+B$, and analogously for $\gg$ and $tag1$.

For relations R and S, the product $R\times S$ and the disjoint sum $R+S$ have, in a way, dual properties. It turns out, however, that the properties of the disjoint sum are far more important in programming. In the following we only present relevant ones, thus obscuring the duality between $\times$ and $+$. We deal with product first.

As mentioned earlier, the product and the sum of two relations can be viewed as an "apply to each element". This will become clear from the interpretations. We introduce the product first, together with its *generator* $\vartriangle$ (called split) , the essence of which is that it doubles arguments and thus generates pairs. (Note that in Chapter 4 the split is denoted by the symbol ∇.) Their interpretations are as follows.

$$x\langle R{\vartriangle}S\rangle(y,z) \quad\equiv\quad x\langle R\rangle y \;\wedge\; x\langle S\rangle z$$
$$(v,x)\langle R{\times}S\rangle(y,z) \quad\equiv\quad v\langle R\rangle y \;\wedge\; x\langle S\rangle z\;.$$

(Note the link between product and parallel composition. A discussion of parallel composition in a relational framework can be found in [Rietman 1995].)

Besides the general properties enjoyed by (binary) relators, a property used quite frequently is the following fusion rule

[$\vartriangle$-$\times$-fusion] $(R{\vartriangle}S);(T{\times}U) \quad=\quad (R;T)\vartriangle(S;U)$

Note that a special instance of this rule is $T{\vartriangle}U = (\mathbb{I}{\vartriangle}\mathbb{I});(T{\times}U)$, a property justifying our earlier remark that the essence of the split is its doubling arguments: the interpretation shows that $\mathbb{I}{\vartriangle}\mathbb{I}$ *is* the doubling function [2].

Unlike $\mathbb{I}{\vartriangle}\mathbb{I}$, however, $R{\vartriangle}R$ when "applied" to an argument x, does not in general yield a pair of equal values (it does if R is a function) : by the nondeterminacy of relations, first applying R and then doubling is not the same as first doubling the argument and then applying R to each ; we only have the *inclusion* $R;(\mathbb{I}{\vartriangle}\mathbb{I}) \subseteq R{\vartriangle}R$. As a consequence, in general composition does not distribute over $\vartriangle$ from the left; we just have an inclusion.

Analogously to the product, the disjoint sum comes with a generator $\blacktriangledown$ (co-junc), which in essence boils down to a test. It also comes with a co-generator, the conditional denoted $\triangledown$ and called "junc" (for "junction"). Operators junc and co-junc are each other's $\smile$-conjugate, i.e. $R{\blacktriangledown}S = (R^{\smile}{\triangledown}S^{\smile})^{\smile}$. As a result, properties of junc and co-junc come in pairs. Operator $\triangledown$ "joins" relations that have different left domains and the same right domain :

$$R{\triangledown}S \in A{+}B \leftrightarrow C \quad \text{for } R \in A{\leftrightarrow}C \text{ and } S \in B{\leftrightarrow}C\;.$$

Informally, the interpretation of $R{\triangledown}S$ is : first test how the input is tagged, untag the input, then apply either R or S. So,

$$x\langle R{\triangledown}S\rangle tag0(y) \equiv x\langle R\rangle y \qquad \text{and} \qquad x\langle R{\triangledown}S\rangle tag1(y) \equiv x\langle S\rangle y\quad.$$

Formally, $R{\triangledown}S$ is defined by

$$R{\triangledown}S \quad=\quad tag0^{\smile};R \cup tag1^{\smile};S\;.$$

Disjoint sum $R{+}S$, with typing

$$R{+}S \in A{+}B \leftrightarrow C{+}D \quad \text{for } R \in A{\leftrightarrow}C \text{ and } S \in B{\leftrightarrow}D\;,$$

is to be interpreted as "apply to each element of the input". In this case, an input is one of the two shapes $tag0(y)$ or $tag1(y)$; such an input contains the single

[2]We render program specifications by relations with input from the left and output to the right. As a consequence, the relational rendering of a function has the function argument on the left.

element y. Therefore, applying to each element is: first untag, then apply the proper argument of the sum, then tag again. Formally, using $\triangledown$, this is rendered by

$$R{+}S \quad = \quad (R;tag0) \triangledown (S;tag1) \ .$$

While product is related to parallel execution, disjoint sum is related to selection. To show this we shall exploit both the junc and the co-junc and, again, introduce some fusion rules. Note that $R \blacktriangledown S$ is equal to $R;tag0 \cup S;tag1$; in particular, for monotypes A and B we have $A \blacktriangledown B = A;tag0 \cup B;tag1$.Thus, $A \blacktriangledown B$ can be interpreted as a test on A versus B, together with a tagging operation to distinguish the outcome of the test (which may be nondeterministic).

The fusion rules to be introduced are:

[**$\blacktriangledown$-+-fusion**]	$(P \blacktriangledown Q);(R+S)$	$=$	$(P;R) \blacktriangledown (Q;S)$
[**+-$\triangledown$-fusion**]	$(R+S);(P \triangledown Q)$	$=$	$(R;P) \triangledown (S;Q)$
[**$\blacktriangledown$-$\triangledown$-fusion**]	$(R \blacktriangledown S);(P \triangledown Q)$	$=$	$R;P \ \cup \ S;Q$
[**$\triangledown$-;-fusion**]	$(R \triangledown S);T$	$=$	$(R;T) \triangledown (S;T)$
[**;-$\blacktriangledown$-fusion**]	$T;(R \blacktriangledown S)$	$=$	$(T;R) \blacktriangledown (T;S)$.

Note, again, the special instance of $\blacktriangledown$-+-fusion $R \blacktriangledown S \ = \ \mathbb{I} \blacktriangledown \mathbb{I} \,;\, R{+}S$, a property justifying our earlier remark that, in essence, the co-junc boils down to a selection: The relation $\mathbb{I} \blacktriangledown \mathbb{I}$ is a nondeterministic selection of either a left or a right tag. Depending on which is chosen, $R{+}S$ subsequently executes either R or S. Note also that, in contrast with the split, composition does distribute over junc (from the right, that is).

Now we are ready to explore the connection of all the above with selection. We consider the selection statement

$$\text{if } A \rightarrow R \ [] \ B \rightarrow S \text{ fi} \ ,$$

where A and B are conditions (monotypes), and R and S are statements (relations). One way of rendering this selection relationally is, of course, as the union $A;R \ \cup \ B;S$ of R restricted to A and S restricted to B . By $\blacktriangledown$-$\triangledown$-fusion, this is equal to

$$(A \blacktriangledown B);(R \triangledown S) \ , \qquad (10.1)$$

which expresses more explicitly that an if-statement is a test followed by a statement. Applying some fusion rules to the latter formula, we find it to be equal to

$$(\mathbb{I} \blacktriangledown \mathbb{I});(A+B);(R{+}S);(\mathbb{I} \triangledown \mathbb{I}) \ . \qquad (10.2)$$

And thus, we hope, we have clarified the connection between disjoint sum and selection. There is one more connection to be mentioned: we consider (10.1) and compose it from the right with relation T; distributing the latter over the junc yields the familiar

$$\text{if } A \rightarrow R \ [] \ B \rightarrow S \text{ fi} \ ; \ T \quad = \quad \text{if } A \rightarrow R;T \ [] \ B \rightarrow S;T \text{ fi} \ .$$

For the subsequent discussion on the shape of programs note that formula (10.2) has shape

$$U;F(X);V$$

where F is a relator (in (10.2) above, the relator is the binary relator $+$ and X is the pair (R, S)) .

Recursive datatypes and initial algebras

As an example of how datatypes are defined by recursion we consider the natural numbers. A natural number is either the constant zero or the successor of a natural number and "nothing else is a natural number". The constant zero can be modelled as a functional relation, call it *zero*, from the unit type to the naturals: $zero \in \mathbf{1} \leftrightarrow \mathrm{N}$. The successor function *succ* has type $\mathrm{N} \leftrightarrow \mathrm{N}$. These two functions are known as the constructors of the natural numbers. All natural numbers are the result of applying a constructor, this is called the "no-junk property [3]", and may be formulated as: the union of the right domains of the constructors is the natural numbers. Given a natural number, there has to be a test to find out by which constructor it was formed and from which argument. In datatype theory terms this is called: the "no-confusion property". In terms of relation algebra this means that the constructors are injective functions with disjoint right domains.

Now consider the type of the constructors: they both have target type N. As a consequence they can be combined into one function using the junc operator: $zero \triangledown succ \in \mathbf{1}+\mathrm{N} \leftrightarrow \mathrm{N}$. This is an example of an *algebra*. In general if we have a number of relations with the same target A (in other words a number of operations on set A), they can be combined into one relation with the junc operator. This will result in a relation of type $F(A) \leftrightarrow A$ for some relator F. In the example the relator is $(\mathbf{1}+)$. It is common practice to call *any* relation of type $F(A) \leftrightarrow A$ an *algebra*, even if it is not formed by combining a number of operations with the $\triangledown$ operator. The type A is called the *carrier* of the algebra.

Next we return to the definition of the natural numbers. The requirements we have thus far, no junk and no confusion, can be reformulated now as: $zero \triangledown succ$ is a bijection between $\mathbf{1}+\mathrm{N}$ and N. This can be shown in some simple calculations. Of course, this is not enough to define the naturals: there are many types A and functions f such that f witnesses the fact $A \cong \mathbf{1}+A$. But there is one more requirement: induction over the natural numbers is valid. It boils down to N being in a certain sense the least type A such that $A \cong \mathbf{1}+A$. (We want to stress that this will turn out to be the only place in this chapter where it really matters that a least solution is taken.)

A problem of the form "find a bijection $A \cong F(A)$" is called "solving a domain equation". The solution of such an equation consists of an algebra; the solution for A can then be recovered as the carrier of this algebra. Of course often a special kind of solution is required: the "least" one, which is called the initial algebra. This is in general a tricky problem: first we have to find out what least means in this context and, second, it is not at all clear for which class of relators such equations do have solutions.

[3] The use of the word "junk" here should not be confused with the name "junc" given to the co-generator of disjoint sum. The latter is short for "junction" and the former means waste or trash.

Working in a homogeneous relation algebra has the advantage that domain equations always have a "least solution": an initial F-algebra exists for all relators F. To see this we strengthen the equation $F(A) \cong A$ to be an equality rather than just an isomorphism. Because relators are monotonic functions from monotypes to monotypes and the monotypes form a complete lattice, Tarski's fixed-point theorem (see Chapter 1) guarantees that there is a smallest monotype A such that $A = F(A)$, denoted μF. Then μF, viewed as a partial identity relation, is certainly an isomorphism between $F(\mu F)$ and μF. So this gives the feature: carrier, algebra and its inverse are the same. Of course there is the need to check that our interpretation of "least" is reasonable; this question will be addressed in a later section.

It might not be clear how the constructors can be recovered from μF. As an example, we show how it is done for the naturals:

$$\begin{array}{ll} & \mu(\mathbf{1}+) \\ = & \{ \quad \mu(\mathbf{1}+) \text{ is a fixed point of } (\mathbf{1}+) \quad \} \\ & \mathbf{1} + \mu(\mathbf{1}+) \\ = & \{ \quad \text{definition of } + \quad \} \\ & (\mathbf{1};tag0) \triangledown (\mu(\mathbf{1}+)\,;\,tag1) \end{array}$$

The constructor *zero* is $\mathbf{1};tag0$: the function that maps the element $*$ of the unit type to $tag0(*)$. This then is the number 0. The constructor succ is $\mu(\mathbf{1}+)\,;\,tag1$: the function that maps a natural number n to $tag1(n.)$ So a natural number n has the form $tag1^n(tag0(*))$.

The construction of the natural numbers does not lead to an interesting datatype transformer: it only yields a constant. For parameterised datatypes like *List* we do, however, get something interesting. The definition of *List* yields the two constants *nil* and *cons*, that can be derived in a similar way to what was done for the naturals, now taking domain equation $List(B) \cong \mathbf{1}+(A \times List(B))$ as a starting point.

So for recursive datatypes we do have a set with operations and, thus, a somewhat more classical datatype.

10.3 Programs

Having discussed our choice of datatypes, we can now move on to discussing the programs operating on these datatypes. The question is: what do we mean by "a program"? To say that it is a relation is not good enough: programs should contain some information on how they are to be (or could be) executed. Here we present the view that programs are relational equations, i.e. equations in terms of relations. We do not mean to say that *all* programs are: we discuss a restricted though wide class of equations of a particular shape.

We want to stress that we equate a program with an equation rather than with, say, the least solution of an equation.

Consider, for instance, equations $X = X$ and $X = \bot\!\!\!\bot$: $\bot\!\!\!\bot$ is the least solution of both ; the first equation, however, leads to an infinite recursion, but the second does not. (The first can be viewed as the program "loop forever" or "abort" , the second as "magic": it cannot even start to operate, since it cannot accept input.)

For the larger part of this section we shall, thus, be concerned with equations. We shall also have to deal with the question of the semantics of such equations, i.e. with the question, which of the solutions of an equation is the input-output relation constructed by the program. This discussion is, however, postponed.

In the above, we implicitly interpret the (possible) execution of an equation $X = E(X)$ in the familiar way: occurrences of X in expression $E(X)$ constitute recursive calls. So we can talk about the efficiency of an equation, if so desired.

Here, however, we are more interested in the other computational aspect of programs, viz. in termination. In the subsequent section we shall deal with this aspect quite extensively, showing how we can attach a useful meaning to the "termination of an equation", capturing it formally, and relating it to more familiar notions of termination. Before we can do so, however, we first have to discuss (and justify) the shape of the relational equations to be considered.

Hylo equations

We recall the relational rendering of the selection statement $\mathsf{if}\ A \rightarrow R\ [\!]\ B \rightarrow S\ \mathsf{fi}$ by $A \blacktriangledown B \,;\, R{+}S \,;\, \mathbb{I}\triangledown\mathbb{I}$, and our remark that it is of the shape $U;F(X);V$ where F is a relator. As we shall motivate shortly, the programs we concentrate on are given by the so-called *hylo-equation* [4]

$$X = T;F(X);R \qquad \text{, where } F \text{ is a relator.}$$

The first motivating instance is the repetition $\mathsf{do}\ B \rightarrow S\ \mathsf{od}$, which is defined by the equation $X = \mathsf{if}\ B \rightarrow \mathsf{skip}\ [\!]\ B \rightarrow S;X\ \mathsf{fi}$. In relational terms it, thus, satisfies

$$X \quad = \quad ({\sim}B \blacktriangledown B) \;;\; \mathbb{I} + (S;X) \;;\; (\mathbb{I}\triangledown\mathbb{I}) \qquad ,$$

(where $\sim$ denotes monotype complementation) or, equivalently, showing that in this case relator F is $(\mathbb{I}+)$,

$$X \quad = \quad ({\sim}B \blacktriangledown B) \;;\; \mathbb{I}{+}S \;;\; \mathbb{I}{+}X \;;\; (\mathbb{I}\triangledown\mathbb{I}) \;.$$

As a second motivating example consider the factorial function. Its recursive definition is as follows (we mark each arrow with the function involved).

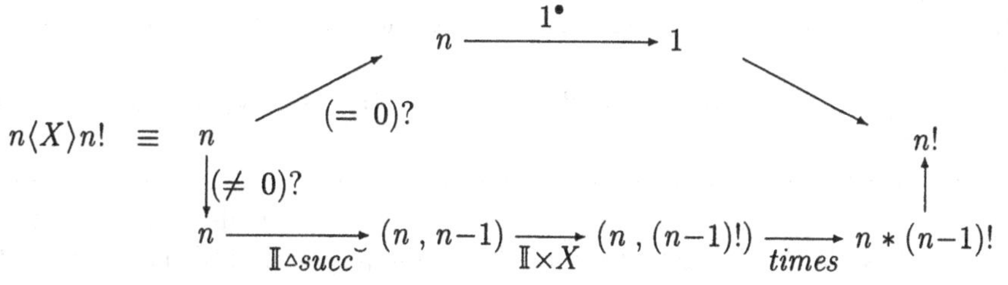

[4]According to Fokkinga [Fokkinga 1992], the term hylo(morphism) was coined by Erik Meijer.

Relationally, with B equal to the test $(\neq 0)?$, and $T = \mathbb{I}\vartriangle succ^{\smile} ; \mathbb{I}\times X ; times$, this is rendered by the equation

$$X \quad = \quad (\sim B \blacktriangledown B) \; ; \; 1^{\bullet} + T \; ; \; (\mathbb{I} \triangledown \mathbb{I}) \; ,$$

where $1^{\bullet}$ is the constant function that maps any n to 1. Thanks to the fact that relator $+$ distributes over composition and by $+$-$\triangledown$-fusion, is equivalent to

$$X \quad = \quad (\sim B \blacktriangledown B) \; ; \; \mathbb{I} + (\mathbb{I} \vartriangle succ^{\smile}) \; ; \; \mathbb{I} + (\mathbb{I} \times X) \; ; \; (1^{\bullet} \triangledown times) \; .$$

The shape of the latter equation shows that for factorial the relator F in the hylo equation is given by $F(X) = \mathbb{I} + (\mathbb{I} \times X)$. (Note that its least fixed point equals $List(\mathbb{I})$.)

The solution of a programming problem typically begins by choosing the relator F. Choosing relator $(\mathbb{I}\, +)$ means choosing a repetition, whereas divide-and-conquer, for example, as used in merge sort, corresponds to the choice of relator F where $F(X) = \mathbb{I} + (X \times X)$. In short, choosing a relator means choosing a solution strategy.

Virtual data structures

There is yet another view on the relator in a hylo equation, which may provide an illuminating insight into programming methodology. The essence is captured by a theorem that states that the least solution of a hylo equation is a composition of two special relations, which we introduce first:

If *in* is an arbitrary initial F-algebra, one can prove that the equation in X

$$X \quad = \quad in^{\smile} ; F(X) ; R$$

has a unique solution; this solution will be denoted $(\!|R|\!)$. Similarly, let $[\![S]\!]$ denote the unique solution of equation in X

$$X \quad = \quad S ; F(X) ; in \; .$$

The theorem now is the following.

Theorem 10.3.1 The least solution of the hylo equation in X

$$X \quad = \quad S ; F(X) ; R$$

is given by the composition $[\![S]\!] ; (\!|R|\!)$. □

(For a proof, note that $(\!|R|\!)$ and $[\![S]\!]$ are *least* solutions as well; using this, the theorem is easily proved by means of residuals and Tarski's fixed-point theorem (Chapter 1).)

Assuming that $S \in A \leftrightarrow F(A)$ and $R \in F(B) \leftrightarrow B$, the so-called *anamorphism* $[\![S]\!]$ constructs an element of the carrier set of *in*, given an input value of type A; the so-called *catamorphism* $(\!|R|\!)$ "destructs" that element into an output value of type B. The composition $[\![S]\!] ; (\!|R|\!)$ makes this construction/destruction process explicit. A standard implementation of the hylo equation hides this process. In other words, it remains invisible and of no concern to the programmer.

For this reason, the relator F is said to specify an intermediate (or *virtual* [de Moor, Swierstra 1992]) data structure. In the case of the factorial function, the relator is $List(\mathrm{N})$ and, indeed, the intermediate data structure is a list of numbers: the theorem states the well-known fact that the factorial of n can be obtained by enumerating the numbers from 1 to n (constructing an element of the intermediate data structure) followed by multiplying them together (destructing the element). In the case of a do-statement, the intermediate data structure is (isomorphic with) $\mathrm{N}\times B$: what is constructed is a count of how many times the loop is executed, together with the output. For an interesting example in which the intermediate data structure is a heap we refer the reader to [de Moor, Swierstra 1992].

10.4 Termination of programs

In this section we discuss termination in the setting of hylo equations. We shall do so in terms of *safe* sets, i.e. sets in which the program can be started without the danger of non-termination. Such sets, of course, do exist: the empty set is always safe; the idea is to find a maximal safe set. To simplify the discussion, we shall also use the familiar Hoare triples and weakest liberal preconditions, cast in our relational framework.

Preconditions

The connection between Hoare triples and the weakest liberal precondition (wlp) is given by

$$\{A\}R\{B\} \quad \equiv \quad A \Rightarrow wlp.R(B) \ . \tag{10.3}$$

The interpretation of $\{A\}R\{B\}$ is that program R when started in a state satisfying condition A only produces output satisfying B. In relation calculus this is expressed by $(A;R)> \ \subseteq B$, where the expression "$(A;R)>$" denotes the relation R restricted to values taken from A. Because the left-hand side, viewed as a function in A, distributes over arbitrary joins, it has an upper adjoint (see Chapt. 1, page 10). This adjoint then is the relational form of the weakest liberal precondition:

$$(A;R)> \ \subseteq B \quad \equiv \quad A \subseteq B/R \ . \tag{10.4}$$

The notation "B/R" indicates the *monotype factor.* From the point-wise interpretation of the left-hand side of (10.4) one can see that the point-wise interpretation of B/R is given by

$$y \in B/R \ \equiv \ \forall(x : \ y\langle R\rangle x : \ x{\in}B) \ .$$

In fact, properties (10.3) and (10.4) together provide four different ways of saying the same thing. The monotype factor enjoys a number of useful properties, all corresponding to well-known properties of the wlp :

- $/R$ distributes over $\cap$ and, hence, is monotonic
- $B/(R;S) \ = \ (B/S)/R$

The main reason for modelling conditions (or sets) by monotypes rather than, say, vectors is, to be able to use annotations: the annotated program $\{A\}R$ is the relation $A;R$, and thus the annotated program $\{A\}R\{B\};S$ is the relation $A;R;B;S$. Expressed in terms of vectors, the relational version would be more complicated.

Also, the fact that $\{A\}R\{B\}$ is correctly annotated, i.e. that program $\{A\}R$ yields program $\{A\}R\{B\}$, can now be expressed as $A;R \subseteq A;R;B$.

Termination and reductivity

In order to derive a definition of termination, we assume that a maximal safe set B, say, exists and our goal is to characterize this set. "Maximal safe set" B is to be interpreted as the set of all states in which the program can be started without the risk of non-termination. All subsets of B can, therefore, be considered safe as well. Given a program

$$X = S;F(X);R \ , \tag{10.5}$$

we shall derive reasonable conditions on such safe sets so as to formalize the notion of termination.

A first requirement on a safe set A (i.e. a set with property $A \subseteq B$) is that, whenever program X is started in it, the recursive calls terminate. So if the program is started in precondition A, the recursive calls should be started in the maximal safe set B. Using program annotation this is expressed as: given precondition A annotation $A;S;F(B;X);R$ is correct. Because relators distribute over composition this boils down to requiring property $\{A\}S\{F(B)\}$ to hold. The requirement on A can thus be formalised as:

$$A \subseteq B \quad \Rightarrow \quad (A\,;S)\!> \ \subseteq F(B) \tag{10.6}$$

Conversely, if property $\{A\}S\{F(B)\}$ holds (i.e. the recursive calls are started in the safe set B whenever the program is started in set A) then set A can be considered safe and should therefore be contained in the maximal safe set:

$$A \subseteq B \quad \Leftarrow \quad (A\,;S)\!> \ \subseteq F(B) \tag{10.7}$$

Now we can use the connection between triples and the wlp: $(A\,;S)\!> \ \subseteq F(B)$ can be rewritten as $A \subseteq F(B)/S$. Therefore (10.6) and (10.7) can be combined:

$$\forall(A :: A \subseteq B \equiv A \subseteq F(B)/S) \tag{10.8}$$

The rule of indirect equality gives that (10.8) is equivalent to $B = F(B)/S$. This leaves us with the choice which solution of this equation in B to take. Basically there are only two obvious choices: the least and the greatest solution. In general, however, the greatest solution is (for S of type $C \leftrightarrow F(C)$ for some C) equal to the identity relation. So if we were to take this one as a safe set, we would adopt the view that the computation can be started safely in any state, regardless of what S is. This would obviously be too optimistic. Therefore we take the least solution as the maximal safe set. (The pessimistic view is a "demonic" view as opposed to an "angelic" view.)

Definition 10.4.1 (*Termination*) The program $X = S;F(X);R$ terminates if it is started in the set $\mu(A \mapsto F(A)/S)$. □

If $\mu(A \mapsto F(A)/S) = \mathbb{I}$, the program can be started safely in any state. This is such a nice property of a program that the condition deserves a name.

Definition 10.4.2 (*Reductivity*) Relation S is F-reductive iff

$$\mu(A \mapsto F(A)_{\!/}S) \;=\; \mathbb{I} \;. \qquad \square$$

Note that reductivity states that equation $B \;=\; F(B)_{\!/}S$ has only one solution. So, if we restrict our attention to reductive relations S, the choice which solution should be taken as the maximal safe set becomes irrelevant.

Having defined termination, we have to check that the definition given here is compatible with the programmer's definition. A programmer proves termination by using well-founded relations: she has to prove that the argument of every recursive call is smaller than the original argument. For hylo program $X \;=\; S;F(X);R$ this means that all members of an F-structure that is the output of S have to be smaller than the corresponding input of S. More formally, with $y\langle mem\rangle x$ standing for "x is a member of F-structure y", we need for all x and z

$$\forall(y:\;\; y \text{ is an } F\text{-structure}:\; z\langle S\rangle y \wedge y\langle mem\rangle x \Rightarrow z \succ x) \;,$$

for some well-founded ordering $\succ$. Assuming that the output of S always is an F-structure, i.e. that $S{>} \;\subseteq\; F(\mathbb{I})$, this condition can be rewritten as $S;mem \;\subseteq\; \succ$, for some well-founded $\succ$, i.e. as $S;mem$ is well-founded. This, however, means that the old and the new definition are compatible, because it can be proved that $S;mem$ is well-founded iff S is (F-)reductive. (In order to conduct such a proof one should, of course, formalize the membership relation mem associated with a relator F. We shall not do so here. Suffice it to say that, on monotypes, F equals $_{\!/}mem$, and that well-founded is the same as Id-reductive, where Id is the identity relator.) This is where the name "reductive" originates from: relation S reduces its argument according to a well-founded relation.

Reductivity and unicity

We have motivated and formalized the notion of termination connected with programs that are hylo equations. We have not yet dealt with the question, which input-output relation we wish to connect with the equation. The nice thing is, however, that a consequence of termination is that the program has only one solution. Therefore, if a terminating program is executed and an input-output relation is constructed, we can be sure that it is the intended one. The program uniquely defines a relation [Doornbos, Backhouse 1995]:

Theorem 10.4.3 (*Hylomorphism characterization*) For R an arbitrary relation and S an F-reductive relation, equation

$$X \quad :: \quad X \;=\; S;F(X);R \tag{10.9}$$

has a unique solution. $\square$

Towards a calculus of terminating programs

Programming can be made less error-prone by making it into a more calculational activity. Proving termination being so essential in the design, our ultimate goal is the design of a calculus of terminating programs. This calculus should include a calculus of reductive relations, by means of which proving termination can become a more syntactic activity.

In addition, one might think of program transformations that preserve or even introduce termination. Especially the latter might constitute a strong tool, in combination with other techniques like program inversion: designing, for instance, a program that flattens a parse tree, is simpler than designing a parser; the converse of a flattening program is a parser, but it may very well be that by inversion of the flattener termination is lost. Transformations that reduce nondeterminism might then restore termination.

10.5 Specification, programming, and execution

The design of a program starts with a specification (relation) S, often expressed using predicates. Relational, i.e. pointfree, formulae can be far more compact than predicates, so S is rewritten into relational form. In this process, however, one should not be pedantic by insisting that all points be removed. One can, for instance, avoid making the plumbing of rearrangements of arguments and other natural transformations explicit by introducing a name for the plumbing relation and just define it pointwise; (using the name *swap*, say, in the relational form of S and just defining it pointwise as the swap of the elements in a pair is much to be preferred to explicitly having $\gg;\ll^{\cup} \cap \ll;\gg^{\cup}$ in the rewrite of S).

Once specification S has been moulded into relational form, program design starts. We have to determine a relator F and relations R and T such that:

$$R;F(S);T \subseteq S \quad \text{, where } R \text{ is } F\text{-reductive.} \tag{10.10}$$

The choice of F corresponds to the choice for structural recursion, repetition, divide-and-conquer, etc.. That relation R has to be reductive should be of great help in its design: somewhere in R there has to be a reductive part. The typing and properties of the other parts are then almost dictated. (In the standard construction of repetitions this is a well-known phenomenon.) If R and T are not directly implementable, they serve as new specifications to be dealt with separately.

This is what constitutes the core of programming. (Sometimes additional coding can be necessary to overcome the limitations of a target programming language. Such obligations might be considered an indication of what features a programming language should have so as to make it a more convenient language.)

The program designed by solving (10.10) is the terminating program

$$X = R;F(X);T \ .$$

If it is executed by a computer, the input-output relation constructed is a —and, hence, *the*— solution of the equation. That this relation does, indeed, satisfy the specification, i.e. is contained in S, is an immediate consequence of the Tarski fixed-point theorem (1.3.3) and (10.10), since a unique solution is also a least solution.

10.6 Conclusion

In the above our primary goal has been to show how core concepts in programming, like the notions of datatype, specification, program, and termination, can be captured compactly in the relational framework. Although important, compactness was, however, not our main concern. Our main concern was to formalize the concepts in such a way that one can work with them; that is to say that they are useful for programming.

In our framework, datatypes are formed by relators, i.e. special functions from relations to relations. Specifications are relations and programs are equations in relations. Termination is formalized by means of the concept of reductivity. Unicity of the solution of a hylo-program is a byproduct of termination, i.e. not a goal in itself. Of course, it is an extremely useful byproduct, for the following reason.

Traditionally one has to guarantee that the relation constructed satisfies the specification by *assuming* that it is the least solution. This enforces the choice for initial semantics and in addition it restricts the freedom of the implementor of the language. Our discussion of hylo-programs shows, however, that if we are interested in terminating programs only, these restrictions are quite unneccesary.

We would like to stress that unicity is a genuine byproduct of termination: all by itself unicity is not enough to guarantee termination. In this sense, reductivity seems an important generalization of the (categorical) notion of initiality.

In the above, we have confined our attention to terminating programs. The study of programs that do not necessarily terminate is a topic outside the current scope. (A possible way of attaching a semantics to such programs is by means of pairs (R, A), where R denotes an input-output relation and A denotes a safe set [Doornbos 1996].) Another issue that was not discussed here is the problem of totality. In the case of a parsing program, for instance, not every input string has a parse tree; capturing this is an additional difficulty not studied here.

The examples given in the above might suggest that the relational framework is suitable for imperative and functional programs only. This is not the case. Logic programs, for instance, can be incorporated equally well. Normally, a logic program is viewed as a predicate, but of course it is a relation just as well. The relational semantics that can be given to logic programs may turn out to be simpler and more useful than the traditonal semantics. Investigation of parallel programming in a relational framework is an interesting area of current development.

As we have tried to demonstrate, relation algebra is an excellent vehicle for concise expression of fundamental notions in computing. It should not be supposed, however, that it will ever completely replace the pointwise predicate calculus. A degree of good taste is essential to deciding where and when point-free reasoning is to be preferred.

Chapter 11

Refinement and Demonic Semantics

Jules Desharnais, Ali Mili[1]*, Thanh Tung Nguyen*

In Chapter 8, it was shown how functional programs can be regarded as elements of a relation algebra. In this chapter, we consider imperative programs, which we view as computing an input-output relation on a set of states. We are interested here in programs that are meant to terminate, not in reactive programs. Our programming language is Dijkstra's language of guarded commands [Dijkstra 1976], which allows the expression of nondeterminism, thus making a relational approach very natural.

The central notion of this chapter is that of *refinement* of specifications. Refining a specification is an iterative process leading to increasingly refined specifications until one obtains a specification that is simple enough to be implemented by a program that is then, by construction, totally correct with respect to the initial specification. We will describe this concept by means of abstract relation algebra, but the specifications used in the examples are given as concrete relations defined by a predicate on a set of states; however, to simplify the presentation, the algebraic notation is used all along, even in the examples. These examples are very simple and have been designed to illustrate the concepts rather than to present the resolution of a moderate-size problem.

Two hypotheses underlying the approach described here are the following:

- If the specifier considers that an input x might be submitted to the program, then there is at least one pair (x, x') in the specification (relation) showing that the output should be x'. There could be many such pairs, since nondeterministic programs are allowed, but there can be no pair $(x, \bot)$, where $\bot$ means nontermination. That is, we assume that no specifier will require a program that might either produce a result or not terminate when started in state x.
- If the specifier knows that an input x is not going to be submitted to the program, then x is not in the domain of the specification; also, the specifier does not care what the program does when started in state x.

Since we are interested in the total correctness of programs, we consider that nondeterministic programs behave as badly as they can and loop forever whenever

[1]The first and second authors acknowledge the financial support of the NSERC (Natural Sciences and Engineering Research Council) of Canada.

they have the possibility to do so. This is the *demonic* approach to the semantics of nondeterministic programs.

11.1 Refinement and demonic operators

If one views a relation s as a specification of the input-output behavior of a program p, then one is naturally led to consider p to be totally correct with respect to s if (i) for any input x in the domain of s, x' is a possible output of p only if $(x, x') \in s$, and (ii) p always terminates for any input belonging to the domain of s [Mili 1983]. Note that for an input that does not belong to the domain of specification s, program p may return any result or return no result; that is, the specifier does not care what happens following the submission of such an input. Whether it is wise for a specifier to write partial specifications can be debated. However, such specifications may arise from the decomposition of higher-level specifications. For example, a total specification r could be decomposed as $s;t$, where t is partial and $ran(s) \subseteq dom(t)$. How an input value outside the domain of t is handled does not matter, since s does not return such values.

This is the rationale behind the following definition of refinement.

Definition 11.1.1 [Mili, Desharnais+ 1987] We say that a relation r *refines* a relation s, denoted by $r \sqsubseteq s$, iff $r \sqcap s;\mathbb{T} \sqsubseteq s$ and $s;\mathbb{T} \sqsubseteq r;\mathbb{T}$.

Thus, for instance, $\{(0,0),(1,2)\} \sqsubseteq \{(0,0),(0,1)\}$. However, on the other hand, $\{(0,2)\} \not\sqsubseteq \{(0,0),(0,1)\}$ and $\{(1,2)\} \not\sqsubseteq \{(0,0),(0,1)\}$.

The relation $\sqsubseteq$ is a partial ordering. After making the connection between predicate transformers and relations (see [Maddux 1996]), it can be seen that this notion of refinement is the same as that of Back and von Wright [Back 1981; Back, von Wright 1992], Morgan and Robinson [Morgan, Robinson 1987] and Morris [Morris 1987].

The refinement ordering $\sqsubseteq$ induces a join semilattice, called a *demonic semilattice*. The operations on this structure include demonic join ($\sqcup$), demonic meet ($\sqcap$), and demonic composition ($\Box$). We now introduce these operations, give some of their properties and illustrate them with simple examples. For more details on relational demonic semantics and demonic operators, see [Backhouse, van der Woude 1993; Berghammer, Schmidt 1993; Berghammer, Zierer 1986; Boudriga, Elloumi+ 1992; Desharnais, Belkhiter+ 1995; Frappier, Mili+ 1995; Nguyen 1991].

The least upper bound of relations r and s with respect to $\sqsubseteq$, denoted by $r \sqcup s$ and called the *demonic join*, is

$$r \sqcup s = (r \sqcup s) \sqcap r;\mathbb{T} \sqcap s;\mathbb{T}. \tag{11.1}$$

Here is an example of this operation:

$$\{(0,0),(0,1),(1,0)\} \sqcup \{(1,1),(2,1),(2,2)\} = \{(1,0),(1,1)\}.$$

This operation corresponds to a demonic nondeterministic choice, since the possibility of failure (e.g., 2 is not in the domain of the first relation) is reflected in the result (2 is not in its domain). For the input value 1, failure is not possible, and the set of allowed results is the union of the results of the two operands.

The greatest lower bound of relations r and s with respect to $\sqsubseteq$, denoted by $r \sqcap s$ and called the *demonic meet*, exists provided that $r;\mathbb{T} \sqcap s;\mathbb{T} \sqsubseteq (r \sqcap s);\mathbb{T}$. Under this condition, its value is

$$r \sqcap s = (r \sqcap s) \sqcup (r \sqcap \overline{s;\mathbb{T}}) \sqcup (\overline{r;\mathbb{T}} \sqcap s). \tag{11.2}$$

The condition $r;\mathbb{T} \sqcap s;\mathbb{T} \sqsubseteq (r \sqcap s);\mathbb{T}$ simply means that for each value in the intersection of their domains, r and s have to agree on at least one result. For example, consider

$$\{(0,0),(0,1),(1,0),(1,2)\} \sqcap \{(1,1),(1,2)\} = \{(0,0),(0,1),(1,2)\}.$$

On the intersection of their domains ($\{1\}$), the operands agree on the result 2 and thus the meet is defined. This is not the case for

$$\{(0,0),(0,1),(1,0)\} \text{ and } \{(1,2)\},$$

because they contradict each other on the intersection of their domains (again $\{1\}$). Note that the domain of the result is the union of the domains of the operands. We call $\sqcap$ a *demonic meet*, because it is associated with the demonic join, even though, the domain of $r \sqcap s$ being larger than that of $r \sqcap s$, $\sqcap$ appears more demonic than $\sqcap$.

We assign to $\sqcup$ and $\sqcap$ the same binding power as that of $\sqcup$ and $\sqcap$.

There are special cases where the expressions for $\sqsubseteq, \sqcap$ and $\sqcup$ (Definition 11.1.1, and Equations 11.1 and 11.2) take a simple form. Before describing them, we recall the notion of partial identity.

A *partial identity* (also called *monotype* in Chapt. 10) is a relation $r \sqsubseteq \mathbb{I}$. The set of partial identities is a Boolean lattice isomorphic to the lattice of domain elements (i.e., relations r such that $r = r;\mathbb{T}$, see Sect. 2.4), with ordering, meet and join given by $\sqsubseteq, \sqcap$ and $\sqcup$, respectively. It is easy to verify that a partial identity r satisfies $r = \mathbb{I} \sqcap r;\mathbb{T}$. We denote the complement of a partial identity r relative to $\mathbb{I}$ by $r^\sim$; that is, $r^\sim = \mathbb{I} \sqcap \overline{r;\mathbb{T}}$.

We are now ready to present the special cases alluded to above:

- If r and s are domain elements or partial identities, then $r \sqcap s$ is defined, and

$$r \sqsubseteq s \iff s \sqsubseteq r, \qquad r \sqcup s = r \sqcap s, \qquad r \sqcap s = r \sqcup s. \tag{11.3}$$

- If $r;\mathbb{T} = s;\mathbb{T}$ (in particular, if r and s are total), then

$$r \sqsubseteq s \iff r \sqsubseteq s, \qquad r \sqcup s = r \sqcup s, \qquad r \sqcap s = r \sqcap s \text{ (if } \sqcap \text{ defined)}. \tag{11.4}$$

Figure 11.1 shows the general structure of $\sqsubseteq$.

If the base algebra $(A, \sqcup, \overline{}, \perp\!\!\!\perp, ;, \breve{\ }, \mathbb{I})$ is complete, then $(A, \sqsubseteq)$ is a complete join semilattice, except for $\bigsqcup \emptyset$, which does not exist (since there is no $\sqsubseteq$-minimal element). Let f be an isotonic function (with respect to $\sqsubseteq$) having at least one fixed point. If $(A, \sqsubseteq)$ is a complete join semilattice (except for the case just mentioned), the greatest fixed point of f exists and is given by $\bigsqcup\{x : f(x) = x\}$ (see Sect. 1.3).

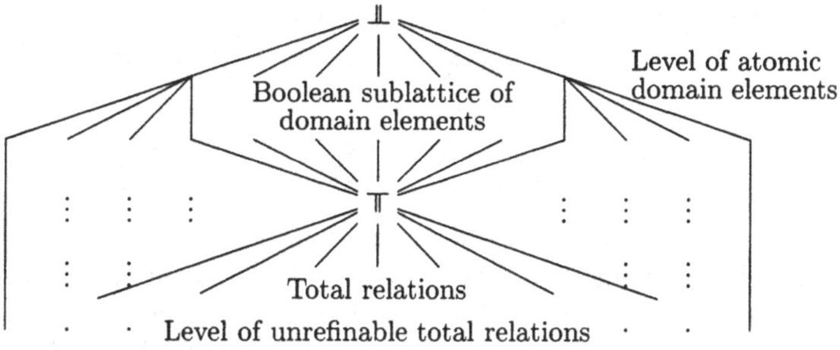

Fig. 11.1 General structure of a semilattice ordered by $\sqsubseteq$

The *demonic composition* of relations r and s is

$$r \Box s \triangleq r ; s \sqcap \overline{r ; \overline{s ; \top\!\!\!\top}}.$$

For concrete relations, this definition is simply read as follows: a pair (x, y) belongs to relation $r \Box s$ iff it belongs to $r ; s$ and it is not possible, from x, by r, to reach a z that is not in the domain of s. This indeed reflects demonic behavior: if something bad can happen (reaching a point z outside the domain of s), then it happens (x is not in the domain of the result).

The following gives an example where $;$ and $\Box$ differ:

$$\begin{aligned}&\{(0,0),(0,1),(1,1),(1,2)\};\{(1,0),(2,2)\} = \{(0,0),(1,0),(1,2)\},\\ &\{(0,0),(0,1),(1,1),(1,2)\}\Box\{(1,0),(2,2)\} = \{(1,0),(1,2)\}.\end{aligned}$$

We assign to $\Box$ the same binding power as that of $;$. Demonic composition is associative and reduces to the usual composition in two particular cases:

$$\begin{array}{lll} \text{(a)} & (r \Box s) \Box t = r \Box (s \Box t), & \\ \text{(b)} & s \text{ total } \implies r \Box s = r ; s, & \\ \text{(c)} & r \text{ univalent } \implies r \Box s = r ; s. & \end{array} \tag{11.5}$$

The demonic operators have a number of additional properties, such as

$$\begin{array}{llll} \text{(a)} & \mathbb{I} \Box r = r \Box \mathbb{I} = r, & \text{(j)} & r \Box (s \mathbin{\dot\sqcap} t) \sqsubseteq r \Box s \mathbin{\dot\sqcap} r \Box t, \\ \text{(b)} & s \sqsubseteq t \implies r \Box s \sqsubseteq r \Box t, & \text{(k)} & (s \mathbin{\dot\sqcap} t) \Box r \sqsubseteq s \Box r \mathbin{\dot\sqcap} s \Box r, \\ \text{(c)} & s \sqsubseteq t \implies s \Box r \sqsubseteq t \Box r, & \text{(l)} & \perp\!\!\!\perp \Box r = r \Box \perp\!\!\!\perp = \perp\!\!\!\perp, \\ \text{(d)} & r \Box (s \mathbin{\dot\sqcup} t) = r \Box s \mathbin{\dot\sqcup} r \Box t, & \text{(m)} & r \mathbin{\dot\sqcap} (s \mathbin{\dot\sqcup} t) = (r \mathbin{\dot\sqcap} s) \mathbin{\dot\sqcup} (r \mathbin{\dot\sqcap} t), \\ \text{(e)} & (s \mathbin{\dot\sqcup} t) \Box r = s \Box r \mathbin{\dot\sqcup} t \Box r, & \text{(n)} & r \mathbin{\dot\sqcup} (s \mathbin{\dot\sqcap} t) = (r \mathbin{\dot\sqcup} s) \mathbin{\dot\sqcap} (r \mathbin{\dot\sqcup} t), \\ \text{(f)} & \multicolumn{3}{l}{r ; \top\!\!\!\top \sqcap s ; \top\!\!\!\top = \perp\!\!\!\perp \implies r \mathbin{\dot\sqcap} s = r \sqcup s,} \\ \text{(g)} & \multicolumn{3}{l}{r ; \top\!\!\!\top \sqcap s ; \top\!\!\!\top = \perp\!\!\!\perp \implies (r \mathbin{\dot\sqcap} s) \Box t = r \Box t \mathbin{\dot\sqcap} s \Box t,} \\ \text{(h)} & \multicolumn{3}{l}{r \text{ univalent } \implies r \Box (s \mathbin{\dot\sqcap} t) = r \Box s \mathbin{\dot\sqcap} r \Box t,} \\ \text{(i)} & \multicolumn{3}{l}{r, s \text{ partial identities } \implies r \Box s = r \sqcap s.} \end{array} \tag{11.6}$$

Note how most of these properties of the demonic operators correspond to those of the usual relational operators. Of course, they hold provided that the necessary meets exist (in laws (f) and (g), they do, because of the hypothesis). One can check the following laws and the fact that the implications are strict:

$$\begin{aligned}(r \sqcap s) \sqcup (r \sqcap t) \text{ defined} &\Longrightarrow r \sqcap (s \sqcup t) \text{ defined},\\ r \sqcup (s \sqcap t) \text{ defined} &\Longrightarrow (r \sqcup s) \sqcap (r \sqcup t) \text{ defined},\\ r \Box (s \sqcap t) \text{ defined} &\Longrightarrow r \Box s \sqcap r \Box t \text{ defined},\\ (s \sqcap t) \Box r \text{ defined} &\Longrightarrow s \Box r \sqcap t \Box r \text{ defined}.\end{aligned} \tag{11.7}$$

The operators $\sqcup, \sqcap$ and $\Box$ are order-preserving in their two arguments; hence, refining an argument yields a refined result. In the case of the demonic meet, however, one must be careful, since refining an argument may lead to an undefined expression. For instance, take $r \triangleq \{(0,0),(0,1)\}, r' \triangleq \{(0,1)\}, r'' \triangleq \{(0,0),(0,1),(1,1)\}$ and $s \triangleq \{(0,0),(1,0)\}$. Using Definition 11.1.1 and Equation 11.2, we find that $r' \sqsubseteq r, r'' \sqsubseteq r$ and $r \sqcap s = \{(0,0),(1,0)\}$; however, neither $r' \sqcap s$ nor $r'' \sqcap s$ is defined.

The next operation that we consider is the *demonic left residual* of s by r, denoted $s /\!\!/ r$ (see Sect. 1.2 and Sect. 1.3 for notions about residuation). By analogy with the relational left residual $/$, we would like $s /\!\!/ r$ to be the largest solution ($\sqsubseteq$-wise) to the inequation (in x) $x \Box r \sqsubseteq s$. However, this largest solution does not always exist. For instance, there is no solution to $x \Box \mathbb{L} \sqsubseteq \mathbb{I}$, since this is equivalent, by 11.6(l), to $\mathbb{L} \sqsubseteq \mathbb{I}$, which does not hold. In fact, $x \Box r \sqsubseteq s$ has a solution precisely when $s ; \mathbb{T} \subseteq (s/r) ; r ; \mathbb{T}$. Under this condition, $s /\!\!/ r$ is defined by

$$x \Box r \sqsubseteq s \iff x \sqsubseteq s /\!\!/ r.$$

When $s /\!\!/ r$ is defined, its value is given by $s /\!\!/ r = s/r \sqcap \mathbb{T} ; r^{\smile}$. In many cases, this expression can be simplified; for example,

$$\text{if } s ; r^{\smile} ; r \subseteq s, \text{ then } s /\!\!/ r \text{ is defined iff } s ; \mathbb{T} \subseteq s ; r^{\smile} ; \mathbb{T}, \text{ and } s /\!\!/ r = s ; r^{\smile}. \tag{11.8}$$

Note that a particular case of the above is when r is univalent ($r^{\smile} ; r \subseteq \mathbb{I}$).

We give a simple example of the use of the demonic left residual. Let $\mathbb{Z}$ be the set of integers. Let s be the relation over $\mathbb{Z}^3$

$$s \triangleq \{((l,m,n),(l',m',n')) : n > 0 \wedge l' \times m'^{n'} = l \times m^n \wedge 0 \leq n' < n\}.$$

We abbreviate the expression of s as follows:

$$s = \{n > 0 \wedge l' \times m'^{n'} = l \times m^n \wedge 0 \leq n' < n\} \tag{11.9}$$

(similar abbreviations are used without mention in the remaining portion of this chapter; unprimed variables denote initial values, and primed variables, final values). We want to refine s by a sequence $t \Box r$, with $r \triangleq \{l' = l \wedge m' = m \wedge n' = n - 1\}$. Since r is univalent, the $\sqsubseteq$-largest t that solves our problem is (by 11.8)

$$t = s /\!\!/ r = s ; r^{\smile} = \{n > 0 \wedge l' \times m'^{n'-1} = l \times m^n \wedge 0 \leq n' - 1 < n\}. \tag{11.10}$$

Note that $s /\!\!/ r$ is well defined, since $s ; \mathbb{T} = \{n > 0\} = s ; r^{\smile} ; \mathbb{T}$.

The following laws are properties of left residuals. Of course, they hold provided that the partial operations ($\sqcap, /\!\!/$) are defined.

$$\begin{array}{llll}\text{(a)} & r \sqsubseteq s \Longrightarrow r /\!\!/ t \sqsubseteq s /\!\!/ t, & \text{(e)} & (r /\!\!/ s) /\!\!/ t = r /\!\!/ (t \Box s),\\ \text{(b)} & r \sqsubseteq s \Longrightarrow t /\!\!/ s \sqsubseteq t /\!\!/ r, & \text{(f)} & (r \sqcap s) /\!\!/ t = (r /\!\!/ t) \sqcap (s /\!\!/ t),\\ \text{(c)} & (r /\!\!/ s) \Box s \sqsubseteq r, & \text{(g)} & r /\!\!/ (s \sqcup t) = (r /\!\!/ s) \sqcap (r /\!\!/ t),\\ \text{(d)} & (r /\!\!/ r) \Box r = r, & \text{(h)} & (r /\!\!/ s) \Box (s /\!\!/ t) \sqsubseteq r /\!\!/ t.\end{array} \tag{11.11}$$

As an important special case, we note that

$$\text{For any relation } r\text{, the left residual } r /\!\!/ r \text{ is defined.} \tag{11.12}$$

In a similar fashion, one defines the *demonic right residual* of s by r, denoted by $r \backslash\!\!\backslash s$, as the largest solution to the inequation (in x) $r \,\square\, x \sqsubseteq s$. In other words, $r \,\square\, x \sqsubseteq s \iff x \sqsubseteq r \backslash\!\!\backslash s$. The demonic right residual has properties dual to those given in (11.11) for the left residual. Like the demonic left residual, it is not always defined. Here are some specific properties of the right residual required later on:

$$\begin{array}{llll} (a) & \mathbb{I} \sqsubseteq r \backslash\!\!\backslash r, & (d) & (r \backslash\!\!\backslash s) \,\square\, (s \backslash\!\!\backslash t) \sqsubseteq r \backslash\!\!\backslash t, \\ (b) & r \,\square\, (r \backslash\!\!\backslash r) = r, & (e) & r ; r^{\smile} ; r \sqsubseteq r \iff r \backslash\!\!\backslash r = r^{\smile} ; r, \\ (c) & \multicolumn{3}{l}{\text{For any relation } r\text{, the right residual } r \backslash\!\!\backslash r \text{ is defined.}} \end{array} \tag{11.13}$$

11.2 Demonic semantics and program derivation

In Chapt. 8, the semantics of a nondeterministic program was given by means of a pair of relations; using such pairs allowed the description of angelic, demonic and erratic nondeterminism. In this chapter, we are interested in demonic semantics, and a single relation suffices, so that the semantics of a program p is given by a relation $\mathcal{D}(p)$, where $\mathcal{D}$ is a function[2] from the set of programs to a suitable relation algebra. For imperative programs, this algebra is usually the full relation algebra *Rel*(X) on a set of states X (a state is simply a function mapping the variables of the program to values). In order to be able to describe loops, we need complete algebras. Hence, from now on, we assume that relation algebras are complete; note that *Rel*(X) satisfies this criterion.

We say that a program p is *correct with respect to* a specification s, or that p *implements* s, if and only if $\mathcal{D}(p) \sqsubseteq s$.

The constructs that we will consider are *assignments*, *sequences*, *alternations* and *loops*. We discuss them in this order. Considerations of semantics and program derivation (refinement) will take place together.

Assignments

The assignment is the base case. Assume that the variables of the program are $x_0, \ldots, x_n$, that they take their values in the sets $X_0, \ldots, X_n$, respectively, and that f is some directly executable function (this could be made precise by considering the full syntax of a specific language). Then, the semantics of the assignment $x_i := f(x_0, \ldots, x_n)$, where $0 \le i \le n$, is the relation

$$\begin{aligned} \mathcal{D}(x_i := f(x_0, \ldots, x_n)) \triangleq \{&((x_0, \ldots, x_n), (x'_0, \ldots, x'_n)) : \\ &\forall j\,(x_j \in X_j \land x'_j \in X_j) \land x'_i = f(x_0, \ldots, x_n) \land \forall j\,(j \neq i \to x'_j = x_j)\}. \end{aligned}$$

For example, if x and y are two integer variables and are the only variables of the program under consideration, then $\mathcal{D}(x := y + 2)$ is the relation over $\mathbb{Z}^2$ that we write under its abbreviated form (see Sect. 11.1) as $\{x' = y + 2 \land y' = y\}$.

[2] As in Chapt. 8, the symbol $\mathcal{D}$ is used to remind of the qualificative *demonic*.

Sequences

The sequential composition of two programs p and q is written $p; q$. Its demonic semantics is simply

$$\mathcal{D}(p; q) \triangleq \mathcal{D}(p) \,\square\, \mathcal{D}(q).$$

Suppose we are given a specification

$$s \triangleq \{n > 0 \wedge l' \times m'^{n'} = l \times m^n \wedge 0 \leq n' < n\}, \tag{11.14}$$

where l, m, n are integer variables (this is the same relation as in Equation 11.9) and that we want to refine it as a sequence $t \,\square\, r$, where $r \triangleq \{l' = l \wedge m' = m \wedge n' = n - 1\}$. This can be done by guessing a suitable t, say $t \triangleq \{l' = l \times m \wedge m' = m \wedge n' = n\}$, then by computing $t \,\square\, r = \{l' = l \times m \wedge m' = m \wedge n' = n - 1\}$ (use either of 11.5(b) or 11.5(c)), and finally by verifying that $t \,\square\, r$ indeed refines s, which it does. In this case, t and r can be implemented by the assignments $l := l \times m$ and $n := n - 1$, respectively; hence, the program $l := l \times m; n := n - 1$ implements s.

Another way of finding t is to calculate it by means of the demonic left residual $s /\!\!/ r$. This was done in the previous section (see Equation 11.10). The result was $t = \{n > 0 \wedge l' \times m'^{n'-1} = l \times m^n \wedge 0 \leq n' - 1 < n\}$. This relation cannot be implemented directly, at least in most current languages. However, it is refined by the relation $t' \triangleq \{l' = l \times m \wedge m' = m \wedge n' = n\}$, which is our first solution.

The relation $s /\!\!/ r$ is also called the *weakest prespecification of r to achieve s* [Hoare, He 1986a; Hoare, He 1986b; Hoare, He 1987]. In programming terms, $s /\!\!/ r$ is the solution to the inequation (in x) $x \,\square\, r \sqsubseteq s$ that is the easiest to refine, in the sense that it leaves most design options open, since it is the least refined solution (the highest in the semilattice). This is quite apparent of the second solution t, given in the previous paragraph.

Alternations

We now turn our attention to the case of alternations. The semantic definition of the alternation **if** $g_0 \rightarrow p_0 \,[\!]\, g_1 \rightarrow p_1$ **fi** is

$$\begin{aligned} \mathcal{D}(\textbf{if } g_0 \rightarrow p_0 \,[\!]\, g_1 \rightarrow p_1 \textbf{ fi}) \triangleq \;& \mathcal{G}(g_0) \,\square\, \mathcal{G}(g_1)^{\sim} \,\square\, \mathcal{D}(p_0) \\ & \sqcap \mathcal{G}(g_0)^{\sim} \,\square\, \mathcal{G}(g_1) \,\square\, \mathcal{D}(p_1) \\ & \sqcap \mathcal{G}(g_0) \,\square\, \mathcal{G}(g_1) \,\square\, (\mathcal{D}(p_0) \sqcup \mathcal{D}(p_1)). \end{aligned} \tag{11.15}$$

This seemingly complex definition, expressed mostly with demonic operators, can be transformed into a more familiar-looking one as follows. The semantics of a guard g is a relation $\mathcal{G}(g)$ that is a partial identity whose domain satisfies the condition of the guard. Note that the domains of the three operands of the demonic meets are pairwise disjoint; in this case the demonic meet acts like $\sqcup$ (see 11.6(f)) and is always defined. Also, partial identities are univalent relations, so that the demonic compositions in Equation 11.15 can be replaced by the usual relational composition (see 11.5(c)). Thus,

$$\begin{aligned} \mathcal{D}(\textbf{if } g_0 \rightarrow p_0 \,[\!]\, g_1 \rightarrow p_1 \textbf{ fi}) = \;& \mathcal{G}(g_0) ; \mathcal{G}(g_1)^{\sim} ; \mathcal{D}(p_0) \\ & \sqcup \mathcal{G}(g_0)^{\sim} ; \mathcal{G}(g_1) ; \mathcal{D}(p_1) \\ & \sqcup \mathcal{G}(g_0) ; \mathcal{G}(g_1) ; (\mathcal{D}(p_0) \sqcup \mathcal{D}(p_1)). \end{aligned}$$

This can be read as

> If g_0 is true and g_1 is false, execute p_0; if g_0 is false and g_1 is true, execute p_1; if both guards are true, so that both p_0 and p_1 could be executed, make a demonic choice between them.

If the guards are mutually exclusive, Equation 11.15 simply becomes

$$\mathcal{D}(\textbf{if } g_0 \to p_0 \mathbin{[\!]} g_1 \to p_1 \textbf{ fi}) = \mathcal{G}(g_0) \mathbin{\Box} \mathcal{D}(p_0) \sqcap \mathcal{G}(g_1) \mathbin{\Box} \mathcal{D}(p_1).$$

The definition of alternations can be generalized to the case of an arbitrary number of guards:

$$\mathcal{D}(\textbf{if } [\!]_{i=0}^{n-1}\ g_i \to p_i \textbf{ fi}) \triangleq \bigsqcap_{\emptyset \neq X \subseteq \{0,\ldots,n-1\}} g_X \mathbin{\Box} g_{\overline{X}}^{\sim} \mathbin{\Box} p_X, \tag{11.16}$$

where

- $n \geq 0$,
- $\overline{X}$ is the complement of X relative to $\{0, \ldots, n-1\}$,
- g_X is the demonic composition of the partial identities $\mathcal{G}(g_i)$, for $i \in X$, and $g_{\overline{X}}^{\sim}$ is the demonic composition of the partial identities $\mathcal{G}(g_i)^{\sim}$, for $i \in \overline{X}$ (by 11.6(i), the order of composition does not matter); for $\overline{X} = \emptyset$, we define $g_{\overline{X}}^{\sim} \triangleq \mathbb{I}$,
- $p_X = \bigsqcup_{i \in X} \mathcal{D}(p_i)$.

It is easily verified that, for $n = 2$, this definition is identical to that given in Equation 11.15. For $n = 1$, the result is $g_0 \mathbin{\Box} p_0$ (only one choice) and, for $n = 0$, is $\perp\!\!\!\perp$ (the demonic meet over an empty set is the largest element in the demonic semilattice).

Consider the alternation A below, where the unique variable n ranges over the set $\mathbb{Z}$ of integers.

$$\begin{array}{llll} A \triangleq & \textbf{if} & n = 1 \to n := 1 & [\!]\ \ n = 1 \to n := -3 \\ & [\!] & n = 3 \to n := 2 & [\!]\ \ n = 3 \to n := -1 \\ & [\!] & n > 3 \to n := n - 4 & \\ & \textbf{fi} & & \end{array} \tag{11.17}$$

Using Equation 11.16, we obtain:

$$\begin{aligned} \mathcal{D}(A) = &\quad \{n = 1 \wedge n' = n\} \mathbin{\Box} (\{n' = 1\} \sqcup \{n' = -3\}) \\ &\sqcap \{n = 3 \wedge n' = n\} \mathbin{\Box} (\{n' = 2\} \sqcup \{n' = -1\}) \\ &\sqcap \{n > 3 \wedge n' = n\} \mathbin{\Box} \{n' = n - 4\}. \end{aligned}$$

Applying 11.6(d) and 11.5(c), this can be simplified as

$$\begin{aligned} \mathcal{D}(A) = &\quad (\{n = 1 \wedge n' = 1\} \sqcup \{n = 1 \wedge n' = -3\}) \\ &\sqcap (\{n = 3 \wedge n' = 2\} \sqcup \{n = 3 \wedge n' = -1\}) \\ &\sqcap \{n > 3 \wedge n' = n - 4\}. \end{aligned} \tag{11.18}$$

Note that, by Equations 11.4 and 11.6(f), both the demonic meets and the demonic joins in this expression could be replaced by $\sqcup$.

Let $s \triangleq c_0 \mathbin{\Box} c_1^{\sim} \mathbin{\Box} r_0 \sqcap c_0^{\sim} \mathbin{\Box} c_1 \mathbin{\Box} r_1 \sqcap c_0 \mathbin{\Box} c_1 \mathbin{\Box} (r_0 \sqcup r_1)$, where $c_0, c_1 \sqsubseteq \mathbb{I}$. If c_0, c_1, r_0, r_1 are simple enough, then s can be implemented by an alternation. Otherwise, s

must be refined. Recall the caveat about refining the operands of a demonic meet (Sect. 11.1). Here, it is possible to refine r_0 and r_1 without restriction, because the operands are pairwise disjoint and remain so if c_0 and c_1 (which we call the *guards*, by an abuse of language) do not change. However, refining the guards may lead to an undefined demonic meet.

Loops

An important problem associated with loops is that of their termination. So, let us introduce the notion of *initial part* of a relation, which is useful for the description of the set of initial states of a program for which termination is guaranteed.

Definition 11.2.1 [Schmidt 1977; Schmidt, Ströhlein 1993] The *initial part* of a relation r, denoted $\mathcal{I}(r)$, is given by $\mathcal{I}(r) \triangleq \sqcap\{x : x = x;\mathbb{T} \text{ and } \overline{r;\overline{x}} = x\}$.

In other words, $\mathcal{I}(r)$ is the least fixed point of the $\sqsubseteq$-isotone (but not continuous) function $g(x) \triangleq \overline{r;\overline{x}}$, restricted to domain elements. Note that $\mathcal{I}(r)$ is itself a domain element. One can show that the restriction to domain elements can be removed, so that

$$\mathcal{I}(r) = \bigsqcap\{x : \overline{r;\overline{x}} = x\}. \tag{11.19}$$

In a concrete setting, $\mathcal{I}(r)$ is the set of points which are not the origins of infinite paths (by r). When $\mathbb{T} \sqsubseteq \mathcal{I}(r)$, we say that r is *progressively finite* (see also Sect. 9.2); in this case, there is no infinite path by r.

Let

$$L \triangleq \mathbf{do}\ g \rightarrow p\ \mathbf{od}$$

be a loop, where g is a condition (the guard) and p is the loop body. The semantics of g is given by a partial identity, as was the case for the guards of an alternation. Then, the demonic semantics of L is the greatest fixed point, with respect to $\sqsubseteq$, of the function

$$L_d(x) \triangleq \mathcal{G}(g)^{\sim} \sqcap \mathcal{G}(g) \,\square\, \mathcal{D}(p) \,\square\, x.$$

That is,

$$\mathcal{D}(L) = \bigsqcup\{x : \mathcal{G}(g)^{\sim} \sqcap \mathcal{G}(g) \,\square\, \mathcal{D}(p) \,\square\, x = x\}.$$

(It is easy to show that there is at least one fixed point, and thus, that $\mathcal{D}(L)$ is well defined, by completeness of the $\sqcup$-semilattice). Other similar definitions of **do** loops can be found in [Hoare, He 1986a; Hoare, He 1986b; Nguyen 1991; Sekerinski 1993]. Note that because $\mathcal{G}(g)^{\sim} \sqcap \mathcal{G}(g) = \mathbb{L}$ and because partial identities are univalent, we can give the following, somewhat more familiar, equivalent expression for L_d (see 11.6(f) and 11.5(c)): $L_d(x) = \mathcal{G}(g)^{\sim} \sqcup \mathcal{G}(g);\mathcal{D}(p) \,\square\, x$.

Calculating $\mathcal{D}(L)$ from this definition is difficult. The following theorem allows one to make a guess and to verify if it is correct. It is a generalization to a nondeterministic context of the *while statement verification rule* of Mills [Mills 1975; Mills, Basili+ 1987]. The theorem shows that the greatest fixed point w of L_d (w is mnemonics for while loop) is uniquely characterized by two conditions, firstly, that w be a fixed point of L_d and, secondly, that there be no infinite path from a state that belongs to the domain of w. Half of this theorem (the $\Leftarrow$

direction) was also proved by Sekerinski (the *main iteration theorem* [Sekerinski 1993]) in a predicative programming set-up [Hehner 1984].

Theorem 11.2.2 Let c be a partial identity and b be a relation. A relation w is the greatest fixed point, with respect to $\sqsubseteq$, of the function

$$L_d(x) \triangleq c^{\sim} \sqcap c \,\square\, b \,\square\, x$$

iff the following two conditions hold:

$$\begin{array}{ll} \text{(a)} & L_d(w) = w, \\ \text{(b)} & w ; \mathbb{T} \sqsubseteq \mathcal{I}(c \,\square\, b). \end{array}$$

□

The following example is an application of this proposition. It is rather contrived, but it is simple and fully illustrates the various cases that may happen. Consider the following loop, where the unique variable n ranges over the set $\mathbb{Z}$ of integers:

$$\begin{array}{llll} L \triangleq \textbf{do}\ n > 0 \rightarrow & \textbf{if} & n = 1 \rightarrow n := 1 & [\!]\ \ n = 1 \rightarrow n := -3 \\ & [\!] & n = 3 \rightarrow n := 2 & [\!]\ \ n = 3 \rightarrow n := -1 \\ & [\!] & n > 3 \rightarrow n := n - 4 & \\ & \textbf{fi} & & \\ \textbf{od} & & & \end{array}$$

Notice how all $n > 0$ such that $n \ \mathsf{mod}\ 4 = 1$ may lead to termination with a final value $n' = -3$, but may also lead to an infinite loop over the value $n = 1$; these initial values of n do not belong to the domain of the relation $w \triangleq \mathcal{D}(L)$ giving the semantics of the loop. Note also that all $n > 0$ such that $n \ \mathsf{mod}\ 4 = 3$ may lead to termination with a final value $n' = -1$, but may also lead to a value $n = 2$, for which the loop body is not defined (by the semantics of **if**...**fi**); these n do not belong to the domain of w. Because they also lead to $n = 2$, all $n > 0$ such that $n \ \mathsf{mod}\ 4 = 2$ do not belong to the domain of w.

The semantics of the loop condition is $c \triangleq \mathcal{G}(n > 0) = \{n > 0 \wedge n' = n\}$, so that $c^{\sim} = \{n \leq 0 \wedge n' = n\}$. The body of the loop is just the alternation A above (11.17), hence the semantics of the loop body is $b \triangleq \mathcal{D}(A)$, where $\mathcal{D}(A)$ is given by Equation 11.18. Using 11.6(h,d) and 11.5(c), it is easy to see that $c \,\square\, b = b$.

Let us now show that

$$w \triangleq \{(n \leq 0 \wedge n' = n) \vee (n > 0 \wedge n \ \mathsf{mod}\ 4 = 0 \wedge n' = 0)\} \tag{11.20}$$

is the abstraction (semantics) of the loop. We have to verify conditions (a) and (b) of Theorem 11.2.2. Condition (a) is $L_d(w) = w$, that is, $c^{\sim} \sqcap c \,\square\, b \,\square\, w = w$. By 11.6(f),

$$\begin{aligned} w &= \{(n \leq 0 \wedge n' = n)\} \sqcap \{(n > 0 \wedge n \ \mathsf{mod}\ 4 = 0 \wedge n' = 0)\} \\ &= c^{\sim} \sqcap \{(n > 0 \wedge n \ \mathsf{mod}\ 4 = 0 \wedge n' = 0)\}. \end{aligned}$$

Since, as was mentioned above, $c \,\square\, b = b$, it suffices to show that

$$b \,\square\, w = \{n > 0 \wedge n \ \mathsf{mod}\ 4 = 0 \wedge n' = 0\}.$$

$$
\begin{array}{ll}
& b \mathbin{\square} w \\
= & \quad \langle\ b = \mathcal{D}(A) \text{ and } (11.18)\ \rangle \\
& ((\{n = 1 \wedge n' = 1\} \sqcup \{n = 1 \wedge n' = -3\}) \\
& \sqcap (\{n = 3 \wedge n' = 2\} \sqcup \{n = 3 \wedge n' = -1\}) \sqcap \{n > 3 \wedge n' = n - 4\}) \mathbin{\square} w \\
= & \quad \langle\ (11.6(\text{g}))\ \rangle \\
& (\{n = 1 \wedge n' = 1\} \sqcup \{n = 1 \wedge n' = -3\}) \mathbin{\square} w \\
& \sqcap (\{n = 3 \wedge n' = 2\} \sqcup \{n = 3 \wedge n' = -1\}) \mathbin{\square} w \sqcap \{n > 3 \wedge n' = n - 4\} \mathbin{\square} w \\
= & \quad \langle\ (11.6(\text{e}))\ \rangle \\
& (\{n = 1 \wedge n' = 1\} \mathbin{\square} w \sqcup \{n = 1 \wedge n' = -3\} \mathbin{\square} w) \\
& \sqcap (\{n = 3 \wedge n' = 2\} \mathbin{\square} w \sqcup \{n = 3 \wedge n' = -1\} \mathbin{\square} w) \\
& \sqcap \{n > 3 \wedge n' = n - 4\} \mathbin{\square} w \\
= & \quad \langle\ \text{All left operands of } \square \text{ are univalent, } (11.5(\text{c})) \text{ and } (11.20)\ \rangle \\
& (\perp\!\!\!\perp \sqcup \{n = 1 \wedge n' = -3\}) \sqcap (\{\perp\!\!\!\perp \sqcup \{n = 3 \wedge n' = -1\}) \\
& \sqcap \{n > 0 \wedge n \bmod 4 = 0 \wedge n' = 0\} \\
= & \quad \langle\ \perp\!\!\!\perp \text{ is the } \sqsubseteq\text{-maximal element.}\ \rangle \\
& \perp\!\!\!\perp \sqcap \perp\!\!\!\perp \sqcap \{n > 0 \wedge n \bmod 4 = 0 \wedge n' = 0\} \\
= & \quad \langle\ \perp\!\!\!\perp \text{ is the } \sqsubseteq\text{-maximal element.}\ \rangle \\
& \{n > 0 \wedge n \bmod 4 = 0 \wedge n' = 0\}
\end{array}
$$

This shows part (a) of the theorem. Part (b) can be established informally by noting that the domain of w is $\{n \leq 0 \vee n \bmod 4 = 0\}$, and that there is no infinite sequence by b for any n in the domain of w; in other words, $w ; \mathbb{T} \sqsubseteq \mathcal{I}(b) = \mathcal{I}(c \mathbin{\square} b)$.

A more satisfying way to show $w ; \mathbb{T} \sqsubseteq \mathcal{I}(b)$ would be to compute $\mathcal{I}(b)$. However, because $\mathcal{I}(b)$ characterizes the domain of guaranteed termination of the associated loop, there is no systematic way to compute it (this would solve the halting problem). To demonstrate termination of the loop from every state in the domain of w, classical proofs based on variant functions or well-founded sets could be given. But formal arguments based on the definition of initial part (Definition 11.2.1) can also be used. We sketch one such argument.

For $k \geq 0$, let $v_k \triangleq \{n \leq 0 \vee (n \leq 4k \wedge n \bmod 4 = 0)\}$. Obviously, $w ; \mathbb{T} = \bigsqcup_{k \geq 0} v_k$. Also, let $g(x) \triangleq \overline{b ; \overline{x}}$. It is easy to show by induction that for all $k \geq 0$, $v_k \sqsubseteq g^{k+1}(\perp\!\!\!\perp)$, where, as usual, $g^0(x) \triangleq x$ and $g^{k+1}(x) \triangleq g(g^k(x))$. Now, using well-known facts of lattice theory and Equation 11.19,

$$
w ; \mathbb{T} = \bigsqcup_{k \geq 0} v_k \sqsubseteq \bigsqcup_{k \geq 0} g^{k+1}(\perp\!\!\!\perp) \sqsubseteq \bigsqcup_{k \geq 0} g^k(\perp\!\!\!\perp) \sqsubseteq \bigsqcap \{x \mid g(x) = x\} = \mathcal{I}(b).
$$

In this example, Theorem 11.2.2 was used to verify that the guessed semantics w of the loop L was correct, given the semantics c of the loop condition and b of the loop body. The theorem can also be used in the other direction. If we are given a specification w, we can guess c and b, and then apply Theorem 11.2.2 to verify the correctness of the guess. If it is correct, then a loop of the form **do** $c \rightarrow p$ **od**, where p is an implementation of b, is correct with respect to w. For example, with the relation w of Equation 11.20, one can check that a good choice is

$$c \triangleq \{n > 0 \wedge n' = n\} \quad \text{and} \quad b \triangleq \{n > 3 \wedge n' = n - 4\}, \qquad (11.21)$$

which leads to a program much simpler than the one we started from:

$$\textbf{do } n > 0 \rightarrow \textbf{if } n > 3 \rightarrow n := n - 4 \textbf{ fi od}.$$

If the relation b cannot be directly implemented by a program, or is deemed to be still too complex, then it can be refined further. What makes this possible is the fact that the greatest fixed point of L_d with respect to $\sqsubseteq$ (that is, $\sqcup\{x : c^{\sim} \sqcap c \,\Box\, b \,\Box\, x = x\}$) is $\sqsubseteq$-isotonic in b. For example, the b of (11.21) is refined by $\{n' = n - 4\}$, which leads to the still simpler program

$$\textbf{do } n > 0 \rightarrow n := n - 4 \textbf{ od}.$$

One problem with this approach is that Theorem 11.2.2 forces one to guess b and c such that the greatest fixed point of L_d is *exactly* w, whereas it would be enough that it *refines* w. This is where the following theorem can be useful.

Theorem 11.2.3 Let c be a partial identity and b, r be relations. Let w be the greatest fixed point, with respect to $\sqsubseteq$, of the function $L_d(x) \triangleq c^{\sim} \sqcap c \,\Box\, b \,\Box\, x$. Then,

$$L_d(r) \sqsubseteq r \text{ and } r;\mathbb{T} \subseteq \mathcal{I}(c \,\Box\, b) \implies w \sqsubseteq r. \qquad \Box$$

Suppose we are given the specification

$$r \triangleq \{(n \leq 0 \wedge n' = n) \vee (n > 0 \wedge n \bmod 4 = 0 \wedge n' = 0)\}$$

(this is the w of Equation 11.20). Let $c \triangleq \{n > 0 \wedge n' = n\}$ and $b \triangleq \{n' = n - 4\}$ (compare with Equations 11.21). Using (11.5(c)) and (11.6(f)), it is easy to show that

$$L_d(r) = r \sqcap \{0 < n < 4 \wedge n' = n - 4\} \sqsubseteq r.$$

Also, $c \,\Box\, b = c;b = \{n > 0 \wedge n' = n - 4\}$ is progressively finite, so that $r;\mathbb{T} \subseteq \mathbb{T} \subseteq \mathcal{I}(c \,\Box\, b)$. Hence, by Theorem 11.2.3, $w \sqsubseteq r$, whatever w is (Theorem 11.2.2 could be used to show that $w = \{(n \leq 0 \wedge n' = n) \vee (n > 0 \wedge n \bmod 4 = 0 \wedge n' = 0) \vee (n > 0 \wedge n \bmod 4 \neq 0 \wedge n' = (n \bmod 4) - 4)\}$). Thus, the following program is correct with respect to r:

$$\textbf{do } n > 0 \rightarrow n := n - 4 \textbf{ od}.$$

Another example of the use of Theorem 11.2.3 will be given in Sect. 11.3.

So far, so good, but still, b and c must be guessed. The following theorem provides quite a lot of guidance in their choice. Moreover, it gives a necessary and sufficient condition for the existence of a loop refining a given specification r.

Theorem 11.2.4 Let r be a relation.

1. If r satisfies the condition
$$r;\mathbb{T} \sqsubseteq r;(r \sqcap \mathbb{I});\mathbb{T} \sqcup r;\overline{r;\mathbb{T}}, \tag{11.22}$$
then there exist a partial identity c and a relation b such that the greatest fixed point of the function $L_d(x) \triangleq c^\sim \sqcap c \circ b \circ x$ refines r. The relations b and c must satisfy the following constraints:
   (a) $c^\sim \sqsubseteq r \sqcup \overline{r;\mathbb{T}}$ and $r;\mathbb{T} \sqsubseteq r;c^\sim;\mathbb{T}$,
   (b) $b \sqsubseteq t /\!\!/ t \sqcup t$, where $t \triangleq c;r;c^\sim$,
   (c) $r;\mathbb{T} \sqsubseteq \mathcal{I}(c \circ b)$.
2. If r does not satisfy (11.22), then there is no partial identity c and relation b such that the greatest fixed point of $L_d(x)$ refines r [Mili, Mili 1992]. □

Note that $t /\!\!/ t$ is always defined, by (11.12). Also, it is easy to show that the existence of a c satisfying condition (a) is enough to ensure that condition (11.22) holds. Thus, when applying the theorem, one may choose to first exhibit c rather than show (11.22); this also guarantees the existence of b.

Consider a relation algebra where $\bar{\mathbb{I}};\mathbb{T} = \mathbb{T}$ (*e.g.*, an algebra of relations on a set with at least two elements). The relation $\bar{\mathbb{I}}$ does not satisfy condition (11.22) of the theorem. Hence, it cannot be implemented by a loop.

As an example of application of Theorem 11.2.4, let us refine the specification
$$r \triangleq \{n \geq 0 \wedge l' = l \times m^n\},$$
where all three variables l, m, n are of type integer. We assume that addition, subtraction and multiplication are available in the programming language, but that exponentiation is not. Because the final values of m and n are not specified, this specification displays some nondeterminism (i.e., r is not univalent).

We have to choose c such that $c^\sim \sqsubseteq r \sqcup \overline{r;\mathbb{T}}$. The maximal $c^\sim$ is given by:
$$(r \sqcup \overline{r;\mathbb{T}}) \sqcap \mathbb{I} = \{(n \geq 0 \wedge l' = l \times m^n \vee n < 0)\} \sqcap \mathbb{I} = \{l = 0 \vee m = 1 \vee n \leq 0\} \sqcap \mathbb{I}.$$
There are seven (nonempty) combinations of the three predicates $l = 0$, $m = 1$ and $n \leq 0$. It is an interesting exercise to try them all. Some, like $l = 0$, correspond to a c that does not satisfy $r;\mathbb{T} \sqsubseteq r;c^\sim;\mathbb{T}$; choosing the maximal $c^\sim$ (i.e., the minimal c) leads to a loop with a minimal number of iterations. We choose $c \triangleq \{n > 0\} \sqcap \mathbb{I}$ (thus, $c^\sim = \{n \leq 0\} \sqcap \mathbb{I}$). It is a simple matter to verify that $r;\mathbb{T} \sqsubseteq r;c^\sim;\mathbb{T}$ is satisfied.

Next, we calculate t:
$$t = c;r;c^\sim = \{n > 0 \wedge l' = l \times m^n \wedge n' \leq 0\}.$$
Since t is *difunctional*, that is, $t;t^\smile;t \sqsubseteq t$ (see Sect. 9.2 and Chapt. 13 for more on difunctionality), we have, by (11.8),
$$t /\!\!/ t = t;t^\smile = \{n > 0 \wedge l' \times m'^{n'} = l \times m^n \wedge n' > 0\}.$$
This relation carries the invariant "$l \times m^n$ is constant"; it is a partial equivalence relation over positive values of n, whereas t is a way from positive values to nonpositive values of n.

Notice that $t /\!\!/ t$ and t have the same domain, namely $\{n > 0\}$; hence, by (11.4), $t /\!\!/ t \uplus t = t /\!\!/ t \sqcup t$ and the domain of $t /\!\!/ t \uplus t$ is also $\{n > 0\}$. Any relation b refining $t /\!\!/ t \uplus t$ has a larger domain (Definition 11.1.1), hence the domain of $c \mathbin{\Box} b$ is $\{n > 0\}$. The condition $r ; \mathbb{T} \sqsubseteq \mathcal{I}(c \mathbin{\Box} b)$ of Theorem 11.2.4 means that there must be no infinite path by $c \mathbin{\Box} b$ from any state in the domain of r, that is, from any state satisfying $n \geq 0$. Altogether, this is equivalent to saying that $c \mathbin{\Box} b$ must be progressively finite. We cannot take $b = t /\!\!/ t \uplus t$, since $c \mathbin{\Box} b$ $(= b)$ is not progressively finite (obvious from the observation that $t /\!\!/ t$ is a partial equivalence relation). But t is progressively finite and $t \sqsubseteq t /\!\!/ t \uplus t$, hence we can take $b \triangleq t$. This is not peculiar to the present specification: this choice can always be made. But it is a bad choice in the sense that no real progress is made in the refinement process: refining t is as hard as refining r, because they are so similar. Now, looking at $t /\!\!/ t \uplus t$, it is obvious that in order to get progressive finiteness, n must be decreased. This justifies the following refinement.

$t /\!\!/ t \uplus t$

$\sqsupseteq$ $\langle$ $\{n > 0 \wedge l' = l \times m^n \wedge n' = 0\} \sqsubseteq t$ by reduction of the nondeterminism (Definition 11.1.1); this refinement is done in order to get the simplification two transformations below. $\rangle$

$t /\!\!/ t \uplus \{n > 0 \wedge l' = l \times m^n \wedge n' = 0\}$

$=$ $\langle$ Substituting the value of $t /\!\!/ t$ from above and arithmetics. $\rangle$

$\{n > 0 \wedge l' \times m'^{n'} = l \times m^n \wedge n' > 0\}$

$\uplus \{n > 0 \wedge l' \times m'^{n'} = l \times m^n \wedge n' = 0\}$

$=$ $\langle$ Both operands of $\uplus$ have the same domain, (11.4). $\rangle$

$\{n > 0 \wedge l' \times m'^{n'} = l \times m^n \wedge n' \geq 0\}$

$\sqsupseteq$ $\langle$ Refinement by reduction of nondeterminism. $\rangle$

$\{n > 0 \wedge l' \times m'^{n'} = l \times m^n \wedge 0 \leq n' < n\}$

$\triangleq$ $\langle$ This is our choice for b. $\rangle$

b

The relation b is progressively finite; it is not univalent and can be refined in many different ways. For example, the variable n may be brought to zero by

- decreasing it by 1 if it is odd and dividing it by 2 if it is even;
- decreasing it by 1 if its value is 1, otherwise decreasing it by 1 or 2 (thus keeping some nondeterminism);
- decreasing it by 1.

This last possibility was explored in Sect. 11.2 (see Equation 11.14). Combining all results together, we find the following program:

$$\textbf{do } n > 0 \rightarrow l := l \times m; n := n - 1 \textbf{ od}.$$

11.3 Program construction by parts

Program construction by parts [Frappier 1995; Frappier, Mili+ 1995] is a paradigm of program construction from relational specifications that is based on the premise that complex relational specifications are structured as demonic meets of simpler specifications; these simpler specifications are then refined. But, as we indicated in Sect. 11.1, refining the arguments of a demonic meet may lead to an undefined expression, so that some care is needed.

The demonic meet of two relations r and s has two different limit behaviors:

- If $r;\mathbb{T} \sqcap s;\mathbb{T} = \mathbb{L}$, then $r \sqcap\!\!\!\!\sqcap s = r \sqcup s$ (by (11.6(f))).
- If $r;\mathbb{T} = s;\mathbb{T}$, then $r \sqcap\!\!\!\!\sqcap s = r \sqcap s$ (by (11.4)).

In the first case, the refinement has an alternation as a natural target, because the semantics of an alternation is precisely structured as a demonic meet of domain disjoint components (see Equation 11.15). In the second case, the refinement is more difficult. A general strategy is to try to push the demonic meet within its operands, hoping to get more easily refinable components. Two properties of interest in this respect, which are easily derived from (11.6(j,k,m,n)), are

$$\begin{aligned} p_1 \,\Box\, p_2 \sqcap\!\!\!\!\sqcap q_1 \,\Box\, q_2 &\sqsupseteq\!\!\!\!\sqsupseteq (p_1 \sqcap\!\!\!\!\sqcap q_1) \,\Box\, (p_2 \sqcap\!\!\!\!\sqcap q_2), \\ (p_1 \sqcup\!\!\!\!\sqcup p_2) \sqcap\!\!\!\!\sqcap (q_1 \sqcup\!\!\!\!\sqcup q_2) &\sqsupseteq\!\!\!\!\sqsupseteq (p_1 \sqcap\!\!\!\!\sqcap q_1) \sqcup\!\!\!\!\sqcup (p_2 \sqcap\!\!\!\!\sqcap q_2). \end{aligned} \qquad (11.23)$$

These two rules show how, when two components have a common structure, this structure commutes with the demonic meet to become the overall structure of the specification. Using (11.7), one can see that, if the refining expression (the one on the right) is defined, then the refined expression is defined, too. This is also a general property of meets: When refining the operands of a meet, it is not necessary to check definedness at every step; if the end result is defined, then the intermediate steps were.

Our example consists in a partial refinement of a specification of a sort program. Assume that $N \geq 0$ is an integer constant. The space of the specification is defined by two variables, f, of type array[0..$N-1$] of integers (i.e., $f : \{0, \ldots, N-1\} \to \mathbb{Z}$) and $n \in \{0, \ldots, N\}$, an index for the array. We define the relation swap, which relates two arrays if one is obtained from the other by swapping two elements, and the relation perm, which relates two arrays if one is a permutation of the other, by

$$\begin{aligned} \mathsf{swap} &\triangleq \{\exists i,j\,(0 \leq i,j < N \wedge f'(i) = f(j) \wedge f'(j) = f(i) \\ &\qquad \wedge \forall k\,(0 \leq k < N \wedge k \neq i \wedge k \neq j \to f'(k) = f(k)))\}, \\ \mathsf{perm} &\triangleq \{\exists \text{bijective } g : \{0, \ldots, N-1\} \to \{0, \ldots, N-1\}\,(f' = g;f)\}. \end{aligned}$$

Here are some properties of swap and perm that are needed later on:

$$\text{(a) } \mathsf{perm} \,\Box\, \mathsf{perm} = \mathsf{perm}, \qquad \text{(b) } \mathsf{swap} \,\Box\, \mathsf{perm} = \mathsf{perm}, \qquad \text{(c) } \mathbb{I} \sqsubseteq\!\!\!\!\sqsubseteq \mathsf{perm}. \qquad (11.24)$$

The predicate pst (stands for *partially sorted*) is

$$\mathsf{pst}(f, n) \triangleq \forall i,j\,(0 \leq i < n \wedge i \leq j < N \to f(i) \leq f(j)).$$

The relation ps is then defined with the help of pst:

$$\mathsf{ps} = \{\mathsf{pst}(f', n')\}.$$

It *establishes* the truth of pst. Note that ps is a range element (i.e., $\mathsf{ps} = \mathbb{T};\mathsf{ps}$, see Sect. 2.4). Like any range element, ps is difunctional, that is, $\mathsf{ps};\mathsf{ps}^\smile;\mathsf{ps} \sqsubseteq \mathsf{ps}$; hence, by (11.13(e)),

$$\mathsf{ps} \backslash\!\!\backslash \mathsf{ps} = \mathsf{ps}^\smile;\mathsf{ps} = \{\mathsf{pst}(f, n) \wedge \mathsf{pst}(f', n')\}. \qquad (11.25)$$

One can say that $\mathsf{ps}\backslash\mathsf{ps}$ *preserves* the truth of pst.

The specification of the sorting program is

$$\mathsf{sort} \triangleq \mathsf{perm} \mathbin{\dot\sqcap} \mathsf{ps} \mathbin{\dot\sqcap} \{n' = N\}$$

(this is not necessarily the top level specification, but may have been arrived at after refining such a specification). The three relations perm, ps and $\{n' = N\}$ are total, so that the demonic meets could be replaced by $\sqcap$ (by (11.4)). Because $\mathsf{perm} \sqcap \mathsf{ps} \sqcap \{n' = N\}$ is total, the demonic meet of these three relations exists (see Equation 11.2), so that the specification sort is well defined.

There are many possible refinements of the operands of sort, some of which quickly lead to a dead end. For example, perm can be refined by $\mathbb{I}$, but then $\mathbb{I} \mathbin{\dot\sqcap} \{n' = N\}$ is not defined. We first decompose the specification as an initialization sequenced with the specification of a loop.

$$\begin{array}{ll}
 & \mathsf{sort} \\
= & \mathsf{perm} \mathbin{\dot\sqcap} \mathsf{ps} \mathbin{\dot\sqcap} \{n' = N\} \\
= & \qquad \langle\ (11.24(a), 11.13(b)) \text{ and } \{n' = N\} = \mathbb{T} \mathbin{\square} \{n' = N\}.\ \rangle \\
 & \mathsf{perm} \mathbin{\square} \mathsf{perm} \mathbin{\dot\sqcap} \mathsf{ps} \mathbin{\square} (\mathsf{ps}\backslash\mathsf{ps}) \mathbin{\dot\sqcap} \mathbb{T} \mathbin{\square} \{n' = N\} \\
\sqsupseteq & \qquad \langle\ (11.23)\ \rangle \\
 & (\mathsf{perm} \mathbin{\dot\sqcap} \mathsf{ps} \mathbin{\dot\sqcap} \mathbb{T}) \mathbin{\square} (\mathsf{perm} \mathbin{\dot\sqcap} \mathsf{ps}\backslash\mathsf{ps} \mathbin{\dot\sqcap} \{n' = N\}) \\
\sqsupseteq & \qquad \langle\ \text{Definition 11.1.1, } \{f' = f \wedge n' = 0\} \text{ refines all operands of the} \\
 & \qquad \text{first factor.}\ \rangle \\
 & \{f' = f \wedge n' = 0\} \mathbin{\square} (\mathsf{perm} \mathbin{\dot\sqcap} \mathsf{ps}\backslash\mathsf{ps} \mathbin{\dot\sqcap} \{n' = N\})
\end{array}$$

The first factor is the initialization and can be refined by the assignment $n := 0$. We refine the second factor by deriving the condition c and the specification b of the body of a loop implementing it. In this derivation, we use the abbreviation $c \triangleq \{n < N \wedge f' = f \wedge n' = n\}$ and its (partial identity) complement $c^{\sim} = \{n = N \wedge f' = f \wedge n' = n\}$ (note that $n = N$ is indeed the negation of $n < N$, because $n \in \{0, \ldots, N\}$).

$$\begin{array}{ll}
 & \mathsf{perm} \mathbin{\dot\sqcap} \mathsf{ps}\backslash\mathsf{ps} \mathbin{\dot\sqcap} \{n' = N\} \\
= & \qquad \langle\ \text{Preparing for a case analysis: } c^{\sim} \mathbin{\dot\sqcap} c = c^{\sim} \sqcup c = \mathbb{I} \text{ (by (11.3)).}\ \rangle \\
 & (c^{\sim} \mathbin{\dot\sqcap} c) \mathbin{\square} (\mathsf{perm} \mathbin{\dot\sqcap} \mathsf{ps}\backslash\mathsf{ps} \mathbin{\dot\sqcap} \{n' = N\}) \\
= & \qquad \langle\ (11.6(g))\ \rangle \\
 & c^{\sim} \mathbin{\square} (\mathsf{perm} \mathbin{\dot\sqcap} \mathsf{ps}\backslash\mathsf{ps} \mathbin{\dot\sqcap} \{n' = N\}) \mathbin{\dot\sqcap} c \mathbin{\square} (\mathsf{perm} \mathbin{\dot\sqcap} \mathsf{ps}\backslash\mathsf{ps} \mathbin{\dot\sqcap} \{n' = N\}) \\
\sqsupseteq & \qquad \langle\ c^{\sim} \text{ is univalent, (11.6(h), 11.5(c)) and definition of } c^{\sim}.\ \rangle \\
 & c^{\sim} \mathbin{\square} \mathsf{perm} \mathbin{\dot\sqcap} c^{\sim} \mathbin{\square} \mathsf{ps}\backslash\mathsf{ps} \mathbin{\dot\sqcap} \{n = N \wedge n' = N\} \mathbin{\dot\sqcap} c \mathbin{\square} (\mathsf{perm} \mathbin{\dot\sqcap} \mathsf{ps}\backslash\mathsf{ps} \mathbin{\dot\sqcap} \{n' = N\}) \\
\sqsupseteq & \qquad \langle\ \text{By (11.24(c), 11.13(a)), Definition 11.1.1 and (11.6(a,b)), } c^{\sim} \text{ re-} \\
 & \qquad \text{fines the first three factors.}\ \rangle \\
 & c^{\sim} \mathbin{\dot\sqcap} c \mathbin{\square} (\mathsf{perm} \mathbin{\dot\sqcap} \mathsf{ps}\backslash\mathsf{ps} \mathbin{\dot\sqcap} \{n' = N\}) \\
\sqsupseteq & \qquad \langle\ (11.24(b), 11.13(d)), \{n' = N\} = \mathbb{T} \mathbin{\square} \{n' = N\} \text{ and } (11.6(b))\ \rangle \\
 & c^{\sim} \mathbin{\dot\sqcap} c \mathbin{\square} (\mathsf{swap} \mathbin{\square} \mathsf{perm} \mathbin{\dot\sqcap} (\mathsf{ps}\backslash\mathsf{ps}) \mathbin{\square} (\mathsf{ps}\backslash\mathsf{ps}) \mathbin{\dot\sqcap} \mathbb{T} \mathbin{\square} \{n' = N\}) \\
\sqsupseteq & \qquad \langle\ (11.23)\ \rangle
\end{array}$$

$$\begin{array}{ll} & c^{\sim} \sqcap c \,\square\, (\mathsf{swap} \sqcap \mathsf{ps}\backslash\!\!\backslash \mathsf{ps} \sqcap \mathbb{T}) \,\square\, (\mathsf{perm} \sqcap \mathsf{ps}\backslash\!\!\backslash \mathsf{ps} \sqcap \{n' = N\}) \\ = & \quad \langle \text{ swap and } \mathbb{T} \text{ are total, (11.4) } \rangle \\ & c^{\sim} \sqcap c \,\square\, (\mathsf{swap} \sqcap \mathsf{ps}\backslash\!\!\backslash \mathsf{ps}) \,\square\, (\mathsf{perm} \sqcap \mathsf{ps}\backslash\!\!\backslash \mathsf{ps} \sqcap \{n' = N\}) \\ \sqsupseteq & \quad \langle \text{ The expression is still defined after adding } \{n' = n + 1\} \text{ (11.2) } \rangle \\ & c^{\sim} \sqcap c \,\square\, (\mathsf{swap} \sqcap \mathsf{ps}\backslash\!\!\backslash \mathsf{ps} \sqcap \{n' = n + 1\}) \,\square\, (\mathsf{perm} \sqcap \mathsf{ps}\backslash\!\!\backslash \mathsf{ps} \sqcap \{n' = N\}) \end{array}$$

Define $b \triangleq \mathsf{swap} \sqcap \mathsf{ps}\backslash\!\!\backslash \mathsf{ps} \sqcap \{n' = n + 1\}$ and $r \triangleq \mathsf{perm} \sqcap \mathsf{ps}\backslash\!\!\backslash \mathsf{ps} \sqcap \{n' = N\}$; the above derivation then shows that $L_d(r) \sqsubseteq r$ (see Theorem 11.2.3 for the definition of L_d). Because $c \,\square\, b \sqsubseteq \{n < N \wedge n' = n + 1\}$, it is a progressively finite relation (indeed, getting progressive finiteness was the reason for introducing $\{n' = n + 1\}$). Thus the two premises of Theorem 11.2.3 are satisfied. We conclude that $\mathsf{perm} \sqcap \mathsf{ps}\backslash\!\!\backslash \mathsf{ps} \sqcap \{n' = N\}$ is refined by the greatest fixed point of L_d, whence it is implemented by a loop with a guard and body implementing c and b, respectively.

Assume that p is a program that is correct with respect to b. Assembling the previous results, we see that the following program is an implementation of **sort**:

$$n := 0; \textbf{do } n < N \rightarrow p \textbf{ od}.$$

The refinement of b can be done by swapping $f(n)$ with the minimal element of f in the interval $n, \ldots, N - 1$ and increasing n by 1; we will not present it here.

11.4 Conclusion

The basic concept we have presented in this chapter is that of a relational refinement ordering and some associated principles for deriving correct programs. With this ordering come (semi-)lattice and composition operators, which we have used to give a so-called *demonic semantics* of Dijkstra's language of guarded commands. Many other orderings have been used for this purpose: the Egli-Milner ordering, the Smyth ordering, the *approximates* ordering of Nelson, the relational *restriction-of* ordering of Nguyen, etc. For a survey of these, see [Nguyen 1991].

The partiality of some of the demonic operators (demonic meet and the demonic residuals) may seem to be a hindrance, but, in the context of program derivation, showing that an operator is defined simply amounts to showing that a proposed decomposition of a specification is possible. As a trivial example, it is not possible to refine the specification $\mathbb{I}$ by a sequence of the form $r \,\square\, \mathbb{L}$, since $\mathbb{I} /\!\!/ \mathbb{L}$ is not defined, as noted in Sect. 11.1. Thus, showing that $s /\!\!/ r$ is defined is a very natural proof obligation in the context of program derivation.

The approach to demonic semantics presented here is not the only possible one, far from it. One way to treat infinite looping is to add to the state space a fictitious state $\perp$ to denote nontermination (see [Hoare, He 1986a; Hoare, He 1986b; Hoare, He 1987]). Programs and specifications are then denoted by relations on this extended state space. Another way is to describe a program by a pair (relation, set), or, in an abstract setting, by a pair (relation, domain element). The relation describes the input-output behavior of the program, whereas the set component describes the domain of guaranteed termination. For a sample of this method, see Chapt. 8 and [Berghammer, Zierer 1986; Doornbos 1994; Maddux 1996; Parnas 1983]. In these two approaches, the operators combining the relations or the pairs are total operators, which is an advantage. But some relations or pairs correspond to nonimplementable specifications. The verification of the definedness of an operation, in the formalism presented in this chapter, is replaced in these approaches by the verification that the result of an operation applied to implementable specifications is also implementable. Finally, we note that the preponderant formalism employed until now for the description of demonic nondeterminism and refinement is the wp-calculus; see, *e.g.*, [Back 1981; Back, von Wright 1992; Dijkstra 1976; Dijkstra, Scholten 1990; Morgan, Robinson 1987; Morris 1987].

Chapter 12

Tabular Representations in Relational Documents

Ryszard Janicki, David Lorge Parnas, Jeffery Zucker[1]

In this chapter the use of relations, represented as tables, for documenting the requirements and behaviour of software is motivated and explained. A formal model of tabular expressions, defining the meaning of a large class of tabular forms, is presented. Finally, we discuss the transformation of tabular expressions from one form to another, and illustrate some useful transformations.

12.1 A relational model of documentation

More than 30 years ago, managers of large software projects began to understand the importance of having precise documentation for software products. The industry was experiencing the frustration of trying to get software to work; most of the many "bugs" that delayed completion and led to unreliable products were caused by misunderstandings that would have been alleviated by better documentation. Since that time, hundreds of "standards" have been proposed; each was intended to improve the consistency, precision and completeness of natural language documents. In spite of these efforts, documentation is still inadequate. Because of the vagueness and imprecision of natural languages, even the best software documentation is unclear. Because informal documentation cannot be analysed systematically, it is usually inconsistent and incomplete as well.

Software Engineering, like other forms of Engineering, can benefit from the use of mathematics. Mathematical notation is commonly used in engineering documents. Only through the use of mathematics can we obtain the precision that we need. In computer science, there has been a great deal of discussion about the use of mathematics to verify the correctness of software. Before program verification becomes practical, the use of mathematics in documentation must be well-established. Specifications and design documents state the theorems that should be proven about programs.

The state of the art among software developers is such that there is no agreement on the contents of the various documents. It is quite common to hear de-

[1]Research was supported by funding from the Telecommunications Research Institute of Ontario, and from the Natural Sciences and Engineering Research Council of Canada.

velopers arguing about whether or not a certain fact should be included in some given document. Often information is included in several documents or not found in any. The first step in using mathematics in this context is to find mathematical definitions of the contents of the documents.

In [Parnas, Madey 1995] the contents of a number of standard documents are defined by stating that each document must contain a representation of one or more binary relations. Each relation is a set of pairs. If the document contains enough information to determine whether or not any pair is included in the specified relation, it is complete. No additional information should be included. Below we give two examples of such document definitions, one for a system requirements document, the other for a program specification.

The system requirements document

The first step in documenting the requirements of a computer system is the identification of the environmental quantities to be measured or controlled and the association of those quantities with mathematical variables. The environmental quantities include: physical properties (such as temperatures and pressures), the readings on user-visible displays, etc. The association of these quantities with mathematical variables must be carefully defined, and coordinate systems, signs etc. must be unambiguously stated.

It is useful to characterise each environmental quantity as monitored, controlled, or both. *Monitored* quantities are those that the user wants the system to measure. *Controlled* quantities are those whose values the system is intended to control. For real-time systems, time can be treated as a monitored quantity. We will use $m_1, m_2, \ldots, m_p$ to denote the monitored quantities, and $c_1, c_2, \ldots, c_q$ to denote the controlled ones.

Each of these environmental quantities can be considered as a function of time. When we denote a given environmental quantity by ν, we will denote the time-function describing its value by ν^t. Note that ν^t is a function whose domain consists of real numbers; its value at time τ is written "$\nu^t(\tau)$". The vector of time-functions $(m_1^t, m_2^t, \ldots, m_p^t)$ containing one element for each of the monitored quantities, will be denoted by "$\underline{m}^t$"; similarly $(c_1^t, c_2^t, \ldots, c_q^t)$ will be denoted by "$\underline{c}^t$".

A *systems requirements document* should contain representations of two relations. NAT describes the *environment*. REQ describes the effect of the system when it is installed.

The relation NAT

- domain(NAT) is a set of vectors of time-functions containing exactly the instances of $\underline{m}^t$ allowed by the environmental constraints,
- range(NAT) is a set of vectors of time-functions containing exactly the instances of $\underline{c}^t$ allowed by the environmental constraints,
- $(\underline{m}^t, \underline{c}^t) \in$ NAT if and only if the environmental constraints allow the controlled quantities to take on the values described by $\underline{c}^t$, if the values of the monitored quantities are described by $\underline{m}^t$.

The relation REQ

- domain(REQ) is a set of vectors of time-functions containing those instances of $\underline{m}^t$ allowed by environmental constraints,
- range(REQ) is a set of vectors of time-functions containing only those instances of $\underline{c}^t$ considered permissible, i.e. values that would be allowed by a correctly functioning system.
- $(\underline{m}^t, \underline{c}^t) \in$ REQ if and only if the computer system should permit the controlled quantities to take on the values described by $\underline{c}^t$ when the values of the monitored quantities are described by $\underline{m}^t$.

NAT and REQ are used to guide the programmers, inspectors, and testers.

Program descriptions

We use the term *program* to denote a text describing a set of state sequences in a digital (finite state) machine. Each of those state sequences will be called an *execution* of the program. Often, we do not want to document the intermediate states in the sequence. For each starting state, s, we want to know only:

1. is termination possible, i.e. are there finite executions beginning in s?
2. is termination guaranteed, i.e. are all executions that begin in s finite?
3. if termination is possible, what are the possible final states?

This information can be described by an LD-relation [Parnas 1983; Parnas 1994b]. An LD-relation comprises a relation and a subset of the domain of that relation, called the *competence set*. The competence set, in a description of a program, is the set of starting states for which termination is guaranteed. (This is called the "safe set" for that program in Sect. 10.2, or the initial part in Definition 3.5.3). The set of starting states for which termination is possible is the domain of the relation. An ordered pair (x, y) is in the relation if it is possible that the program's execution would terminate in state y after being started in state x.

12.2 Industrial experience with relational documentation

The need for relational documentation and tabular representations became apparent in attempts to apply the ideas above to describe programs that were in use in military and civilian applications. We now describe three of those experiences.

The A-7 experience

An early version of this relational requirements model was used in 1977 at the U.S. Naval Research Laboratory to write a software requirements document for the Onboard Flight Program used in the U.S. Navy's carrier-based attack aircraft, the A-7E. The software was being redesigned as an experiment on programming methods, but the design team was inexperienced in that application area. They set out to produce a document that could be used as the exclusive source of information for programmers. At the same time, it was essential that the document could be carefully reviewed by people who understood the requirements, i.e. pilots. This document was reviewed by pilots (who found hundreds of detail errors in the first versions) and then guided the programmers for several years. A description of the A-7 requirements document (with samples) can be found in [Heninger 1980]; the complete document was published as [Heninger, Kallander+ 1978]. It has been used as a model for many other requirements documents (cf. [Hester, Parnas+ 1981; Parnas, Asmis+ 1991]). Further discussion of requirements documents can be found in [Schouwen, Parnas+ 1993].

The Darlington experience

The relational requirements model and the program documentation model were used in a multi-million dollar effort to inspect safety-critical programs for the Darlington Nuclear Power Generating Station in Ontario, Canada. A relational requirements document, modelled on the A-7 document [Heninger, Kallander+ 1978], was written by one group after the English document was shown to have dangerous ambiguities. A second group wrote relational descriptions of the programs. The third group compared the two descriptions. A fourth group audited the process. In all, some 60 people were involved. Although the code had been under test for six years, numerous discrepancies were found and many had to be corrected. This experience is discussed in more detail in [Parnas, Asmis+ 1991; Parnas 1994a].

The Bell Labs (Columbus Ohio) experience

An earlier experience with relational documentation is reported in [Hester, Parnas+ 1981]. In this experience, the time invested in producing a precise statement of the requirements was paid back when the system received its first on-site test. The test period was the shortest in the lab's history because most of the misunderstandings that usually become apparent during testing had been eliminated when the relational requirements document was reviewed. Another, unique, characteristic of this project was the use of the formal mathematical documentation in preparing informal user documentation. It was found that the informal documentation could imitate the structure of the formal documentation and that its quality was greatly enhanced by basing it on the formal document.

12.3 Why use tabular representations of relations?

In all of these experiences, we found that conventional mathematical expressions describing the relations were too complex and hard to parse to be really useful. Instead, we began to use two-dimensional expressions, which we call *tables*. Each table is a set of cells that can contain other expressions, including tabular ones.

No single notation is well-suited for describing all mathematical functions and relations. The history of mathematics and engineering shows clearly that when we become interested in a new class of functions, we also invent new notations that are well-suited for describing functions in that class. The functions that arise in the description of computer systems have two important characteristics. First, digital technology allows us to implement functions that have many discontinuities, which can occur at arbitrary points in the domain of the function. Unlike the designers of analogue systems, we are not constrained to implement functions that are either continuous or exhibit exploitable regularity in their discontinuities. Second, the range and domain of these functions are often tuples whose elements are of distinct types; the values cannot be described in terms of a typical element. The use of traditional mathematical notation, developed for functions that do not have these characteristics, often results in function descriptions that are complex, lengthy, and hard to read. As a consequence of this complexity, mathematical specifications are often incorrect, and, even more often, misunderstood. This has led many people to conclude that it is not practical to provide precise mathematical descriptions of computer systems. However, our experience shows that the use of tabular representations makes the use of mathematics in these applications practical.

Discovering the first tables

The need for new notation for software documentation first became apparent when producing the requirements document for the A-7. Attempts to write the descriptions in English were soon abandoned because we realised that our turgid prose was no better than that of others. Attempts to write mathematical formulae were abandoned because they quickly developed into parentheses-counting nightmares. We tried using the notations that are commonly used in hardware design, but found that they would be too large in this application. The present table types evolved to meet specific needs. In this work, the tables were used in an *ad hoc* manner, i.e., without formal definition.

Use of program function tables at Darlington

In the Darlington experience, the requirements document was prepared using the notation developed in the A-7 project. However, for this experience, it was necessary to describe the code. New table formats were introduced for this purpose. Once more the tables were used without formal definition. This time, the lack of formal definition caused some problems. During the inspection process heated discussions showed that some of the tables had more than one possible interpretation. This led to a decision to introduce formal definitions for the tables, but this work had to be postponed until the inspection was completed. The first attempt to formalise the meaning of tabular expressions was [Parnas 1992].

Why we need more than one type of table

The initial experience with tables on the A-7 went smoothly for a few weeks. Then we encountered an example that would not fit on one piece of paper. In fact, there were so many different cases that we used four large pieces of paper, taped to four physical tables, to represent one table. In examining this monster, we found that although there were many cases to consider, there were only a few distinct expressions that appeared in the table. This led us to invent a new form of table ("inverted tables") that represented the same information more compactly. The various types of tables will be discussed more extensively in later sections.

Tables help in thinking

To people working in theoretical computer science or mathematics, the use of tabular representations seems a minor matter. It is fairly easy to see that the use of tabular representations does not extend the expressive power of the notation. Questions of decidability and computational complexity are not affected by the use of this notation.

However, our experience on a variety of projects has shown that the use of this notation makes a big difference in practice. When someone sets out to document a program, particularly when one sets out to document requirements of a program that has not yet been written, they often do not know what they are about to write down. Writing, understanding, and discovery go on at the same time. Document authors must identify the cases that can arise and consider what should be done one case at a time. If the program controls the values of many variables of different types, they will want to think about each of those variables separately. They may have to consult with users, or their representatives, to find out what should be done in each case. Tabular notations are of great help in situations like this. One first determines the structure of the table, making sure that the headers cover all possible cases, then turns one's attention to completing the individual entries in the table. The task may extend over weeks or months; the use of the tabular format helps to make sure that no cases get forgotten.

"Divide and conquer" – how tables help in inspection

Tabular notations have also proven to be of great help in the inspection of programs. Someone reviewing a program will have to make sure that it does the right thing in a variety of situations. It is extremely easy to overlook the same cases that the designers failed to consider. Moreover, the inspection task is often conducted in open review meetings and stretch over many days. Breaks must be taken, and it is easy to "lose your place" when there is a pause in the middle of considering a program.

In the Darlington inspection, the tables played a significant role in overcoming these problems. Inspectors followed a rigid procedure. First, they made sure that the set of columns was complete and that no case was included in two columns. Second, they made sure that the set of rows was complete and that there were no duplications or overlaps. Then we began to consider the table column by column proceeding sequentially down the rows. A break could be taken at the end of any entry's consideration; a simple marker told us where to begin when we returned. For a task that took many weeks, the structure provided by the tabular notation was essential.

12.4 Formalisation of a wide class of tables

The industrial applications of tabular expressions were conducted on an *ad hoc* basis, i.e. without formal syntax or semantics. New types of tables were invented when needed and the semantics was intuitive. The first formal syntax and semantics of tabular expressions (or simply tables) was proposed in [Parnas 1992]. Several different classes of tables, all invented for a specific practical application, were defined and for each class a separate semantics was provided. A different approach was proposed in [Janicki 1995]. Instead of many different classes and separate semantics, one general model was presented. The model covered all table classes in [Parnas 1992] as well as some new classes that had not been considered before. The new model followed from the topology of an abstract entity called "table", and *per se*, was application independent. In this section we will present the basic concepts of this model using very simple examples. For more interesting examples, the reader is referred to [Parnas, Madey$^+$ 1994]. We shall not, here, make a distinction between relations and functions, as we view functions as a special case of relations. (However, in those cases where our relations are, in fact, functions, we shall generally use functional notation.)

Let us consider the two following definitions of functions, $f(x, y)$ and $g(x, y)$.

$$f(x,y) = \begin{cases} 0 & \text{if } x \geq 0 \wedge y = 10 \\ x & \text{if } x < 0 \wedge y = 10 \\ y^2 & \text{if } x \geq 0 \wedge y > 10 \\ -y^2 & \text{if } x \geq 0 \wedge y < 10 \\ x+y & \text{if } x < 0 \wedge y > 10 \\ x-y & \text{if } x < 0 \wedge y < 10 \end{cases} \qquad g(x,y) = \begin{cases} x+y \text{ if } (x < 0 \wedge y \geq 0) \\ \qquad \vee (x < y \wedge y < 0) \\ x-y \text{ if } (0 \leq x < y \wedge y \geq 0) \\ \qquad \vee (y \leq x < 0 \wedge y < 0) \\ y-x \text{ if } (x \geq y \wedge y \geq 0) \\ \qquad \vee (x \geq 0 \wedge y < 0) \end{cases}$$

If we were to describe $f(x, y)$ using classical predicate logic, we would write an expression like:

$$\begin{array}{l} (\forall x, (\forall y, ((x \geq 0 \wedge y = 10) \rightarrow f(x,y) = 0) \wedge ((x < 0 \wedge y = 10) \rightarrow f(x,y) = x) \wedge \\ ((x \geq 0 \wedge y > 10) \rightarrow f(x,y) = y^2) \wedge ((x \geq 0 \wedge y < 10) \rightarrow f(x,y) = -y^2) \wedge \\ ((x < 0 \wedge y > 10) \rightarrow f(x,y) = x + y) \wedge ((x < 0 \wedge y < 10) \rightarrow f(x,y) = x - y))). \end{array}$$

Such classical mathematical notation is not very readable, even though the functions are very simple ones. The description becomes much more readable when tabular notation, even without a formal semantics, is used (see Figures 1 and 2).

We must now compare these two examples. What are the differences and similarities between the two tables? They both have the same *raw skeleton*, as illustrated in Fig. 3, which in both cases consists of two *headers*, H_1 and H_2, and the grid, G. However, in Fig. 1 the formulae giving the final value is in the main grid, while in Fig. 2 the final formula to be evaluated is in header H_1.

Formally, a *header* is an indexed set of cells, say $H = \{h_i | i \in I\}$, where $I = \{1, 2, ..., k\}$ (for some k) is an index set. We treat *cell* as a primitive concept that does not need to be explained. A grid G indexed by headers $H_1, \ldots, H_n$, with $H_j = \{h_i^j | i \in I^j\}$, $j = 1, \ldots, n$, is an indexed set of cells G, where $G =$

$\{g_\alpha | \alpha \in I\}$, and $I = I^1 \times \cdots \times I^n$. The set I is the index set of G. A *raw table skeleton* is a collection of headers plus a grid indexed by this collection.

		$y = 10$	$y > 10$	$y < 10$	H_1
H_2	$x \geq 0$	0	y^2	$-y^2$	G
	$x < 0$	x	$x + y$	$x - y$	

Fig. 1: A (normal) table defining f

		$x + y$	$x - y$	$y - x$	H_1
H_2	$y \geq 0$	$x < 0$	$0 \leq x < y$	$x \geq y$	G
	$y < 0$	$x < y$	$y \leq x < 0$	$x \geq 0$	

Fig. 2: An (inverted) table defining g

$H_1 = \{h_i^1 | i = 1, 2, 3\}$

	h_1^1	h_2^1	h_3^1
h_1^2	g_{11}	g_{21}	g_{31}
h_2^2	g_{12}	g_{22}	g_{32}

$H_2 = \{h_i^2 | i = 1, 2\}$ $\quad$ $G = \{g_{ij} | i = 1, 2, 3 \text{ and } j = 1, 2\}$

Fig. 3: A raw table skeleton of tables from Fig. 1 and Fig. 2

The first step in expressing the semantic difference between the two types of tables is to define the *Cell Connection Graph* (CCG in short), which characterises *information flow* ("where do I start reading the table and where do I get my result?"). A CCG is a *relation* that could be interpreted as an *acyclic directed graph* with the grid and all headers as the nodes. The only requirement is that *each arc must either start from or end at the grid G*. There are four different types of CCGs. They are all illustrated in Fig. 4 for $n = 3$. When the number of headers is smaller than 3, type 3 disappears. Type 1 is called *normal*, and type 2 is called *inverted*. These are the two most popular types in practice. The CCG divides the cells into relation cells and predicate cells. A cell is a *relation cell* if no arc starts from it, otherwise it is a *predicate cell*. In Fig. 4 (as well as in Fig. 1) all relation cells are represented by double boxes. Headers and a grid, together with a CCG graph, define a *medium table skeleton*. Each predicate cell defines the domain of a subset of the relation defined; the relation cells define possible values within each domain. However, we *still* do not have enough semantics. How do we determine the domain and values? Let us take the upper table from Figure 3, and the cells h_1^1, h_1^2, g_{11}. If this table represents the function $f(x, y)$, then the CCG (together with the contents of the cells) restricted to these cells should represent the expression $(x \geq 0 \wedge y = 10) \Rightarrow f(x, y) = 0$. But why $(x \geq 0 \wedge y = 10)$? Why not for example: $(x \geq 0 \vee y = 10)$, or $(\neg(x < 0 \vee y = 10))$ etc.?

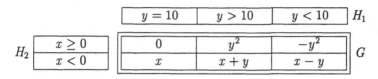

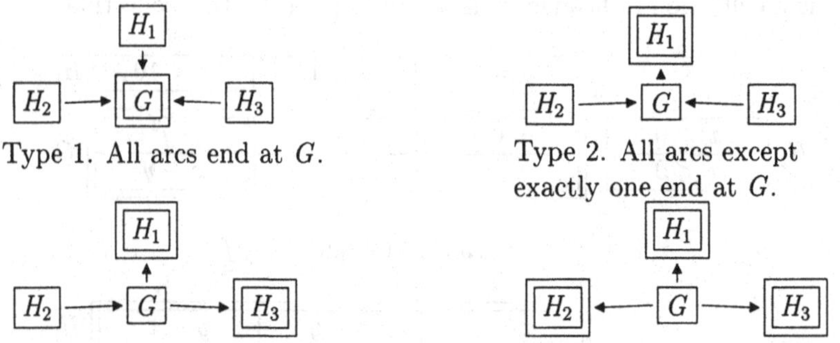

Type 1. All arcs end at G.

Type 2. All arcs except exactly one end at G.

Type 3. At least two arcs start from G and at least one arc ends at G.

Type 4. No arc ends at G.

Fig. 4: Four different types of cell connection graphs

There is no explicit information in the table that indicates conjunction. A medium table skeleton does not provide any information on how the domain and values of the relation (function) specified are determined; such information must be added. The domain is determined by a *table predicate rule*, P_T, and the value is determined by a *table relation rule*, r_T. A medium table skeleton together with table predicate and relation rules is called a *well-done table skeleton*. Figure 5 illustrates two well-done table skeletons. The left-hand one is of type 1, the domain is defined as the conjunction of the predicates contained in appropriate cells of the headers H_1 and H_2, while the value is defined just by an expression held in the appropriate cell of the grid G. The right-hand one is of type 4; the domain is defined by the predicate held in the appropriate cell of the grid G, while the value of the defined relation is the set union of the relations defined by appropriate expressions held in the headers H_1 and H_2.

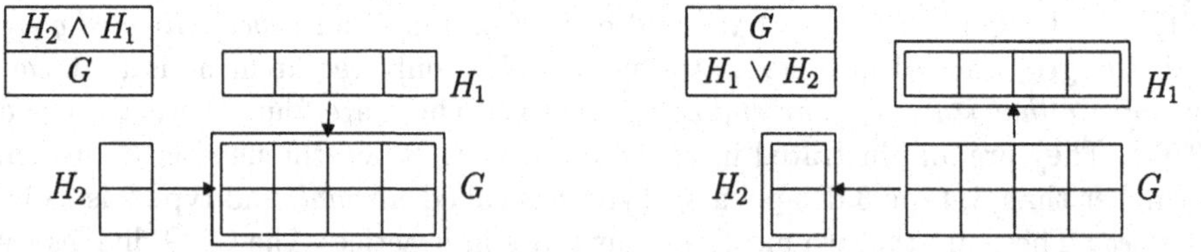

$$P_T(H_1, H_2) = H_1 \wedge H_2, \quad r_T(G) = G \qquad P_T(G) = G, \quad r_T(H_1, H_2) = H_1 \vee H_2$$

Fig. 5: Two examples of well done table skeletons

$H_1 \wedge H_2$			
G	$y = 10$	$y > 10$	$y < 10$

H_2				
$x \geq 0$		0	y^2	$-y^2$
$x < 0$		x	$x + y$	$x - y$

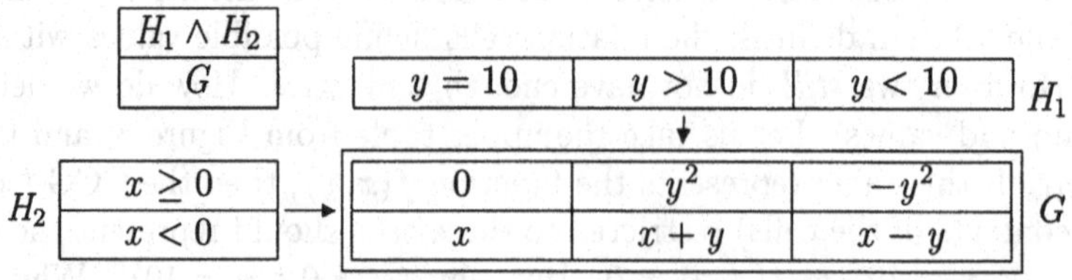

Fig. 6: A tabular expression describing the function f

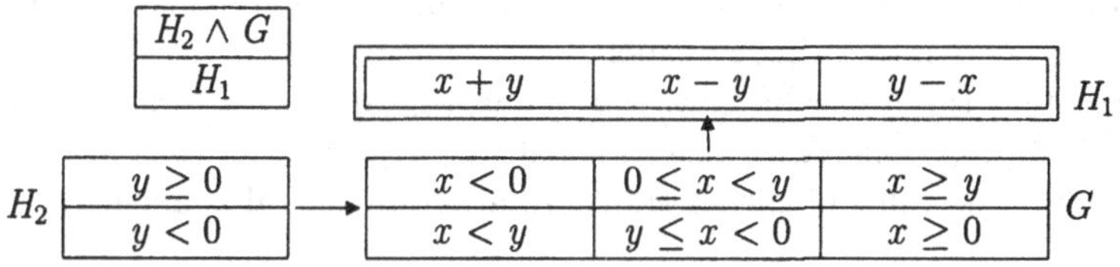

Fig. 7: A tabular expression describing the function g.

To get the *full tabular expression* we need to enrich a well-done table skeleton by a mapping which assigns a predicate expression to each predicate cell, and a relation expression to each relation cell. Figures 6 and 7 show tabular expressions describing the functions f and g.

Summing up, the indexing scheme determines a *raw table skeleton*, where a table is a set of sets of cells. The *cell connection graph* shows "information flow". A *table predicate rule* and a *table relation rule* provide the remaining information.

12.5 Transformations of tables of one kind to another

As we have seen above, many kinds of tables have been found to be useful. Many questions arise naturally: Given a function, what is the best kind of table to represent it? Given two tables (of the same or different kinds), can we see whether they define the same function? Can we (or under what conditions can we) perform certain useful manipulations on tables, such as transforming a table to a "simplest" form of the same kind, or transforming one kind of table to another?

These questions are the subject of ongoing research by the authors and their collaborators at McMaster University. The present section focuses on the last question, on which significant progress has been made [Shen 1995; Zucker 1996]. We consider two of the kinds of tables discussed above, namely *normal function tables* and *inverted function tables*, with conjunction as the predicate rule. (We also revert from the relational to the functional formalism.)

We study various methods for effectively transforming one kind of table to the other. This section gives an informal overview; see also [Zucker 1996].

We are interested in transforming tables to other, semantically equivalent tables, which may be easier to work with. We will consider transformations φ of tables from one kind to another, which satisfy the following two properties: (i) φ is semantics preserving, and (ii) φ is computable. We will consider three examples of such transformations: *changing the dimensionality* of a table, *inverting* a normal table, and *normalising* an inverted table.

Changing the dimensionality of a table

Note first that any n-dimensional (normal or inverted) table can be trivially transformed to an $(n+1)$-dimensional table, by adding an $(n+1)^{\text{th}}$ coordinate header with a single entry, "**true**".

More interestingly: given an n-dimensional table, we can transform it to an $(n-1)$-dimensional table, by "combining" two of the dimensions, i.e., combining two of the headers into a single header.

By iterating this procedure, we can transform any table to a 1-dimensional table — albeit very long (with length equal to the number of cells in the original table), so that the value of such a transformation in general is unclear.

Inverting a normal table

We illustrate inversion with a simple example. Consider the case of a 2-dimensional 3×3 normal table, as in Fig. 8. (The header entries c_i are conditions, and the grid entries t_i^j are terms.)

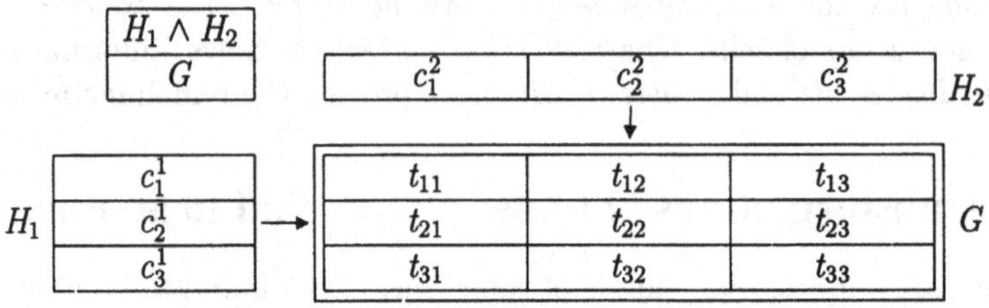

Fig. 8: A two dimensional 3×3 normal table

This can be "inverted along dimension 2" (say), to produce the table in Fig. 9.

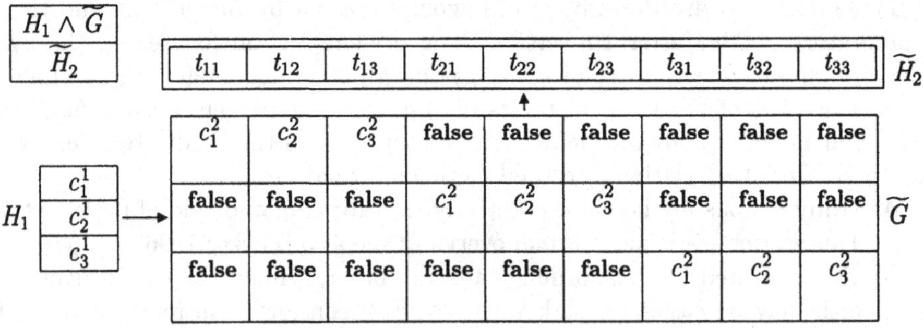

Fig. 9: Inversion of the table in Fig. 8

In general, a normal table can be inverted along any dimension k to produce an inverted table with value header $\widetilde{H}_k$, and the other headers unchanged. The practical value of this transformation is, however, dubious, since the new table is much bigger than the original. (The length of the value header in the new table is equal to the number of cells in the original table!)

In [Zucker 1996] a second method for inversion is also considered, which leads to better (i.e., smaller) inverted tables assuming there are not many distinct terms.

Normalising an inverted table

Here the situation is less satisfactory. We can transform an inverted table to a 1-dimensional normal one, but not (in general) to a many-dimensional one (apart from the trivial many-dimensional version of a 1-dimensional table described in 12.5). As a simple example, consider the 2-dimensional 2×3 inverted table T shown in Fig. 10, with value header H_2.

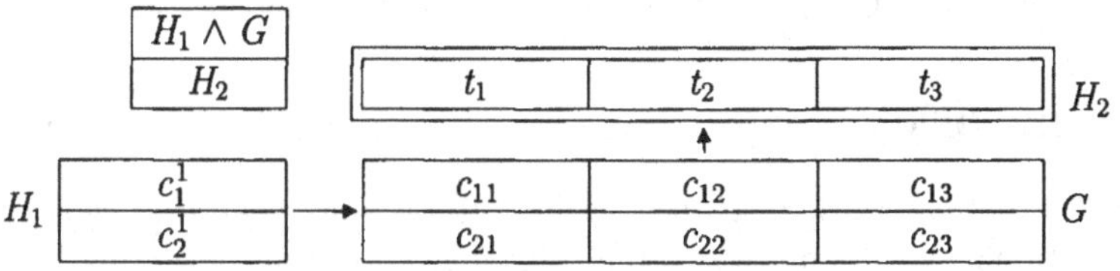

Fig. 10: 2-dimensional inverted table

This can be normalised "along dimension 2" to a 1-dimensional table:

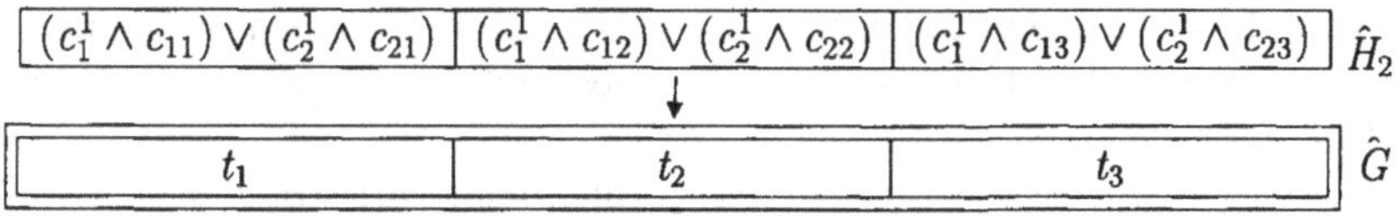

Fig. 11: Normalisation of the table in Fig. 10

Note that the header of the transformed table of Fig. 11 has the same length as H_2, but the conditions in this header are quite complicated. We can, however, effect a trade-off between complexity of conditions and header length by "splitting" disjunctions, as in Fig. 12.

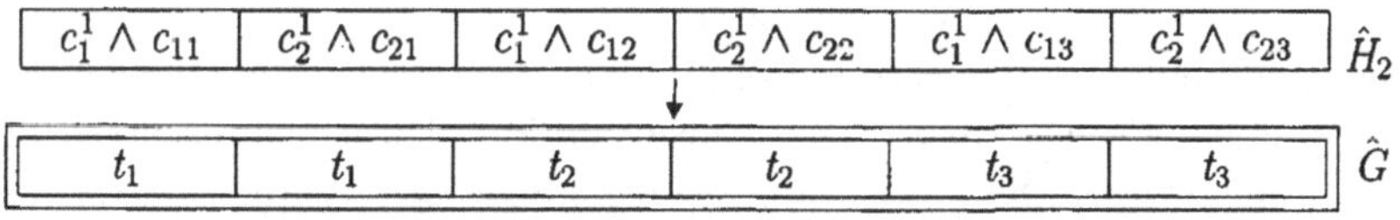

Fig. 12: Another normalisation of the table in Fig. 10

Another (more complex) normalisation algorithm for inverted tables, which preserves dimensionality, was reported in [Shen 1995].

Interrelationship between transformations

We can find a connection between the three types of transformation considered above (changing dimensionality, inverting and normalising), as follows. First we need some definitions.

An *elementary transformation* is any one of the following operations: permuting components of a conjunction or disjunction (in a cell), distributing a conjunction over a disjunction, simplifying conditions '$c \wedge$ **false**' to '**false**' and '$c \vee$ **false**' to 'c', repeating rows by "splitting disjunctions", permuting rows, and deleting rows with '**false**' in the header. (Note that by "rows" we mean rows in any dimension, i.e., rows or columns for two-dimensional tables.) *Elementary equivalence* of tables is the equivalence relation generated by the class of elementary transformations.

Theorem 12.5.1 (*See* [Zucker 1996]) i) If T is a normal table, then the result of an inversion of T followed by a normalisation is elementarily equivalent to the 1-dimensional transform of T.
ii) If T is an inverted table, then the result of a normalisation of T followed by an inversion is elementarily equivalent to the the 1-dimensional transform of T. □

12.6 Conclusion

During the practical work discussed earlier, it became clear that working with tabular relational documents was often dull, mechanical and exacting. The work discussed in this paper provides the basis for tools to assist people when using these notations. We are now engaged in a project to produce a set of prototype tools. The kernel of our system is a "table holder" that creates objects representing uninterpreted tables. Other programs can use this kernel to store and communicate tabular expressions. The tables are stored as abstract data structures and without formatting information.

Separate tools will assist a user to create tables, which are stored in table-holder objects, and to print tables. The printing tool is designed for use by people who do not necessarily understand the expressions. Although the operator of this tool has control over the appearance of the table when printed, the contents cannot be changed by this tool. We are also designing tools to perform basic checks on tabular expressions.

A tool for evaluating and simplifying tabular expressions is being designed. We are also studying the problem of finding a tabular representation of the composition of two relations represented by given tables.

A tool that generates test oracles from tabular program specifications has been produced [Peters, Parnas 1994]. The transformation algorithms discussed above have been implemented [Shen 1995] to provide a prototype tool that assists designers in transforming tables. Much work remains to be done in finding transformation algorithms which are simple and also produce compact transformed tables.

Related work is going on at the University of Quebec (Hull) [Bojanowski, Iglewski+ 1994; Desrosiers, Iglewski+ 1995], Warsaw University [Iglewski, Madey 1996] and Swansea University [Wilder, Tucker 1995].

We believe that the idea of tabular representations of relations is an area deserving attention from both theoretical and practical computer specialists. Although tabular notations were useful before they were formalised [Heninger, Kallander+ 1978; Parnas 1994a], the formalisation discussed in this paper has eliminated misunderstandings and made tools possible.

Chapter 13

Databases

Ali Jaoua, Nadir Belkhiter, Habib Ounalli, Théodore Moukam

Since Codd introduced the relational database model and functional dependencies [Codd 1970], the semantics of relationships between the attributes of a database has been extensively studied by many computer scientists. In this chapter we introduce an application of the calculus of binary relations to the (n-ary) relational database model. This consists of using the notion of *difunctional relations* to define *difunctional dependencies* in the relational database model. Difunctional dependencies are weaker than functional dependencies, but more regular than the other well-known dependencies.

Originally, a study of particular problems on forward program recovery [Jaoua 1987] showed that most of the existing applications can be explained by a difunctional binary relation. Difunctional relations were studied in [Jaoua, Boudriga+ 1991], and more recent and similar works [Jaoua, Beaudry 1989] have been compared to those of Everett [Everett 1944], Riguet [Riguet 1948] and Ore [Ore 1942]. Le Thanh [le Thanh 1986] has shown the practical interest of the concept of "iso-dependencies", which is nothing other than a difunctional relation between the attributes of a database. From a data point of view, the study of difunctional relations is appropriate for a better understanding of semantic relationships between attributes, for data structuring, for maintaining data consistency, and even for reducing data redundancy in a relational database.

In Sect. 13.1 we introduce some mathematical background on equivalence between the two formalisms respectively used for binary relations (due to Tarski) and for n-ary relations (due to Codd). In Sect. 13.2, we state and prove the most interesting properties of difunctional dependencies, for which we propose a set of inference rules. In Sect. 13.3 we show how these results can be applied in practice to obtain a canonical decomposition of a relational schema from its difunctional dependencies. Finally, we conclude this chapter by briefly presenting possible applications of the relational approach in some other computer science fields and by mentioning some research work in progress.

13.1 Mathematical background

In this section, we briefly recall some classical definitions of the relational database model [Codd 1970; Maier 1983; Ullman 1982] and of Tarski's relational algebra

[Tarski 1941]. We propose an equivalent representation of a database instance by a binary relation and we state some equivalent definitions of functional and difunctional dependencies.

Codd's relational formalism

In the relational data model, data are represented as a collection of relations. Informally, each relation resembles a table, where each row represents a collection of related data values. For example, in a database *University*, a table *STUDENT* is used to keep track of each student's name (*SNAME*), student number (*SNUMBER*), social security number (*SSN*), current address (*ADR*), and home phone (*HPHONE*). Each row of this table represents facts about a particular student. The table named *STUDENT* and the column names *SNAME*, *SNUMBER*, *SSN*, *ADR*, and *HPHONE* are used to interpret the meaning of the values in each row of the table. In relational model terminology, a row is called a tuple, a column name is called an attribute, and the table name is called a relation. The data type of values appearing in each column is called a domain. We now formally define these terms.

Attributes are denoted by symbols from the set $U = \{A_1, A_2, \ldots, A_n\}$. To each attribute A_i we associate a domain denoted by $domA_i$ which represents the possible values of this attribute. The elements of $domA$, $domB$ and $domC$ are denoted by a, b, and c respectively. Let X be a set of attributes, then an X-value is an association of values to the different attributes of X on their domain. We use the notation A, $B, \ldots$ for elementary attributes and $X, Y, \ldots$ for subsets of attributes of U. The union of two subsets X and Y is denoted by XY. Furthermore, we do not make any distinction between the elementary attribute A and the set $\{A\}$.

A relation s defined on the set of attributes $U = \{A_1, A_2, \ldots, A_n\}$ is a subset of the Cartesian product $domA_1 \times domA_2 \times \ldots \times domA_n$. A relational schema defined on $U = \{A_1, A_2, \ldots, A_n\}$ is denoted by $S(A_1, A_2, \ldots, A_n)$, $S(U)$ or S. A relation s is an instance of a relational schema S. Elements of an n-ary relation are called tuples or rows. A tuple is usually represented by a sequence of values which are associated to the attributes; for example, (a, b, c) is a tuple of the relation s defined on $S(A, B, C)$. If t is a tuple of a given relation s defined on $S(U)$ and if A is an attribute in U, then $t[A]$ is the component A of t. Similarly, for a subset of attributes $X \subseteq U$, $t[X]$ denotes the restriction of t to X; for example, if $t = (a, b, c)$, then $t[AC] = (a, c)$.

We say that there is a *functional dependency* (abbreviated fd) between X and Y, denoted by $X \to Y$, if and only if, for any instance s of $S(U)$: if for any two tuples t_1 and t_2 of s, if $t_1[X] = t_2[X]$ then $t_1[Y] = t_2[Y]$.

For example, from the semantics of the attributes of the relation schema

STUDENT(SNAME, SNUMBER, SSN, ADR, HPHONE),

we know that the fd *SSN* $\to$ *SNAME* should hold. This fd specifies that the value of the student's social security number (*SSN*) uniquely determines the student's name (*SNAME*).

Functional dependencies have been the most widely used dependencies in practice since their introduction by Codd [Codd 1970]. Functional and multivalued dependencies have been studied by Armstrong [Armstrong 1974], Beeri and Bernstein [Beeri, Bernstein 1979], Biskup [Biskup 1978; Biskup 1980], and Mendelzon [Mendelzon 1979]. Multivalued dependencies have been extended to the notion of generalized multivalued dependencies and first-order hierarchical decomposition by Delobel [Delobel 1978]. Join dependencies have been originally introduced by Rissanen [Rissanen 1978], followed by similar works on the concept of lossless join dependencies [Aho, Beeri+ 1979; Beeri, Bernstein 1979; Maier 1983]. Mutual dependencies, originally introduced by Nicolas [Nicolas 1978], provide a more general decomposition of a database schema. Paredaens [Paredaens 1980] presented a new integrity constraints type called transitive dependencies, which can be seen as a generalization of other well-known dependencies. More recently, Raju and Majumdar have examined the implication problem of fuzzy functional dependencies and have proposed a set of sound and complete inference axioms [Raju, Majumdar 1988].

In the design of relational databases, we are generally interested in lossless join decomposition. [Aho, Beeri+ 1979]. In this chapter, we study another kind of decomposition, called *binary decomposition*, which is a lossless relative product decomposition. This kind of decomposition is particularly interesting in the sense that it provides a data redundancy reduction as well as better data consistency.

Tarski's relational formalism

Binary relations are appropriate for the specification of problems without internal states, for which we can deduce the output from the input. Many mathematical problems can be specified by binary relations, as well as many applications in different computer science fields. In particular, under the assumption of a judicious choice of output and input spaces, any relational schema can be explained in several manners by a binary relation, as we show later.

Informally, the less deterministic a relation is, the more outputs it associates to a given input. Formally [Mili 1985], if R and R' are two relations over a set E, we say that R is *more deterministic* than R' if and only if $R^{\smile};R \subseteq R'^{\smile};R'$.

Example 13.1.1 $R = \{(a,0),(b,1),(c,2)\}$,

$R' = \{(a,0),(a,1),(b,1),(b,2),(c,2),(c,0)\}$,

$R^{\smile};R = \{(0,0),(1,1),(2,2)\}$,

$R'^{\smile};R' = \{(0,0),(0,1),(1,0),(1,1),(0,2),(2,0),(1,2),(2,1),(2,2)\}$;

R is more deterministic than R' since $R^{\smile};R \subseteq R'^{\smile};R'$.

A relation R is called *deterministic* (univalent, or a (partial) function), if and only if it is more deterministic than the identity relation, i.e. if and only if $R^{\smile};R \subseteq I$. A function is called *injective* if and only if $R;R^{\smile} \subseteq I$. We define the *rectangular closure* of a relation R, denoted by R^{++}, as the Cartesian product of its domain and its range: $R^{++} = domR \times ranR$.

Example 13.1.2 Define relation $R = \{(1,2),(2,3),(2,1)\}$ with $domR = \{1,2\}$, and $ranR = \{1,2,3\}$. The rectangular closure of R is $R^{++} = \{1,2\} \times \{1,2,3\}$.

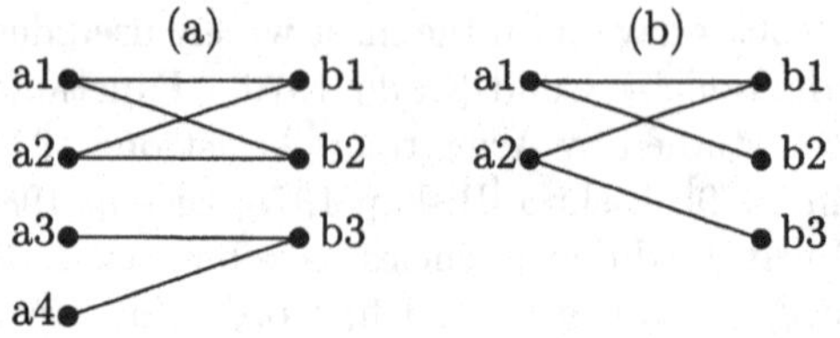

Fig. 13.1 (a) Difunctional relation; (b) non-difunctional relation

A relation is a *rectangle* if it is equal to its rectangular closure, i.e. if there are $A, B \subseteq E$ such that $R = A \times B$. A rectangle $S \subseteq R$ is said to be maximal (in R) if $S \subseteq S' \subseteq R$ implies $S' = S$, for every rectangle S'.

The non-determinism of a relation induces some disorder in the association of images to its antecedents. This disorder results from the fact that any two antecedents can share some particular images and differ on others, as for example in Fig. 13.1(b). Difunctionals are relations (deterministic or non-deterministic) which conserve uniformity in the association of images to antecedents, as for example in Fig. 13.1(a).

Definition 13.1.3 A relation $R : E \leftrightarrow F$ between sets E and F is *difunctional* if and only if $R;R^{\smile};R \subseteq R$, which is equivalent to $R;R^{\smile};R = R$ [Riguet 1948].

The notion and the name of difunctional relations were introduced by Riguet [Riguet 1948]; the study of program specification and forward recovery has brought us to rediscover these relations, which in [Jaoua 1987; Jaoua, Boudriga+ 1991] we named *pseudo-invertible* or *regular*. Everett ([Everett 1944]) used the following equivalent definition of a difunctional relation:

Definition 13.1.4 A relation R from E to F is difunctional if and only if: if $a \in E$ and $b \in E$ are such that $a.R \cap b.R \neq \emptyset$ then $a.R = b.R$.

(Here $x.R = \{y | xRy\}$.)

Example 13.1.5 Define $E = \{a1, a2, a3, a4\}$, $F = \{b1, b2, b3\}$, and $R = \{(a1, b1), (a1, b2), (a2, b1), (a2, b2), (a3, b3), (a4, b3)\}$ (Fig. 13.1(a)). R is non-deterministic and difunctional because $a1.R = a2.R = \{b1, b2\}$ and $a3.R = a4.R = \{b3\}$.

Example 13.1.6 Now, define $E = \{a1, a2\}$, $F = \{b1, b2, b3\}$ and $R = \{(a1, b1), (a1, b2), (a2, b1), (a2, b3)\}$ (Fig. 13.1(b)). We can easily verify that R is not difunctional since $a1.R \cap a2.R = \{b1\}$ but $a1.R \neq a2.R$.

In the following section, we state (in both Codd's and Tarski's formalisms) the definition of a difunctional dependency between two subsets of attributes.

Equivalence between binary and n-ary relations

This kind of equivalence has been proved several times [Delobel 1978; Ullman 1982] for the representation of relational schema as well as for queries. Indeed, for a given binary relation R from X to Y, where X and Y are respectively the Cartesian products of the elementary sets, $SX_1 \times SX_2 \times \ldots \times SX_n$ and $SY_1 \times SY_2 \times \ldots \times SY_m$, we can generate a relational schema S in Codd's formalism and equivalent to the binary relation R:

$$S = \{X_1, ..., X_n, Y_1, \ldots, Y_m\},$$

where SX_i is the name of an attribute X_i (i.e., $SX_i = domX_i$). With relation R, we associate an instance s of S. With an element (e, e') of R, we associate the following tuple t of s:

•$t[X_i]$ = Projection of e on the X_i component, for $1 \leq i \leq n$, and

•$t[Y_j]$ = Projection of e' on the Y_j component for $1 \leq i \leq m$.

Conversely, with any instance s of a relational schema $S(U)$, we can associate a binary relation (denoted by $R[X, Y]$) from $domX$ to $domY$ such that

$$X \cup Y = U, \quad \text{and}$$

$$R[X, Y] = \{(e, e') \mid \exists t \in s[X, Y] : t[X] = e \text{ and } t[Y] = e'\},$$

and where $domX$ (resp. $domY$) is the Cartesian product of domains of attributes of X (resp. of Y).

The choice of the appropriate binary decomposition can be made according to existing functional and difunctional dependencies between the two attribute subsets which are associated respectively with X and Y.

We now turn to specifying a functional dependency by a binary relation. Let X and Y be two attribute subsets of U; we say that there is a functional dependency between X and Y if for any instance s of S, $s[X, Y]$ is a deterministic relation from $domX$ to $domY$, where $s[X, Y]$ is the projection of s on all the attributes of X and Y. In all what follows, we use the following notations:

$X \rightarrow Y$ if there is a functional dependency between X and Y,

$X \leftrightarrow Y$ if there is a functional dependency between X and Y and conversely (i.e. if $\forall s \in S(U)$, $R[X, Y]$ is an injective function).

Example 13.1.7 Let $U = \{A, B, C, D\}, X = \{A, B\}, Y = \{C, D\}$ and $S(U)$ be a relational schema such that $X \rightarrow Y$. Furthermore, if we assume that $domA = domB = domC = domD = \mathbb{N}$, then $domX = domY = \mathbb{N} \times \mathbb{N}$. If s is the following instance of $S(U)$:

s:

A	B	C	D
1	2	4	5
2	1	5	7
2	2	7	7
3	3	4	4

then $s[X, Y]$ can be represented by the following deterministic binary relation:

$$\{((1,2),(4,5)),((2,1),(5,7)),((2,2),(7,7)),((3,3),(4,4))\}$$

defined over $\mathbb{N} \times \mathbb{N}$.

Let S be a relational schema which is defined on an attribute set U, and X, Y and Z be three subsets of U. We can now define the notion of difunctional dependency.

Definition 13.1.8 We say that there is a *difunctional dependency* between X and Y, denoted by $X \rightleftharpoons Y$, if and only if $\forall s \in S(U)$, the binary relation $R[X, Y]$ defined by $s[X, Y]$ is difunctional.

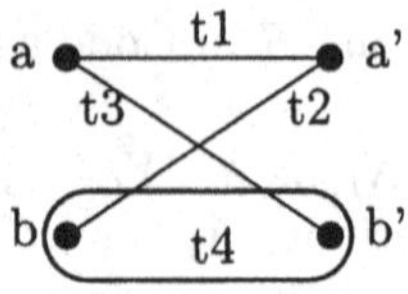

Fig. 13.2 Illustration of a difunctional dependency

The following Theorem 13.1.9 effectively gives an equivalent definition in the Codd formalism. It is also the definition of an *iso-dependency* [le Thanh 1986].

Theorem 13.1.9 There is a difunctional dependency $X \rightleftharpoons Y$ between X and Y if and only if $\forall s \in S(U)$, for three tuples t_1, t_2, and t_3 of s such that

(1) $t_1[X] = t_3[X]$ and $t_1[Y] = t_2[Y]$,

there is always another tuple t_4 of s such that

(2) $t_4[X] = t_2[X]$ and $t_4[Y] = t_3[Y]$.

(See Fig. 13.2.)

Proof: Let us put $a = t_1[X], a' = t_1[Y], b = t_2[X]$ and $b' = t_3[Y]$. Let $R[X, Y]$ be the relation defined by $s[X, Y]$. By assumption, we know (according to definitions 13.1.4 and 13.1.8 and by denoting $R[X, Y]$ by R) that $X \rightleftharpoons Y$ iff R is a difunctional relation, which holds iff

(3) $a.R \cap b.R \neq \emptyset \implies a.R = b.R$

for all $a, b \in X$. Let s be any instance of $S(U)$ for which we assume that the condition (1) above is true.

First case: $(a' = b')$ and $(a = b)$; we can take $t_4 = t_1$ or $t_4 = t_2$.

Second case: $(a' = b')$ and $(a \neq b)$; we can take $t_4 = t_2$.

Third case: $(a' \neq b')$ and $(a = b)$; we can take $t_4 = t_3$.

Fourth case: $(a' \neq b')$ and $(a \neq b)$; as a' and b' are two images of a by $R[X, Y]$, b' must also be an image of b by $R[X, Y]$ (according to condition (3) above); therefore there is at least one tuple t_4 not equal to t_1, t_2, and t_3 such that: $t_4[X] = b$ and $t_4[Y] = b'$.

Conversely, let us assume that if condition (1) is true for any set (t_1, t_2, t_3) of tuples of s, then condition (2) is also true, which implies that condition (3) is true for the four cases above. Therefore, $R[X, Y]$ is difunctional, i.e. $X \rightleftharpoons Y$. □

Example 13.1.10 Consider $U = \{A, B, C\}$, and s the following instance of $S(U)$:

s:

A	B	C
2	3	5
2	4	5
3	3	5
3	4	8

We could easily prove that $A \rightleftharpoons B$ is true for s, because the binary relation $R[A, B]$ is difunctional; on the contrary $B \rightleftharpoons C$ is false.

Example 13.1.11 Consider the relation named FOLLOW with attributes NO-STUD, ACAD-LEVEL and MANDAT-COURSE, that indicates mandatory courses to be followed by a student with a given academic level. If we assume that the same mandatory course never corresponds to two different academic levels, we see that there is a difunctional dependency (amoung others) between (NO-STUD, ACAD-LEVEL) and (MANDAT-COURSE) as illustrated by the following instance of the relation FOLLOW:

NO-STUD	ACAD-LEVEL	MANDAT-COURSE
876543	N1	IFT-10551
876543	N1	IFT-10552
345678	N1	IFT-10551
345678	N1	IFT-10552
123456	N2	IFT-15662
123456	N2	IFT-15662
312786	N3	IFT-17555
312786	N3	IFT-17755

This difunctional dependency could be interpreted as follows: "If two students s1 and s2 with academic levels a1 and a2 follow the same course ck, then the sets C1 and C2 of courses followed respectively by s1 and s2 are the same ones".

Note that, though a difunctional dependency is different from a *multivalued dependency*, there is a relationship between these two kinds of dependencies. If there is a multivalued dependency between X and Y (denoted by $X \rightsquigarrow Y/Z$, where $Z = U-(X \cup Y)$), then there is a difunctional dependency between (X, Y) and (X, Z) (denoted by $(X, Y) \rightleftharpoons (X, Z)$). The proof of this result can be found in [le Thanh 1986]. Note also that a difunctional dependency (DFD) is a particular case of *Generalized Functional Dependencies* (GFD) defined and studied by Sadri and Ullman [Maier 1983; Ullman 1988]. More specifically, it is a particular case of *tuple-generating dependency*. Therefore, the DFD can also be seen as a GFD of the form "$(t_1, t_2, t_3)/t$", where t_i are the assumptions and t is the conclusion. For example, a tabular representation of the DFD $A \rightleftharpoons B$ would be as follows:

a1	b1	c1	d1
a2	b1	c2	d2
a1	b3	c3	d3
a2	b3	c4	d4

The first three rows of the above table are the three tuples t_1, t_2, t_3. The fourth row is the conclusion (i.e. the tuple t_4 to be generated).

13.2 Properties of difunctional dependencies

The properties of difunctional dependencies stated in this section are divided into those which are necessary for a relation (or dependency) to be difunctional, those which are sufficient, and those which are both necessary and sufficient. Proofs of the corresponding theorems (i.e. Theorem 13.2.1 to 13.2.9) can be found in [Jaoua 1987; Jaoua, Boudriga+ 1991] or in [Riguet 1948]. By using Theorem 13.1.9 we then derive a set of inference rules (Theorem 13.2.10 to 13.2.20) under the assumption of functional and difunctional dependencies.

Properties of difunctional relations

Some necessary conditions for a relation to be difunctional are given in the next three theorems.

Theorem 13.2.1 If R is difunctional, and if R_1 and R_2 are two maximal rectangles of R such that $R_1 \neq R_2$, then $domR_1 \cap domR_2 = \emptyset$ and $ranR_1 \cap ranR_2 = \emptyset$, which implies that $R_1 \cap R_2 = \emptyset$.

Theorem 13.2.2 If R is difunctional, then $R;R^{\smile}$ and $R^{\smile};R$ are transitive. In fact, they are equivalence relations on $domR$ and $ranR$ respectively.

Theorem 13.2.3 If R is difunctional, then $K(R) = R;R^{\smile}$.

The next three theorems give sufficient conditions for a relation to be difunctional.

Theorem 13.2.4 If R is deterministic, then R is difunctional.

Theorem 13.2.5 If R is symmetric and transitive, then R is difunctional.

Theorem 13.2.6 If R is a rectangle, then R is difunctional.

Finally, the next three theorems give conditions that are both necessary and sufficient for a relation to be difunctional.

Theorem 13.2.7 A relation is difunctional if and only if it is a union of rectangles whose projections are disjoint. In other words, if and only if, for some index set X, $R = \bigcup_{i=1}^{p}(A_i \times B_i)$, with $A_i \cap A_j = B_i \cap B_j = \emptyset, \forall i \neq j$, and where p is the number of rectangles.

Theorem 13.2.8 A relation R is an equivalence relation if and only if it is reflexive and difunctional.

Theorem 13.2.9 A relation R is difunctional if and only if there exist two functions f and g such that $R = f;g^{\smile}$.

We now turn to inference rules for difunctionality. In the following sequence of Theorems (13.2.10 to 13.2.20) the symbols T, W, X, Y and Z denote attribute subsets of U.

Theorem 13.2.10 (*Reflexivity*) If $Y \subseteq X$ then $X \rightleftharpoons Y$.
Proof: $Y \subseteq X$ implies $X \to Y$, according to the first Armstrong axiom [Armstrong 1974; Ullman 1982]. But $X \to Y$ can be represented by a deterministic relation, since it is a function. According to Theorem 13.2.4, any deterministic relation is difunctional. Therefore, if $X \to Y$, then $X \rightleftharpoons Y$. □

Theorem 13.2.11 (*Symmetry*) If $X \rightleftharpoons Y$ then $Y \rightleftharpoons X$.
Proof: $X \rightleftharpoons Y$ can be represented by the binary relation $R[X, Y]$. Similarly, $Y \rightleftharpoons X$ can be represented by the binary relation $R[Y, X]$. If we denote $R[Y, X]$ by R, then $R[Y, X] = R^{\smile}$. If R is difunctional, then $R;R^{\smile};R \subseteq R$, hence $R^{\smile};R;R^{\smile} \subseteq R^{\smile}$, and so $R^{\smile}$ is difunctional. □

A different proof for each of Theorems 13.2.10 and 13.2.11 can be found in [le Thanh 1986]. Note that the interesting property of symmetry does not apply to functional dependencies.

Theorem 13.2.12 (*Augmentation*) If $X \to Z$ (H1), $Y \to Z$ (H2) and $X \rightleftharpoons Y$ (H3), then $XZ \rightleftharpoons YZ$.

Proof: Let s be an instance of $S(U)$, and assume that there are three tuples t_1, t_2, and t_3 of s such that $t_1[XZ] = t_3[XZ]$ and $t_1[YZ] = t_2[YZ]$. Then $t_1[X] = t_3[X]$ and $t_1[Y] = t_2[Y]$. According to (H3), $\exists t_4 \in s$, such that $t_4[X] = t_2[X]$ and $t_4[Y] = t_3[Y]$. Then this fact, with (H1) and (H2), imply that $t_4[Z] = t_2[Z]$ and $t_4[Z] = t_3[Z]$. Hence $t_4[XZ] = t_2[XZ]$ and $t_4[YZ] = t_3[YZ]$. □

Theorem 13.2.13 (*Mixed transitivity*) If $X \leftrightarrow Y$ (H1), and $Y \rightleftharpoons Z$ (H2), then $X \rightleftharpoons Z$.

Proof: Let s be an instance of $S(U)$, and assume that there are three tuples t_1, t_2, and t_3 of s such that $t_1[X] = t_3[X]$ and $t_1[Z] = t_2[Z]$. Then this fact, and (H1), imply that $t_1[Y] = t_3[Y]$. From this and (H2) we can then deduce that $\exists t_4 \in s$ such that $t_4[Y] = t_2[Y]$ and $t_4[Z] = t_3[Z]$. Hence $t_4[X] = t_2[X]$ (according to (H1)) and $t_4[Z] = t_3[Z]$. □

(Similarly, we could easily prove that if $X \rightleftharpoons Y$ and $Y \leftrightarrow Z$, then $X \rightleftharpoons Z$.)

Theorem 13.2.14 (*Decomposition*) If $Y \to Z$ (H1) and $X \rightleftharpoons YZ$ (H2), then $X \rightleftharpoons Y$.

Proof: Let s be an instance of $S(U)$, and assume that there are three tuples t_1, t_2, and t_3 of s such that $t_1[X] = t_3[X]$ and $t_1[Y] = t_2[Y]$. According to (H1), $t_1[YZ] = t_2[YZ]$. According to (H2), we can deduce that $\exists t_4 \in s$ such that $t_4[X] = t_2[X]$ and $t_4[YZ] = t_3[YZ]$. Hence $t_4[X] = t_2[X]$ and $t_4[Y] = t_3[Y]$. □

(Similarly, we could easily prove that if $Z \to Y$ and $X \rightleftharpoons YZ$, then $X \rightleftharpoons Z$.)

Example 13.2.15 With $U = \{A, B, C\}$ consider the following instance of $S(U)$:

A	B	C
2	3	4
2	5	8
6	3	4
6	5	8

Here $B \to C$, $A \rightleftharpoons BC$ and $A \rightleftharpoons B$.

Theorem 13.2.16 (*Generalized union*) If $X \rightleftharpoons Y$ (H1) and $Y \to Z$ (H2), then $X \rightleftharpoons YZ$.

Proof: Let s be an instance of $S(U)$, and assume that there are three tuples t_1, t_2, and t_3 of s such that $t_1[X] = t_3[X]$ and $t_1[YZ] = t_2[YZ]$. Then we can deduce that $t_1[Y] = t_2[Y]$. According to (H1), $\exists t_4 \in s$ such that $t_4[X] = t_2[X]$ and $t_4[Y] = t_3[Y]$. Hence by (H2) we get $t_4[Z] = t_3[Z]$. It then follows that $t_4[X] = t_2[X]$ and $t_4[YZ] = t_3[YZ]$. □

(Similarly, we could easily prove that if $X \to Y$ and $X \rightleftharpoons Z$, then $X \rightleftharpoons YZ$.)

Example 13.2.17 With $U = \{A, B, C\}$ consider the following instance of $S(U)$:

A	B	C
2	3	6
2	1	8
5	3	6
5	1	8

Here $A \rightleftharpoons B$, $B \to C$ and $A \rightleftharpoons BC$.

Theorem 13.2.18 (*Pseudo-transitivity*) If $X \leftrightarrow Y$ (H1), and $WY \rightleftharpoons Z$ (H2), then $WX \rightleftharpoons Z$.
Proof: According to (H1), we have $X \to Y$ and $Y \to X$, hence $WX \to WY$ and $WY \to WX$ (by applying the augmentation rule). From these together with (H2), and by mixed transitivity (cf. Theorem 13.2.13) we deduce that $WX \rightleftharpoons Z$.□

Example 13.2.19 With $U = \{A, B, C, D\}$ consider the following instance of $S(U)$:

A	B	C	D
4	2	8	1
4	2	8	9
6	7	2	1
6	7	2	9

Here $B \leftrightarrow C$, $AC \rightleftharpoons D$, and $AB \rightleftharpoons D$.

Theorem 13.2.20 (*Pseudo-augmentation*) If $X \rightleftharpoons Y$ (H1), $X \to Z$ (H2) and $Y \to T$ (H3), then $XZ \rightleftharpoons YT$.
Proof: Let s be an instance of $S(U)$, and assume that there are three tuples t_1, t_2, and t_3 of s such that $t_1[XZ] = t_3[XZ]$ and $t_1[YT] = t_2[YT]$. Then $t_1[X] = t_3[X]$ and $t_1[Y] = t_2[Y]$. According to (H1), we can say that $\exists t_4 \in s$ such that $t_4[X] = t_2[X]$ and $t_4[Y] = t_3[Y]$. Hence, according to (H2) and (H3), we can deduce respectively that $t_4[Z] = t_2[Z]$ and $t_4[T] = t_3[T]$, and finally that $t_4[XZ] = t_2[XZ]$ and $t_4[YT] = t_3[YT]$. □

Example 13.2.21 With $U = \{A, B, C, D\}$ consider the following instance $S(U)$:

A	B	C	D
1	3	8	7
1	4	8	6
2	3	5	7
2	4	5	6

Here $A \to C$, $AC \rightleftharpoons D$, and $AB \rightleftharpoons D$. Also $A \rightleftharpoons B$, $A \to C$, $B \to D$ and $AC \rightleftharpoons BD$.

13.3 Binary decomposition of n-ary relations

In this section, we show how the properties of difunctional dependencies can be applied in practice to obtain a binary decomposition of an n-ary relational schema based on its difunctional dependencies. This decomposition is more suitable when there are no classical dependencies (functional or others) between the attributes of the n-ary relational schema.

The basic principle

Let $S(U)$ be a relational schema defined on the set U of attributes, and X and Y two subsets of U such that $X \cup Y = U$. Furthermore, let us assume that the difunctional dependency $X \rightleftharpoons Y$ is true on $S(U)$. If s is any instance of $S(U)$, then the associated binary relation $R[X, Y]$ defined by $s[X, Y]$ is difunctional. The basic principle of the binary decomposition consists in replacing $s[X, Y]$ by two relations R_1 and R_2 associated with $R[X, Y]$. (These are defined below.)

Indeed, according to Theorem 13.2.7, $R[X, Y]$ is the union of rectangles whose projections are disjoint. I.e:

$$R[X, Y] = (A_1 \times B_1) \cup \ldots \cup (A_i \times B_i) \cup \ldots \cup (A_n \times B_n),$$

with $A_i \cap A_j = B_i \cap B_j = \emptyset, \forall i \neq j$. Let C be a new attribute (which we call a *pseudo-attribute* in what follows) whose domain is the set $\mathbf{N}$ of natural numbers. Each of the subsets A_i and B_i (which we call a *class*) is identified by the integer value i of C. With each element (x, y) of $R[X, Y]$, we can associate the couple (i, i) where i stands for the rectangle number i of $R[X, Y]$ and such that $x \in A_i$ and $y \in B_i$. Then, the binary decomposition consists in associating to $R[X, Y]$ the two following relations:

- $R_1[X, C] = \{(x, i) | x \in A_i\}$, and
- $R_2[C, Y] = \{(i, y) | y \in B_i\}$.

From Theorem 13.2.9, we can deduce that $R[X, Y] = R_1[X, C];R_2[C, Y]$. Note that by analogy with the lossless join decomposition, this result shows that the binary decomposition is a "lossless relative product decomposition". In order to simplify notations, $R_1[X, C]$ and $R_2[C, Y]$ will respectively be denoted by s_1 and s_2 in what follows.

Example 13.3.1 Define $U = \{A, B, E, F\}$, $X = \{A, B\}$, $Y = \{E, F\}$ and let s be the instance illustrated by Fig. 13.3 defined on relational schema $S(U)$. We assume that the difunctional dependency $X \rightleftharpoons Y$ is true on $S(U)$. Let $R[X, Y]$ be the equivalent representation associated with s. We see that $X \cup Y = U$, and that $R[X, Y] = (A_1 \times B_1) \cup (A_2 \times B_2)$, where $A_1 = \{(1, 0), (2, 8)\}$, $A_2 = \{(3, 7), (4, 2)\}$, $B_1 = \{(6, 2), (6, 3), (6, 4)\}$ and $B_2 = \{(1, 5), (1, 9)\}$. In fact:
$R[X, Y] = \{((1, 0), (6, 2)), ((1, 0), (6, 3)), ((1, 0), (6, 4)), ((2, 8), (6, 2)), ((2, 8), (6, 3)), ((2, 8), (6, 4)), ((3, 7), (1, 5)), ((3, 7), (1, 9)), ((4, 2), (1, 5)), ((4, 2), (1, 9))\}$.
For example, with the element $\{((3, 7), (1, 9))\}$ of $R[X, Y]$ we can associate the couple $(2, 2)$, since $(3, 7) \in A_2$ and $(1, 9) \in B_2$. The relation $R[X, Y]$ associated with the initial instance s of $S(U)$ can be decomposed in a canonical manner by associating it with the two relations R_1 and R_2 respectively illustrated by Figs. 13.4(a) and 13.4(b). Here we see that:

$R_1 = \{((1, 0), 1), ((2, 8), 1), ((3, 7), 2), ((4, 2), 2)\}$,
$R_2 = \{(1, (6, 2)), (1, (6, 3)), (1, (6, 4)), (2, (1, 5)), (2, (1, 9))\}$, and
$R_1;R_2 = \{((1, 0), (6, 2)), ((1, 0), (6, 3)), ((1, 0), (6, 4)), ((2, 8), (6, 2)), ((2, 8), (6, 3)), ((2, 8), (6, 4)), ((3, 7), (1, 5)), ((3, 7), (1, 9)), ((4, 2), (1, 5)), ((4, 2), (1, 9))\}$.

The interest of the binary decomposition is threefold. First, it provides the decomposition of a relational schema which could not be decomposed using the traditional dependencies (functional dependencies, multivalued dependencies, join dependencies, etc.). Second, it ensures better data consistency after the update operations on the database according to the data classification in each of the instances s_1 and s_2 associated with R_1 and R_2. Third, it reduces data redundancy. In the following subsections we consider each of these three cases.

s:

A	B	E	F
1	0	6	2
1	0	6	3
1	0	6	4
2	8	6	2
2	8	6	3
2	8	6	4
3	7	1	5
3	7	1	9
4	2	1	5
4	2	1	9

Fig. 13.3 Instance s of $S(U)$

A	B	C
1	0	1
2	8	1
3	7	2
4	2	2

(a)

C	E	F
1	6	2
1	6	3
1	6	4
2	1	5

(b)

Fig. 13.4 Binary decomposition of R
(a) $R_1[X, C]$; (b) $R_2[C, Y]$

A	B
1	2
1	3
1	4
2	2
2	3
2	4
3	5
3	9
4	5
4	9

(a)

A	C
1	1
2	1
3	2
4	2

C	B
1	2
1	3
1	4
2	5
2	9

(b)

Fig. 13.5 (a) Instance s of $S(U)$ nondecomposable on the classical approach
(b) binary decomposition of $S(U)$

Relational schemas which cannot be decomposed

The traditional decomposition consists of splitting a given relational schema into several smaller schemas, with some desired properties from existing dependencies between attributes of this schema. To obtain a normalized structure, this decomposition must have two additional properties: preservation of the initial dependencies, and the lossless join property. Difunctional dependencies and the notion of binary decomposition provide a more suitable decomposition of a relational schema, when there are no classical dependencies between the attributes of this schema.

Example 13.3.2 Define $U = \{A, B\}$, with $X = \{A\}$ and $Y = \{B\}$ and let s be the instance illustrated by Fig. 13.5(a) defined on the relational schema $S(U)$, on which we assume that only the difunctional dependency $X \rightleftharpoons Y$ is true. Without this difunctional dependency, the relational schema $S(U)$ cannot be decomposed by the traditional approach. The use of this difunctional dependency allows us to decompose $S(U)$ in a canonical manner to obtain the equivalent representation illustrated by Fig. 13.5(b).

Data consistency and update errors

The translation of an instance s of a relational schema $S(U)$ by a difunctional binary relation $R[X, Y]$ (with $X \cup Y = U$) allows us to represent $R[X, Y]$ by two relations R_1 and R_2 whose respective instances s_1 and s_2 are organized into classes. This kind of representation ensures better data consistency by eliminating some update errors. To illustrate this advantage on operations such as delete, insert and update of tuples, we will observe (below) their respective effects with reference to example 13.3.1, disregarding the time factor. We study the effect of these operations executed on s_1 and s_2, as given by the canonical representations illustrated in Figs. 13.4(a) and 13.4(b).

For the *delete operation*, assume that we want to delete all the tuples for which the value of attribute A is equal to 1 and the value of attribute B is equal to 0. If this operation is executed on s (illustrated by Fig. 13.3), three tuples must be deleted (such that the difunctional dependency $X \rightleftharpoons Y$ remains true, in order to maintain data consistency). However, if this same delete operation is to be made on the equivalent canonical representation (Fig. 13.4), it is sufficient to delete only the tuple t of s_1 (Fig. 13.4(a)) such that $t[X] = (1, 0)$.

For the *insert* operation, assume that we want to insert the tuple $(9, 5, 6, 2)$ in s (Fig. 13.3). To maintain data consistency after this operation, it is important that we do not forget to insert also the tuples $(9, 5, 6, 3)$ and $(9, 5, 6, 4)$ in s. However, if this same operation is to be made on the equivalent canonical representation (Fig. 13.4), it is sufficient to insert only one new element. $\{(9, 5), 1)\}$, in the relation s_1 (Fig. 13.4(a)) to obtain the same result with no risk of update errors. Indeed, by construction of R_1 and R_2, the association of values 9 and 5 of attributes A and B with the values $(6, 2)$, $(6, 3)$ and $(6, 4)$ of attributes E and F (of R_2) is made automatically.

For the *update* operation, assume that we want to replace the tuple $(3, 7, 1, 9)$ by the tuple $(3, 7, 2, 8)$ in s (Fig. 13.3). Since we have to preserve the difunctional dependency $X \rightleftharpoons Y$ after this update operation, we should modify simultaneously also the tuples $(3, 7, 1, 9)$ and $(4, 2, 1, 9)$ in s (to be replaced by $(3, 7, 2, 8)$ and $(4, 2, 2, 8)$, respectively), which implies two operations to avoid any update error. However, if this same operation is to be made on the equivalent canonical representation (Fig. 13.4), it is sufficient merely to replace in s_2 (Fig. 13.4(b)) the element $\{2, (1, 9)\}$ by the element $\{2, (2, 8)\}$, in order to obtain the same result.

Redundancy reduction

Consider a relational schema $S(U)$, any instance s of S and the difunctional binary relation $R[X, Y]$ associated with s (with $X \cup Y = U$), which is the union of rectangles whose projections are disjoint. So:

$$R[X, Y] = (A_1 \cup B_1) \cup \ldots \cup (A_i \cup B_i) \cup \ldots \cup (A_n \cup B_n),$$

with $A_i \cap A_j = B_i \cap B_j = \emptyset, \forall i \neq j$. Now define the following notations:

- p = number of rectangles of $R[X, Y]$,
- $a_i = |A_i|$, i.e. the cardinality of Ai,
- $b_i = |B_i|$, i.e. the cardinality of Bi,
- t_x = sum of attribute sizes of each component of X,
- t_y = sum of attribute sizes of each component of Y,
- t_c = size of the pseudo-attribute C.

(Note: t_x, t_y, t_c could for example be explained in bytes.) Let T_s be the space occupied by s and T_r the space occupied by s_1 and s_2. These two quantities are defined as follows:

$$T_s = (t_x + t_y)\sum_{i=1}^{p} a_i b_i, \text{ and } \quad T_r = \sum_{i=1}^{p} t_x a_i + t_y b_i + t_c(a_i + b_i).$$

Hence,

$$T_s - T_r = \sum_{i=1}^{p}(t_x + t_y)a_i b_i - (t_x(a_i + t_y b_i + t_c(a_i + b_i)),$$

i.e. the total gain is the sum of gains of each rectangle. Now assume that $\forall i$, $a_i = b_i = h$ (i.e. all classes A_i and B_i have the same number of elements). The gain for one rectangle is:

$$G(h) = h^2(t_x + t_y) - h(t_x + t_y + 2t_c).$$

We can now make the following observations:

a) The space gain is effective when $h \geq 2$. Note that $G(h) = 0$ for $h_0 = 0$ (empty rectangle), or for $h_1 = \frac{t_x+t_y+2t_c}{t_x+t_y} = 1 + \frac{2t_c}{t_x+t_y}$, and $G(h) < 0$ when $0 \leq h \leq h_1$. Generally, however, $\frac{2t_c}{t_x+t_y} \leq 1$.

b) Asymptotic behaviour: when h increases sufficiently,

$$G \approx h^2(t_x + t_y),$$

(i.e G is $\mathcal{O}(h^2)$) and therefore greater rectangles make for greater economy and compensate for the loss of several small rectangles.

13.4 Conclusion

We have demonstrated the utility of binary decomposition based on difunctional dependencies in database management by showing that there is a straightforward way of decomposing in a rectangular manner an n-ary relation [Ounalli, Jaoua$^+$ 1994], and mapping any n-ary relation to an equivalent binary one [Jaoua, Ounalli$^+$ 1994]. The study of difunctional dependencies in a relational database has the advantage of reflecting the use of a rectangular relations as atomic components that may be used for information decomposition.

In real databases, the binary relation R that links two sets of objects is not always as uniform as difunctional relations, but we can describe R by an "economical" subset of non-necessarily disjoint rectangles that cover R. This observation has led to further research on rectangular decomposition heuristics that have been used for several computer science applications, e.g. database organization [Belkhiter, Bourhfir$^+$ 1994], cooperative answering to database's queries [Belkhiter, Desharnais$^+$ 1993], or learning and data compression. These studies have shown that it is useful to look for data clusters, using the rectangular clustering method, to discover data dependencies for any (relational or object) database model. We can also use it to derive software or information system architecture. This intuition has been confirmed in [Ounalli, Jaoua$^+$ 1994], where we use a rectangular decomposition of n-ary relations to extract database entity types. We believe that rectangular database clustering may be used further for discovering unknown classes and adequate sets of attributes that best describe objects of such classes.

Chapter 14

Logic, Language, and Information

Patrick Blackburn, Maarten de Rijke, Yde Venema[1]

The rapidly evolving interdisciplinary field of Logic, Language and Information (LLI) treats a variety of topics, ranging from knowledge representation to the syntax, semantics and pragmatics of natural language. Moreover, it does so from a variety of perspectives. However, one word more than any other gives the flavour of much contemporary work in LLI: *dynamics*. The purpose of this chapter is twofold. First, we give an impression of what LLI is and why dynamics plays such a fundamental role there. Second, we relate the study of dynamics to relation algebra. The essential point that will emerge is that many LLI approaches to dynamics can be naturally viewed as explorations of *fragments* of relation algebra via their *set-theoretic representations*.

We proceed as follows. In Sect. 14.1 we sketch the developments that lead to the current focus on dynamics in LLI. The idea of *logics of transitions* emerges naturally from this discussion, and provides the bridge to the world of relation algebra. In Sect. 14.2 we discuss the syntax and semantics of a number of transitional logics in detail, emphasising the variety of options this essentially simple idea offers. In Sect. 14.3 we turn to more general technical themes. Issues discussed include the key model theoretic notion of a *bisimulation*, various metatheoretic properties of these logics and the idea of *relativisation*, and recent work on *dynamic modes of inference*. We conclude with a discussion of newly emerging themes, and the limitations of the relational perspective.

14.1 Dynamics in logic, language and information

In broad terms, research in LLI aims to give abstract models of high-level information processing. It is reasonably simple to explain what is meant by "abstract": in principle it means any mathematical or computational model, though in practice it has tended to mean tools drawn from mathematical logic, theoretical computer science, or the logical and functional programming paradigms. Explaining what is meant by "high-level information processing" is less straightforward. There is a core intuition that many cognitive abilities such as language understanding, planning and spatial visualisation can be usefully thought of in terms of information

[1]The research of the third author has been made possible by a fellowship of the Royal Netherlands Academy of Arts and Sciences.

processing. Much research in LLI is concerned either with analysing such abilities (usually at a fairly high degree of abstraction) or with developing general models of information and information flow. Emphasis has tended to be placed on those aspects of high-level information processing that readily lend themselves to symbolic analysis, hence the ability of humans to work with beliefs and to cope with language have been the focus of attention. In summary: LLI borders such disciplines as theoretical computer science, linguistics and cognitive science, and its practitioners are theoreticians inspired by problems drawn from these fields.

Stated at this level of generality, it is probably unsurprising that "dynamics" has emerged as a key concept. After all, we talk of *forming* beliefs, *amending* them, *changing our mind*, and *learning something new*; the idea implicit in these folk psychological descriptions is of creating, updating or discarding some kind of belief structure. Nonetheless, a little more historical background is necessary to appreciate just why it is that contemporary LLI treatments of dynamics take the form they do — and in particular, why they lead to the study of relational fragments.

Current LLI and its notion of dynamics have been influenced by three developments: first, the growing realisation in linguistics, computational linguistics and artificial intelligence (AI) that process-based explanations at too low a level of abstraction are counterproductive, and the consequent call for more "logical" or "declarative" analyses; second, the highly influential insight (in semantics of natural language and belief revision) that logics equipped with more procedural interpretations can be useful analytic tools, and third, the work in theoretical computer science on logics for reasoning about program behaviour. This last influence has proved particularly important. It has given rise to the idea of general *logics of transitions* or *dynamic logics*, and provides the link with relation algebra. Let us consider these developments in a little more detail.

Early Chomskyan linguistics placed heavy emphasis on procedural explanation: the linguist's task was to explain how grammatical syntactic structures were built up via chains of structure manipulating transformations. By the early 1970s, the Chomskyan program had degenerated into uninhibited programming over trees that yielded an ever-decreasing return of linguistic insight. Chomsky redirected the field: the task of the *Principles and Parameters* approach which emerged was to discover the general principles that govern whether or not sentences are well-formed. The question of how the syntactic structure was actually built up now had secondary status.

Over roughly the same period, interest in "logical approaches" developed in both computational linguistics and AI. Pioneering work on large scale grammars made it clear that processing procedurally specified grammars was difficult. Attention turned to declarative grammar formalisms, in which well-formedness conditions on structure could be described with little or no procedural commitment; the availability of PROLOG, and the development of unification and constraint solving techniques, ensured that such formalisms also had straightforward procedural realisations. With their combination of declarative clarity and ease of implementation, such approaches swiftly became dominant in computational linguistics. Much the same development occurred in mainstream AI; again, higher level methods which abstracted away from procedural details seemed called for, and were developed.

In short, many developments in linguistics, computational linguistics and AI can be seen as a move towards logical (or declarative) analyses of problems that retained a (high-level) procedural content. Although the terminology was not used in these fields, it is natural to sum this up as a quest for *dynamic logics*.

Over roughly the same period, many researchers in natural language semantics and belief revision embarked on a similar quest. Although their ultimate goal — dynamic logic — was similar, these researchers approached it from the opposite direction: from logic to computation. That is instead of thinking of logic as an essentially static tool (the traditional view), they wanted to infuse it with computational content. Let us consider why.

If one is concerned to give the semantics of a single natural language sentence, then the static truth conditions provided by classical logic (perhaps extended with various modal operators) probably suffice. For example, "A man walks in the park" can be represented as $\exists x\,(Man(x) \wedge \textit{Walks-in-park}(x))$. The standard first-order semantics makes this formula true in a model $\mathcal{M}$ if there is an assignment g of values to variables such that

$$\mathcal{M}, g \models Man(x) \wedge \textit{Walks-in-park}(x).$$

That is, the meaning of this sentence is adequately modeled in terms of the existence of a satisfying assignment. As soon as one considers multi-sentence discourses, however, the limits of classical logic start to show. Consider the following discourse: "A man walks in the park. He whistles." One possible representation of this in classical logic is:

$$\exists x\,(Man(x) \wedge \textit{Walks-in-park}(x) \wedge \textit{Whistles}(x)).$$

But this "analysis" assumes too much. The two sentences are presented sequentially, and we understand them in real time; presumably we built up the semantic representation incrementally. However, the natural incremental representation is

$$\exists x\,(Man(x) \wedge \textit{Walks-in-park}(x)) \wedge \textit{Whistles}(x).$$

This representation is more honest — it reflects the sentential structure of the discourse — but it does *not* capture its content. In particular, because the final occurrence of x is free, the anaphoric link between "He" and "A man" is lost.

The (now standard) response in LLI is to add a procedural dimension to logical semantics. The meanings of sentences are no longer thought of in terms of static truth conditions. Rather, they are thought of as *context transformers*. A sentence is uttered in a certain context. Its utterance tranforms that context, thus altering the context in which subsequent utterances will be uttered. These successive transformations of context provide the mechanism by which discourse phenomena, such as anaphoric links, are captured. Intuitively, when we utter "A man walks in the park" in some context, we *introduce a new discourse referent* — let us call it x — and assert that it picks out a man that walks in the park. To put it another way, we instruct our listeners to make a new memory location available, and to associate with this location the information "is a man" and "walks". Given this enriched context, the subsequent utterance of "he walks" makes perfect sense: it can be viewed as an instruction to update the new location with additional information.

In the following section we will consider one way of formalising this idea, viz. dynamic predicate logic; here we'll simply remark that the key idea of adding a procedural dimension to logical semantics underlies other important approaches to natural language semantics, including discourse representation theory [Kamp, Reyle 1993] and file change semantics [Heim 1983]. The reason for the widespread acceptance of such systems of dynamic semantics is their intuitive appeal combined with their applicability to many important semantic phenomena, such as the interaction of tense and temporal reference. Dynamic semantics has even been successfully applied to problems usually relegated to "the dustbin of pragmatics". In particular, it has provided convincing accounts of presupposition; see [Beaver 1995] and [Eijck 1994] for further discussion.

Similar computational metaphors underlie LLI treatments of belief revision. For example, [Gärdenfors 1988] re-examines propositional logic using the idea that a proposition is function taking epistemic states to epistemic states, and Veltman's update logic [Veltman] gives an *eliminative* semantics to a uni-modal language (the effect of evaluating formulas is essentially to discard inconsistent information). This view models belief states as something rather like data-bases, and views reasoning about beliefs as the process of adding new information, querying the database, and retracting or replacing information. The *dynamic modal logic* of [van Benthem 1989] and [de Rijke 1992; de Rijke 1994b] is a powerful tool for exploring the database metaphor and we will examine it in Sect. 14.2.

We are now ready to make the abstraction that will lead us, via logics of transition, to relation algebra. The dynamic analyses of natural language discourse and belief change sketched above revolve around one central idea: viewing logical interpretation as a process of navigating through a network of states. Rather than thinking of the logical connectives as tools for describing a fixed situation, we view them as instructions which take us from one state to another. We are naturally lead to the idea of *transition systems* (these are the entities we navigate) and various *logics of transition* (corresponding to the various means we choose for moving between states). This view raises many questions that need answering, but let us defer them to the following section and push on to relation algebra.

A transition system is simply a set equipped with a collection of relations (i.e., a relational structure). When we fix a choice of connectives we are essentially selecting a number of possibilities for manipulating (or combining) these relations. That is, we are essentially choosing a subset of the combinatoric possibilities offered by relation algebra. Fixing some transition logic corresponds to fixing a reduct of relation algebra.

The way we have approached relation algebra via transition systems should make it clear that we are interested in the set-theoretic representations as much, if not more, than the abstract algebras: it is the transition systems that give us the intuitive "fit" with applications. But of course, once the link with relation algebras is made, transfer of results between the concrete and algebraic domains becomes possible, and fruitful. We will consider such issues later, but first, let us take a closer look at some logics of transition.

14.2 Logics of transition

By a *transition logic* we will mean a formalism that is designed to describe transitions. In this section we give an impressionistic sketch of some transition logics that have been considered in the LLI literature. Our starting point will be the immediate ancestor of today's transition logics, propositional dynamic logic (PDL). We then consider various dimensions of variation. In particular we consider the *choice of connectives*, which leads to a discussion of dynamic modal logic and Lambek calculus; the issue of *states versus transitions*, which leads to a discussion of arrow logic; and the *nature of states and transitions* which leads to a discussion of dynamic predicate logic. As a final example, we discuss evolving algebras.

Propositional dynamic logic

The language of propositional dynamic logic (PDL) (see [Pratt 1976; Harel 1984]) contains modal operators $\langle\alpha\rangle$ and $[\alpha]$ for every regular expression α over some set of atomic programs. The intended reading of a formula $\langle\alpha\rangle\varphi$ is that there is an execution of the program α that terminates in a state satisfying φ, and the intended reading of $[\alpha]\varphi$ is that every execution of the program α terminates in a state satisfying φ. PDL expressions are built up from proposition letters p_0, p_1, ..., atomic programs a_0, a_1, ... using the following production rules:

$$\begin{aligned} \varphi &::= p \mid \mathsf{false} \mid \mathsf{true} \mid \neg\varphi \mid \varphi \wedge \varphi \mid \langle\alpha\rangle\varphi \mid [\alpha]\varphi \\ \alpha &::= a \mid \alpha \cup \alpha \mid \alpha;\alpha \mid \alpha^* \mid \varphi? \end{aligned}$$

The intended reading of $\alpha \cup \beta$ is 'do either α or β non-deterministically'; the program $\alpha;\beta$ stands for "do α, then do β"; α^* stands for 'iterate α a finite number of times'; and $\varphi?$ is the program that tests for φ and succeeds if φ is true, and fails otherwise. PDL allows one to express various "safety" and "liveness" properties of programs:

- $\varphi \to [\alpha]\psi$: the precondition φ implies the postcondition ψ after every terminating execution of α
- $\langle\alpha\rangle\mathsf{true}$: there exists a terminating execution of α.

Being a modal logic, PDL is interpreted on Kripke structures. Let Prog be the collection of all programs in the language. Then, a Kripke structure for PDL has the form $(W, R_\alpha)_{\alpha\in\mathsf{Prog}}$, where each R_α is a binary relation on W, and

$$\mathcal{M}, w \models \langle\alpha\rangle\varphi \text{ iff for some } v,\ wR_\alpha v \text{ and } \mathcal{M}, v \models \varphi.$$

To reflect the intended readings of PDL's program constructions, we require that our models satisfy the requirements $R_{\alpha\cup\beta} = R_\alpha \cup R_\beta$, $R_{\alpha;\beta} = R_\alpha;R_\beta$, $R_{\alpha^*} = (R_\alpha)^* = \bigcup_n (R_\alpha)^n$, and $R_{\varphi?} = \{(w,v) \mid w = v \text{ and } \mathcal{M}, w \models \varphi\}$. To get a complete deductive system for PDL, start with the **K** axioms (see Sect. 1.6) for every modality $\langle\alpha\rangle$, and add the following axioms, most of which are decompositions of the program constructions in terms of booleans and simpler programs:

$$\langle\alpha \cup \beta\rangle\varphi \leftrightarrow \langle\alpha\rangle\varphi \vee \langle\beta\rangle\varphi$$
$$\langle\alpha\,;\,\beta\rangle\varphi \leftrightarrow \langle\alpha\rangle\langle\beta\rangle\varphi$$

$$\langle\varphi?\rangle\psi \leftrightarrow \varphi \wedge \psi$$
$$\langle\alpha^*\rangle\varphi \leftrightarrow \varphi \vee \langle\alpha\rangle\langle\alpha^*\rangle\varphi$$
$$\varphi \wedge \langle\alpha^*\rangle\neg\varphi \rightarrow \langle\alpha^*\rangle(\varphi \wedge \langle\alpha\rangle\neg\varphi).$$

The final two axioms are called the Segerberg axioms; they reflect the more complex infinitary behaviour of iteration. We refer to [Goldblatt 1987] for an accessible completeness proof for PDL.

In addition to finite axiomatizability, PDL also enjoys a second important advantage over full relation algebra, namely decidability. To be precise, the satisfiability problem for PDL is EXPTIME-complete (see [Fischer, Ladner 1979; Pratt 1979]).

The modal algebras of PDL are two-sorted algebras in the style of the Peirce algebras mentioned in Sect. 1.3. These algebras — called *dynamic algebras* — differ from Peirce algebras in that their relational component is based on a so-called Kleene algebra rather than a relation algebra. A *Kleene algebra* is a structure $(K, \sqcup, \mathbb{L}, ;, ^*, \mathbb{I})$, where * is the reflexive, transitive closure operator of Sect. 1.2. We refer the reader to [Kozen 1981] for the axioms governing the interaction between the two sorts in a dynamic algebra.

Let us briefly list the ingredients that we have at hand now, as most of the transition logics found in the literature may be perceived as variations on PDL. The most important dimensions along which variations have been considered are

- *Connectives*. PDL has the usual boolean operations to combine formulas, the regular operators to build programs, the test operation taking formulas to programs, and the modalities that take a program and a formula to return a formula.
- *Site of evaluation*. The models for PDL have both states and transitions, but PDL formulas are evaluated only at states.
- *The nature of states and transitions*. In PDL no assumptions are made about the nature of the states (although in practical applications they are usually memory structures of some kind), and the transitions are taken to be ordered pairs.

The above dimensions have been varied widely in the literature. The two main (and often conflicting) motivations for these variations are the need for descriptively adequate systems that are able to express all the key features of the phenomena being studied, and the need for computationally well-behaved calculi that have tractable, or at least decidable satisfiability problems. We refer the reader to [van Benthem, Muskens$^+$ to appear] for an extensive overview; here we will only give some representative examples.

Choice of connectives

Dynamic modal logic (DML) differs from PDL in that its relational component allows all the usual operations from relation algebra in addition to converse operation. Moreover, for every formula φ DML has two special relations: transitions along an abstract information order to states where φ holds ($exp(\varphi)$), and transitions backward along the information order to states where φ fails ($con(\varphi)$). On the formula side it has three constructs taking relations to propositions: $dom(\alpha)$, $ran(\alpha)$ and $fix(\alpha)$ are true in a state if it is in the domain, range or set of fixed points of α, respectively. All in all, we have

$$\begin{array}{rcl} \varphi & ::= & p \mid \mathsf{false} \mid \mathsf{true} \mid \neg\varphi \mid \varphi \wedge \varphi \mid dom(\alpha) \mid ran(\alpha) \mid fix(\alpha) \\ \alpha & ::= & con(\varphi) \mid exp(\varphi) \mid \overline{\alpha} \mid \alpha \cup \alpha \mid \alpha;\alpha \mid \alpha^{\smile} \mid \varphi? \end{array}$$

DML is interpreted on *information structures* $(W, \sqsubseteq)$ where W is a non-empty set, and $\sqsubseteq$ is a pre-order on W, called the *information order*; intuitively, $x \sqsubseteq y$ if y is at least as informative as x.

DML expressions are interpreted on information structures by means of a valuation V (to take care of the formulas) and an interpretation $[\![\cdot]\!]$ (to take care of the relational expressions). Some of the clauses in the truth definition are

$$\begin{array}{rcl} \mathcal{M}, w \models dom(\alpha) & \text{iff} & \text{there exists } v \text{ with } (w,v) \in [\![\alpha]\!] \\ \mathcal{M}, w \models ran(\alpha) & \text{iff} & \text{there exists } v \text{ with } (v,w) \in [\![\alpha]\!] \\ \mathcal{M}, w \models fix(\alpha) & \text{iff} & (w,w) \in [\![\alpha]\!] \\ [\![exp(\varphi)]\!] & = & \{(x,y) \mid x \sqsubseteq y \wedge y \models \varphi\} \\ [\![con(\varphi)]\!] & = & \{(x,y) \mid x \sqsupseteq y \wedge y \not\models \varphi\} \\ [\![\varphi?]\!] & = & \{(x,y) \mid x = y \wedge y \models \varphi\}. \end{array}$$

So, the original PDL diamonds can be recovered as $\langle\alpha\rangle\varphi = dom(\alpha;(\varphi?))$, and the *dom* operator can be expressed in PDL as $dom(\alpha) = \langle\alpha\rangle\mathsf{true}$.

DML was designed as a general framework for reasoning about information change. Technical results covering expressive power, undecidability and a complete axiomatisation for DML may be found in [de Rijke 1992]. Uses of DML as a framework for dynamic phenomena may be found in [van Benthem 1989; Jaspars, Krahmer 1995; de Rijke 1992]; examples include theory change, update semantics, knowledge representation, and dynamic semantics for natural language.

Site of evaluation

Transitions are commonly depicted as arrows. In a mathematical setting such arrows might denote vectors, functions or morphisms; for a computer scientist they may represent steps in a computation, and to a linguist they may denote the dynamic meaning of a chunk of text. *Arrow logic* is the basic modal logic of arrows. That is, in arrow logic arrows are the sites of evaluation. And so propositions denote sets of arrows, rather than preconditions or postconditions of transitions as in traditional modal logics such as PDL. This doesn't imply that arrows are primitive entities. On the contrary, in the semantics of arrow logic an important role is played by two-dimensional models in which arrows are pairs (a, b) of which a is the start of the arrow, and b is its end.

Following the familiar operations from relation algebra, natural ways of endowing collections of arrows with structure suggest itself. One can think of a ternary relation of *composition* C, where $Cabc$ denotes the fact that the arrow a can be decomposed into two arrows b and c. Or one can designate a subset as the set I of *identity arrows*. The addition of a binary relation R of reverse is slightly more debatable: aRb if the arrow b is a reverse of the arrow a. These ingredients make up an *arrow frame* (W, C, I, R). The *arrow language* is given by the following production rule

$$\varphi ::= p \mid \neg\varphi \mid \varphi \wedge \varphi \mid \varphi \circ \varphi \mid \otimes\varphi \mid \delta.$$

Here, $\circ$ is interpreted as composition, $\otimes$ as converse, and δ as identity (this is the standard notation of the arrow logic community). The language is interpreted on *arrow models* $\mathcal{M} = (\mathcal{F}, V)$ where $\mathcal{F}$ is an arrow frame, and V is a valuation. The interesting clauses in the truth definition are

$$\begin{aligned}
\mathcal{M}, a \models \varphi \circ \psi \;&\text{ iff }\; \text{there are } b, c \text{ with } Cabc \text{ and } \mathcal{M}, b \models \varphi \text{ and } \mathcal{M}, c \models \psi \\
\mathcal{M}, a \models \otimes\varphi \;&\text{ iff }\; \text{there is } b \text{ with } Rab \text{ and } \mathcal{M}, b \models \varphi \\
\mathcal{M}, a \models \delta \;&\text{ iff }\; Ia.
\end{aligned}$$

Two-dimensional frames or *relativised squares* are an important and much-studied class of arrow frames. They are built up using a base set U such that the domain W is a subset of $U \times U$, C satisfies $Cabc$ if $a_0 = b_0$, $a_1 = c_1$, and $b_1 = c_0$; R satisfies aRb iff $a_0 = b_1$ and $a_1 = b_0$; and Ia holds iff $a_0 = a_1$. If W is a reflexive and symmetric binary relation, then the arrow frame is called *locally square*; if W is the full Cartesian square over U, then the frame is called a *square*.

For axiomatic aspects of arrow logic this has the following important consequences. An arrow formula φ is valid on all squares iff $\tau\varphi = 1$ is valid on all representable relation algebras, and φ is valid on all relativised squares iff $\tau\varphi = 1$ is valid on all relativised representable relation algebras. These equivalences form the basis for transferring techniques and results from algebraic logic to arrow logic, and vice versa. We refer the reader to [Venema 1995] for details and further results.

Axiomatic questions for arrow logic have been studied extensively. For the class of all arrow frames the *minimal normal arrow logic* (which is the straightforward generalization to the similarity type $\{\circ, \otimes, \delta\}$ of the minimal normal modal logic **K** in the similarity type with just one diamond $\Diamond$, as defined in Sect. 1.6) is complete. For the (relativised) square frames the situation is far more complicated. For instance, the squares are not finitely axiomatizable (this follows from the well-known non-finite axiomatizability result for representable relation algebras).

To conclude our discussion of arrow logic we briefly mention some systems closely related to it. First, the *Lambek calculus* is a substructural Gentzen-style derivation system involving a fragment of arrow logic; its connectives are $/$, $\backslash$ and $\circ$, and in relativised square models the slashes are interpreted as the *residuals* of $\circ$ (see Sect. 1.3). When interpreted on such models, the Lambek Calculus receives a dynamic, procedural interpretation, based on the paradigm that parsing a sentence is performing a logical deduction. A second example concerns *two-dimensional modal logic*; this is an example of a modal system whose intended semantics is based on domains with objects that are tuples over some base set (see [Marx, Venema] for the general picture). Among others, two-dimensional logics arise in formal analyses of temporal discourse when both a *point of reference* and a *point of utterance* needs to be accounted for. Arrow logic is an example of a two-dimensional logic, as part of its intended semantics is formulated in terms of (relativised) squares.

Finally, there are hybrid systems, languages with two sorts of formulas — one interpreted on states, another interpreted on transitions —, and with a rich set of operators relating the two sorts. [van Benthem 1994] presents an abstract approach, [Marx] has results on concrete interpretations of sorted transition systems, and [de Rijke 1994a] presents a completeness result for the full square case of Peirce algebras.

States and transitions

To see an example of a calculus in which a very natural choice is made as to the units that carry semantic information, let us return to the example text in Sect. 14.1: "A man walks in the park. He whistles." To formalise the idea explained in Sect. 14.1, that meanings of sentences in such a discourse are to be thought of as context transformers rather than in terms of static truth conditions, [Groenendijk, Stokhof 1991] introduces a system of *dynamic predicate logic* (DPL). Its syntax is the same as that of first-order logic, and so are its models. What is novel is that it is interpreted with respect to *ordered pairs* of assignments of values to variables; thus, these become the basic units of semantic information. The first assignment is viewed as the "input". Evaluating a formula may have the effect of altering this assignment. This new assignment is returned as the second component, the "output".

To make this more concrete we give some formal details. First, a model $\mathcal{M}$ is a pair (D, F), where D is a non-empty set, and F is an interpretation function mapping constants and predicates to elements and subsets (of tuples) of the model in the usual way. An assignment is a function assigning individuals to variables; we write $g \approx_x h$ if g and h agree on all variables except possibly x. Then, given a model $\mathcal{M}$ and an interpretation function F, some of the key clauses in the definition of the semantics for DPL are

$$\begin{aligned}
[\![Rt_1 \dots t_n]\!] &= \{(g,h) \mid h = g \text{ and } ([\![t_1]\!] \dots [\![t_n]\!]) \in F(R)\} \\
[\![\varphi \wedge \psi]\!] &= \{(g,h) \mid \exists k\, ((g,k) \in [\![\varphi]\!] \text{ and } (k,h) \in [\![\psi]\!])\} \\
[\![\exists x\, \varphi]\!] &= \{(g,h) \mid \exists k\, (k \approx_x g \text{ and } (k,h) \in [\![\varphi]\!])\}.
\end{aligned}$$

So, an atomic statement admits those assignments which satisfy it, and blocks those that don't; conjunction is simply interpreted as composition, and when an existential quantifier is evaluated, a satisfying assignment to the bound variable is recorded.

In the little discourse above, the effect of evaluating

$$\exists x\, (\mathit{Man}(x) \wedge \mathit{Walks\text{-}in\text{-}park}(x))$$

with respect to an input assignment g would be to produce an output assignment f just like g, save that $f(x)$ is a value satisfying $\mathit{Man}(x)$ and $\mathit{Walks\text{-}in\text{-}park}(x)$. This output assignment f is then used as the *input* to the second part of the discourse, $\mathit{Whistles}(x)$. As f fixes a value for x that satisfies $\mathit{Man}(x)$ and $\mathit{Walks\text{-}in\text{-}park}(x)$, the fact that this occurrence of x is free is unproblematic; subsequent interpretation can take place straightforwardly, and the anaphoric link is accounted for.

Interpreting formulas at pairs of assignments instead of single assignments brings with it various choices for the notions of truth and consequence; [van Benthem 1989] explores a large number of them, and some are discussed in the following section. In addition, the relational perspective can be used to explore the behaviour of DPL's propositional connectives; [Blackburn, Venema 1993] examines the behaviour of dynamic implication in both the presence and absence of the boolean operations. Finally, we should mention that there are close similarities between quantified versions of the logic PDL discussed above and DPL; [Groenendijk, Stokhof 1991] studies various embeddings of the latter into the former.

Evolving algebras

To conclude our impressionistic tour of transition logics, we briefly discuss *evolving algebras*. The subject of evolving algebras was introduced by Yuri Gurevich around 1985 as a means of specifying operational aspects of computation at a level appropriate for reasoning about the system in question. Evolving algebras can be viewed as vast generalization of PDL (see [Gurevich 1991]). It uses many-sorted first-order structures as the *states* of its structures, and in contrast to standard practice, the constant and function symbols are dynamic: they might change according to a set of *transition rules*. Although we can't give any formal details here, the evolving algebra framework seems to provide some general tools for useful specification, and the recent literature indicates that these tools might be fruitfully applied in LLI as well (see for example [Moss, Johnson 1995]).

14.3 General themes

The aim of this section is to acquaint the reader with some examples of questions that the dynamic perspective on semantics brings to the fore. First, we discuss a connection between relation algebra and the important notion of a bisimulation between transition systems; second, we show a few ways to overcome the negative properties of the relation-algebraic approach; and finally, we discuss the notion of semantic consequence in the dynamic way of thinking.

Bisimulations

The simplest models for dynamics are the (labeled) transition systems that we encountered in the previous sections; let us define here, for a given set A of atomic actions or programs, a *transition system* for A as a structure $\mathcal{X} = (X, R_a)_{a \in A}$, where each R_a is a binary relation on X. Such a transition system is supposed to model a process; a relation R_a holding of two states s and t signifies the possibility that an action a is performed in s and leads to t. In many applications, for instance in the semantics of process algebra, cf. [Baeten, Weijland 1990], a transition system has a designated state called its *root*, which is supposed to represent the initial state of the process.

An important question is: when do two transition systems represent the same process? In other words, we are looking for a natural notion of equivalence between transition systems. In the literature on process theory, one sees many different options. For instance, one might be interested only in the various *traces* of a rooted transition system, i.e., sequences consisting of atomic actions corresponding to the paths one can take through the transition system (starting from its root). In this approach, two transition systems are equivalent iff they generate the same set of traces. We will concentrate here on a different notion of equivalence between processes, viz. that of a bisimulation. When compared to trace equivalence, the crucial point in the definition of a bisimulation is that two states are bisimilar only if one has precisely the same choice of actions enabled in each state.

Formally, a *bisimulation* between two transition systems $\mathcal{X} = (X, R_a)_{a\in A}$ and $\mathcal{X}' = (X', R'_a)_{a\in A}$ is a non-empty relation $Z \subseteq X \times X'$ that satisfies the following *back* and *forth* conditions (see also Sect. 3.4):

(*forth*) xR_ay and xZx' imply the existence of a y' such that $x'R'_ay'$ and yZy',
(*back*) $x'R'_ay'$ and xZx' imply the existence of a y such that xR_ay and yZy'.
In a picture, this means the following:

Algebraically, these conditions amount to '$R_a^{\smile};Z \subseteq Z;R_a'^{\smile}$' and '$R'_a;Z^{\smile} \subseteq Z;R_a$.'

There is another reason why bisimulations have received attention and that is their intimate relation with modal logics. This was put into a slogan in de Rijke's dissertation [de Rijke 1993]: bisimulations are to modal logic what partial isomorphisms are to first-order logic. For instance, it is quite easy to prove that bisimilar states satisfy the same poly-modal formulas (in a language with a modality $\langle a\rangle$ for each action a). A deeper result, due to [van Benthem 1976], characterises the "modal fragment" of first-order logic as the set of those formulas that are invariant under bisimulations.

Then, relational *algebra* may come into the picture in a number of ways; we have to confine ourselves to the following example. Suppose that Z is a bisimulation with respect to the transition systems (X, R, S) and (X', R', S'). It is easy to show that Z is also a bisimulation for the union of the relations, i.e., between the systems $(X, R\cup S)$, $(X', R'\cup S')$, or for the (reflexive) transitive closure of one of them, e.g., between (X, R^+) and (X', R'^+). On the other hand, one can give an easy counterexample showing that Z does not always have to be a bisimulation with respect to the intersections $R\cap S$ and $R'\cap S'$. In other words, the following definition is meaningful.

A relational operation $O(R_1,\ldots,R_n)$ is *safe for bisimulation* if a relation Z is a bisimulation between the structures $(X, O(R_1,\ldots,R_n))$ and $(X', O(R'_1,\ldots,R'_n))$ whenever Z is a bisimulation between $(X, R_1,\ldots,R_n)$ and $(X', R'_1,\ldots,R'_n)$.

Apart from the aforementioned union and (reflexive) transitive closure, examples of safe operations for bisimulation include the diagonal relation (i.e., seen as a constant operation), relation composition and dynamic negation '$\sim$', given by

$$\sim R = \{(x,y) :\ \text{not } \exists z\,(xRz)\}$$

In fact, among the relational operations that can be defined in first-order logic, these operations generate the clone of safe operations for bisimulations, as the following theorem shows (for a proof we refer to [van Benthem 1993b]).

Theorem 14.3.1 A first-order definable relational operation $O(R_1, \ldots, R_n)$ is safe for bisimulation if and only if it can be defined from the atomic relations R_i and the diagonal relation, using just the three operations ;, $\cup$ and $\sim$.

The notion of a bisimulation was introduced by van Benthem in [van Benthem 1976] (under the name of a p-relation); in the literature on process theory, bisimulations go back to [Park 1981]. Further references can be found in [Ponse, de Rijke+ 1995].

Taming logics

In earlier chapters we saw that the standard Tarskian framework of relation algebras, although very elegant from the mathematical perspective, has two properties that make it less attractive from the applicational point of view: compared to for instance boolean algebras, the equational theory of the class RRA of representable relation algebras does not have a "clean" equational axiomatization, and it is highly undecidable as well. A number of results recently obtained in the field of LLI, can be grouped together, under the common denominator that the authors all try to fine-tune or modify the standard approach to get around these negative results. The aim is, to use a slogan of [Mikulás 1995], to *tame* transition logics; here, we will mention a few ideas and results that have arisen in the (recent) literature[2]. The various modifications that can tame a logic's undecidability can be divided into the following three areas: studying *reducts* of the original language, interpreting the language in so-called *non-square* models — this is the modal-logic counterpart of *relativizing* relation algebras, or *restricting* the admissible valuations on the full square models.

To start with the second approach, recall from Chapt. 2 that the class RRA is generated by the full relation algebras, and that the full relation algebra on a set U is based on the power set of the *full* cartesian square U^2 over U. In other words, the assumption is that all elements of U^2 are available as transitions, or, possible worlds (in the arrow logic perspective). In the non-square or relativised approach, this assumption is dropped: any subset W of U^2 can serve as the top set of the algebra. Formally, the *W-relativised full relation algebra* on U is defined as the structure

$$Re^W(U) = (\mathcal{P}(W), \cup^W, {}^{-W}, \emptyset, ;^W, {}^{\smile W}, I_U^W),$$

i.e., all relational algebraic operations are *relativised to* W. For instance, the operation $;^W$ is given by

$$R;^W S = (R;S) \cap W.$$

[2]Our presentation is not justified from the historical perspective. For instance, the idea of a relational interpretation of Lambek's Calculus is due to [van Benthem 1989], 30 years after the introduction of the system by Lambek.

Obviously, the equational theory of such non-square algebras is weaker than that of the square ones; for instance, the operation $(\cdot)^{\smile W}$ is not idempotent unless W is symmetric.

Now an interesting landscape of algebras arises if we start from various classes of algebras $Re^W(U)$ for which W meets some constraints. Formally, let R stand for reflexive, S for symmetric and T for transitive; let H be a subset of {R, S, T} and W a binary relation. We use " W is an H-relation" to abbreviate that W has the properties mentioned in H. We define an algebra to be in $\mathsf{RRA}_{\mathrm{H}}$ if it can be embedded in an algebra of the form $Re^W(U)$, with W an H-relation. Then, the effect of taming RRA can be summarised by the following theorem:

Theorem 14.3.2 For every $H \subseteq \{\mathrm{R}, \mathrm{S}, \mathrm{T}\}$, the class $\mathsf{RRA}_{\mathrm{H}}$ is a variety. This variety is finitely axiomatizable and decidable if and only if $\mathrm{T} \notin \mathrm{H}$.

For lack of space, we cannot give proper references to all results covered by this theorem; for a more extensive overview and references, the reader may consult [Marx, Venema]. The proof that $\mathsf{RRA}_{\mathrm{RST}}$ (which happens to be the same variety as RRA) cannot be finitely axiomatised is due to Monk, the other negative results are due to Andréka, Németi and Sain. Positive results concerning finite axiomatizability are due to Maddux (namely, that $\mathsf{RRA}_{\mathrm{RS}}$ is identical to the equationally defined variety WA of so-called weakly associative relation algebras), Kramer ($\mathsf{RRA}_{\emptyset}$) and Marx (the remaining varieties). Concerning computational properties, the undecidability results are due to Tarski (RRA), and Andréka and Németi ($\mathsf{RRA}_{\mathrm{H}}$ with $\mathrm{T} \in \mathrm{H}$), the positive ones to Németi ($\mathsf{RRA}_{\mathrm{RS}}$) and Marx (the remaining ones).

The upshot of Theorem 14.3.2 is that *transitivity* of the top element W is the malefactor. It may therefore come as a surprise that by restricting the language of arrow logic (or algebraically, considering reducts of the relation algebras), there is the following positive result. Theorem 2.1 of [Andréka, Mikulás 1994] states that the original Lambek Calculus (mentioned in the previous section) is sound and complete with respect to a relativised relational interpretation with a transitive top set. This result is further strengthened to obtain an answer to an old question posed in [Schein 1970], viz. to give an axiomatic characterisation of the class of algebras isomorphic to sets of binary relations with the operations of multiplication and two residuations.

A different "taming strategy", originating with two-dimensional temporal logic, is the following. Let $<$ be a designated ordering relation on the base set U, i.e., in the algebraic language there is a special constant referring to this relation, just like the $\perp\!\!\!\perp$ that refers to the empty relation. Now consider the subalgebra of $Re(U)$ that is generated by $<$ and arbitrarily many left-ideal elements (essentially, unary relations in disguise). It follows from results in [Venema 1994], that the set of equations valid in such algebras is decidable, and finitely axiomatizable if we consider only algebras in which $<$ is a well-ordering or the ordering of the natural numbers.

Finally, if one is only interested in axiomatizations of the equational theory of the full relation algebras, the only solution to Monk's negative result on finite axiomatizability is to be more liberal concerning the format of a derivation system.

Results in [Marx, Venema] and [Mikulás 1995] show that RRA can be finitely axiomatised if one allows *unorthodox derivation rules*.

Dynamic consequence

As pointed out in Sect. 14.2, whenever we turn the meaning of a formula into a dynamic notion (in this contribution represented as a set of transitions, hence, as a binary relation), the question arises as to what the proper notion of semantic *consequence* is. In other words, what does it mean that a formula is a consequence of some (finite) bunch of other formulas, in this new semantical perspective? Or, more formally, if φ and $\varphi_1, \ldots, \varphi_n$ are formulas, what is an appropriate definition of the dynamic consequence relation $\models_d$ in (14.1) below?

$$\varphi_1, \ldots, \varphi_n \models_d \varphi. \tag{14.1}$$

To see the point of this question, the reader could try to look at the φ_i's as consecutive updates; it will then become clear that the *order* of the premisses $\varphi_1, \ldots, \varphi_n$ may influence the outcome of the semantic consequence. In other words, the "static" interpretation

$$[\![\varphi_1]\!] \cap \cdots \cap [\![\varphi_n]\!] \subseteq [\![\varphi]\!] \tag{14.2}$$

does not seem the most appropriate choice for (14.1). We briefly discuss two dynamic alternatives to (14.2). The relevance of relational algebra lies in the fact that these interpretations can both be expressed in quite a simple fragment of the language of Tarski's relation algebras, and, hence, that meta-theoretic notions such as dynamic consequence can be studied at the object-level using relational methods.

In the *dynamic* style of inference we interpret (14.1) by saying that in all models, each transition for the sequential composition of the premises must be admissible for the conclusion:

$$[\![\varphi_1]\!]; \cdots ;[\![\varphi_n]\!] \subseteq [\![\varphi]\!]. \tag{14.3}$$

In a formal framework for update semantics, the natural interpretation for (14.1) seems to be the following *update* or *mixed* style of inference, "first process all premises consecutively, then test if the conclusion is satisfied by the remaining state":

$$ran([\![\varphi_1]\!]; \cdots ;[\![\varphi_n]\!]) \subseteq fix[\![\varphi]\!], \tag{14.4}$$

where $x \in fix R$ iff $(x, x) \in R$ as in Sect. 14.2.

Obviously, these new styles of inference do not satisfy all standard structural properties of static inference, such as:

(*Monotonicity*) $X, Z, Y \models \varphi$ if $X, Y \models \varphi$
(*Reflexivity*) $\varphi \models \varphi$
(*Cut*) $Y, X, Z \models \varphi$ if $X \models \psi$ and $Y, \psi, Z \models \varphi$.

(Here X, Y and Z stand for arbitrary sequences of formulas.) For instance, the dynamic inference style (14.3) does not satisfy (Monotonicity), while the mixed style (14.4) satisfies none of the three mentioned properties. However, there are modified versions of these rules which the mixed style does satisfy.

Finally, it is shown by van Benthem that various styles of inference are *characterised*, in some sense, by the structural rules that they admit. For instance, the dynamic interpretation (14.3) is completely determined by the rules (Reflexivity) and (Cut). For reasons of space limitations we cannot go into detail; let us just mention here that these results amount to finite axiomatizations of various very simple *generalised reducts* of representable relation algebras. The interested reader is referred to [van Benthem 1993a], or to [Kanazawa 1994] or [Groeneveld, Veltman 1994] for more recent results.

14.4 Conclusion

In this chapter we have tried to give the reader a taste of the use of relational methods in Logic, Language and Information, both by giving specific examples of logics of transition and their motivation, and by identifying some general themes. As a result of the interaction between theory and application, the use of relational methods in LLI is undergoing rapid changes; in some cases these changes lead to a more intimate connection between relational methods and LLI, in other cases LLI seems to move away from relational methods. To conclude the chapter we sketch examples of both phenomena.

One of the classical themes in relation algebra, and indeed in algebraic logic, has been the study of restricted or finite variable fragments of first-order logic (see [Tarski, Givant 1987]). As we have seen in Sect. 14.2, the use of relational methods in LLI gives rise to new ways of interpreting well-known systems such as first-order logic. Procedural interpretations of the latter turn out to have an unexpected impact on its expressive power: more can be said with fewer variables, as becomes clear from the detailed case study [Hollenberg, Vermeulen 1994]. That work shows that relational methods provide the tools for dealing with familiar issues of expressive power in the new setting of the logics arising in LLI, and that such logics can behave in quite unexpected ways.

A natural question at this point is: how far do relational methods get us in understanding dynamic phenomena in LLI? At this stage it is impossible to answer the question, but it is clear that several researchers feel the limitations of relational methods. For example, in their search for a precise specification of the dynamic interpretation process of natural language texts in humans and machines, [Visser, Vermeulen 1995] take the radical position that one must be able to interpret any chunk of text, and that the interpretation of larger chunks is a function of the interpretations of the smaller chunks. They argue that category theory is the proper format for the description of the flow of interpretation, and for studying the monoidal manner in which interpretations of small chunks of text interact with each other. In a similar vein, [Moshier 1995] uses category theory to provide a better meta-theory for investigating independence of syntactic principles and enforcement of interactions between principles. And, finally, the paper [Moss, Johnson 1995] cited in Sect. 14.2 on using evolving algebras for analysing grammar formalisms is another example of research in LLI that goes beyond relational methods. To conclude, then, it is not clear how far relational methods will take us, but their limitations are being felt by a number of researchers.

Chapter 15

Natural Language

Michael Böttner[1]

The purpose of this chapter is to show how relational algebra can be applied to the semantics of natural language. The use of relational algebra for natural language semantics was first proposed in [Suppes 1976] under the heading of *relational grammar* in the context of model theory. It was extended to various areas of English such as modification of nouns by adjectives [Suppes, Macken 1978], intonation [Suppes 1979b], rules of natural language inference [Suppes 1979a], anaphoric pronouns [Böttner 1992b] and [Böttner 1996], and coordination [Böttner 1994]. An extension to a procedural semantics was proposed in [Böttner 1992a]. Most of Suppes' articles have now become easily accessible in [Suppes 1991].

The organization of this chapter is as follows. In Sect. 15.1 the use of algebraic methods in logic and linguistics is outlined. In sections 15.2 and 15.3 an introduction is given to relational grammar in the context of model theory, and in Sect. 15.4 the notion of relational grammar is extended to procedural semantics.

15.1 Algebraic semantics for natural language

From the beginning of the development of algebra in mathematics there have been attempts to apply algebra to the analysis of natural language. One goal of these attempts was to be able to calculate with natural language expressions very much in the same way as one could already do with numbers in arithmetic, or to put it in modern terms, to do automated theorem-proving. Another goal was to reduce the grammar of a language to a "rational" grammar. Both projects have usually been attributed to Leibniz under the names of *characteristica universalis* and *calculus ratiocinator*. Characteristica universalis is a language regimented to allow to express every possible thought or concept. Calculus ratiocinator is a system of operations that allows to check arguments and logical inferences by computing. The main ideas for these projects had already been present in his dissertation published in Leipzig 1666 but can also be found in many of his later works. For a good introduction to both projects and further reference see [Parkinson 1966], [Burkhardt 1980], or [Ishiguro 1990]. But it was not until two centuries later that progress was made in the area of algebraic logic by Boole, De Morgan, Peirce, and Schröder. Traditional logic was intimately connected with the structure of

[1]Supported by Max Planck Institute for Psycholinguistics, Nijmegen.

natural language and this tradition was also alive in 19th century logic. With the rise of quantifier logic, however, from the *Principia Mathematica* of Russell and Whitehead onward ([Whitehead, Russell 1910]), logic became more and more removed from natural language. But in somewhat the same way as algebraic logic again came to the fore in the second half of this century, an algebraic approach to language has also been re-established.

The operations of Boolean algebra have been illustrated by natural language examples from its very beginnings. An example used by Boole (1854) is

$$z(x+y), \tag{15.1}$$

where z stands for *European*, x stands for *men*, y stands for *women*, $+$ stands for *and*, and juxtaposition of terms stands for modifying a noun by an attributive adjective. With this interpretation (15.1), translates into

European men and women. (15.2)

Applying the law of distributivity to (15.1) yields

$$zx+zy, \tag{15.3}$$

and with the interpretation chosen for (15.1), (15.3) translates into

European men and European women. (15.4)

The algebraic equality between (15.1) and (15.3) then guarantees the semantic equivalence of the expressions (15.2) and (15.4). Algebraic symbolism like (15.1), (15.3) may thus be used as an instrument to compute inferences and equivalences of natural language expressions. Not just phrases like (15.2) but whole sentences can be treated in Boolean terms. Standard examples are sentences occurring in Aristotelian syllogisms. A *syllogism* is an argument with two premises and a conclusion, each of which is a sentence that has one of the following four forms:

$$\begin{array}{ll} S\,\mathbf{a}\,P: & \textit{All } S \textit{ are } P \\ S\,\mathbf{i}\,P: & \textit{Some } S \textit{ are } P \\ S\,\mathbf{e}\,P: & \textit{No } S \textit{ are } P \\ S\,\mathbf{o}\,P: & \textit{Some } S \textit{ are not } P \end{array} \tag{15.5}$$

A sentence exhibiting one of these forms is called *categorical* and is usually identified by a letter indicating whether the structure is affirmative universal (**a**), affirmative particular (**i**), negative universal (**e**), or negative particular (**o**). A truth condition in terms of Boolean operations can be spelt out for each of these structures as follows:

$$\begin{array}{lll} S\,\mathbf{a}\,P & \text{is true iff} & S \subseteq P \\ S\,\mathbf{i}\,P & \text{is true iff} & S \cap P \neq \emptyset \\ S\,\mathbf{e}\,P & \text{is true iff} & S \subseteq \overline{P} \\ S\,\mathbf{o}\,P & \text{is true iff} & S \cap \overline{P} \neq \emptyset. \end{array} \tag{15.6}$$

The semantics of a language is studied by model theory. Originally, model theory was applied to languages of logic, but once rigorous methods had become available for the syntax of natural language, model theory was successfully applied in this domain as well. A major breakthrough in natural language semantics was the work by

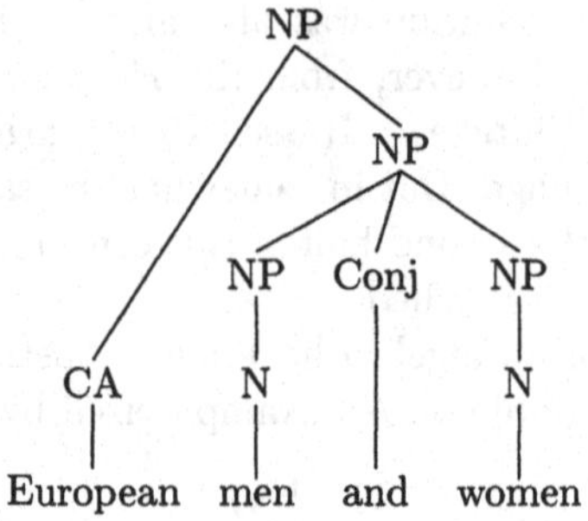

Fig. 15.1 Derivation tree for *European men and women*

Richard Montague [Montague 1974]. He proposed to view any natural language as a pair of two homomorphic algebras: an algebra $\langle A, F\rangle$ of expressions and an algebra $\langle B, G\rangle$ of meanings. The algebra of expressions is defined as an absolutely free algebra, cf. Sect. 1.4, with the finite lexicon of this language as its base set. Other than requiring that it be an absolutely free algebra, not many specific constraints on the algebraic structure of meanings were given by Montague. A first step in defining a structure of the semantic algebra was given in [Keenan, Faltz 1978; Keenan, Faltz 1985]. Keenan & Faltz start from the observation that the Boolean operators *and, or, not* are not limited to sentences as a category of operands, but can have operands of a variety of other syntactic categories like, e.g., common nouns, noun phrases, determiners, verb phrases, transitive verb phrases, adjective phrases, prepositions, prepositional phrases, and adverbial phrases. This led to a characterization of each of these categories as a Boolean algebra, i.e. as sets of objects closed under the Boolean operations. Under this approach, there is no algebra corresponding to the language as a whole in the sense of standard algebraizations for languages of logic. In particular, some categories are interpreted as homomorphisms, i.e. as mapping from one algebra to another algebra. A much more straightforward approach to an algebraization of a formal language was achieved in [Suppes 1976]. In the following, our focus will be on this approach.

15.2 Relational grammar

In this section, we give an introduction to relational grammar in the context of model theory.

Denoting grammar

There is a tradition in linguistics to represent the structure of an expression by a tree. The tree for expression (15.2), for instance, would appear as in Fig. 15.1. This tree represents two kinds of information about (15.2): the grouping of its constituents and the categorization of its constituents. The tree tells us that *European* is a classificatory adjective (CA), that *men* and *women* are common nouns (N), and that *and* is a conjunction ($Conj$). Moreover, it tells us that the substring *men and women* forms a constituent, i.e. that the constituents *men*, *and*, and *women* belong closer together than *European* and *men*. Constituents of this kind are conventionally called noun phrases (NP).

A tree for an expression of a given language is conventionally derived by a grammar for that language. This is usually a so-called context-free grammar. The tree is therefore called a *derivation tree.* A context-free grammar which is able to derive the tree in Fig. 15.1 is the following one:

$$\begin{array}{rrcl} i) & NP & \to & CA + NP \\ ii) & NP & \to & NP + Conj + NP' \\ iii) & NP & \to & N \\ iv) & N & \to & men \\ v) & N & \to & women \\ vi) & CA & \to & European \\ vii) & Conj & \to & and \end{array} \qquad (15.7)$$

The symbol $\to$ means that any symbol of the left-hand side can be replaced by a symbol of the right-hand side. NP is called the *start* symbol of the grammar, *men*, *women*, *European*, and *and* are called *terminal* symbols, N, CA, and $Conj$ as *non-terminal* symbols. The important rules are i) and ii), which allow any phrase of category NP to be replaced by either an adjective and a phrase of category NP, or by two strings of the same category NP combined by a conjunction of category $Conj$. They are important because they are recursive, i.e. any result of their application can be used as input to a new application.

A phrase structure grammar like (15.7) accounts only for the shape of a string of words but not for its meaning. Meanings are taken care of by *model theory.* Model theory provides semantic interpretations for a language with respect to a *model structure.* A model structure for a language is a pair of some non-empty set D and a mapping v from the terminal symbols of that language. D is called the *domain* of the model structure. The mapping v is called the *valuation function* of the model structure. The semantic interpretation assigns to each well-formed expression of the language a certain object of the model structure. This object is called the *denotation* of that expression with respect to that model structure. The stock of denotations is determined by the *hierarchy* over the model-structure.

If we assume that each denoting word of the string has a denotation we would then like to know the denotation for the whole string. The composition of meanings is effected by *semantic functions.* Semantic functions are therefore attached to the grammar rules. One way to attach semantic rules to our grammar (15.7) could be:

$$\begin{array}{rlcl@{\qquad}rlcl} i) & [NP] & = & [CA] \cap [NP] & iv) & [N] & = & [men] \\ ii) & [NP] & = & [NP] \cup [NP'] & v) & [N] & = & [women] \\ iii) & [NP] & = & [N] & vi) & [CA] & = & [European] \end{array} \qquad (15.8)$$

The semantic functions are set-theoretical intersection in Rule i), set-theoretical union in Rule ii), and identity in the Rules iii) through vi). There is no semantic

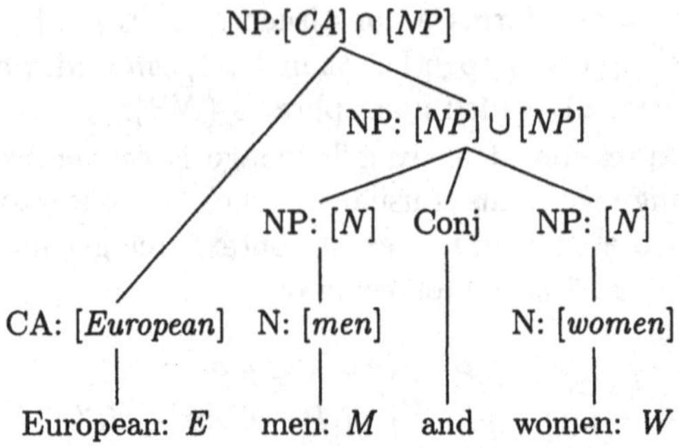

Fig. 15.2 Semantic tree for *European men and women*

function attached to Rule vii). The reason for this is that the word introduced by this rule does not have a denotation.

A phrase structure grammar with semantic functions attached to its production rules is called a *denoting grammar*. The notion of a denoting grammar was proposed in [Suppes 1973b]. Our denoting grammar ((15.7), (15.8)) allows us to compute the denotation of the expression (15.2) from its component denotations for every model structure of the grammar. As domain D we may adopt the set of persons living on this planet in 1996. A quite different domain might have been picked by Boole in 1854. The grammar of (15.7) has the terminal symbols *men*, *women*, *European*, and *and*. We may then pick subsets of our domain D as valuations for the words *men*, *women*, and *European*, namely the set of all male persons as valuation for *men*, the set of all female persons as valuation for *women* and the set of all citizens of some European country for *European*. No set is picked as valuation for the word *and*. The hierarchy used for the grammar ((15.7), (15.8)) is therefore any subset of

$$\mathcal{P}(D) \tag{15.9}$$

that is closed with respect to Boolean operations.

The computation of compound expressions can be represented by a *semantic tree*. The semantic tree for (15.2) is given in Fig. 15.2. If a leaf of this tree is labelled by a denoting word it will be assigned a denotation by the valuation function of the model structure under consideration. Any mother gets its denotation as the result of a certain semantic operation applied to the denotations of the respective daughters. This operation is determined by the particular rule by which the daughters are produced. The procedure is iterated down to the root of the tree. Since the root dominates every leaf of the tree, the root denotation is the denotation of the entire expression.

It should now be straightforward to write down a denoting grammar for the set of categorical sentences of (15.5). If this grammar contains the denoting grammar ((15.7), (15.8)) as a subgrammar it will derive categorical sentences with complex noun phrases like, e.g., *No European women are European men*.

Relational extension

The hierarchy (15.9) provides denotations for expressions involving absolute terms like, e.g., *dog* or *pregnant*. Not all terms of a natural language are absolute though. There is a large and important class of relative terms. Standard examples are kinship terms like, e.g., *sister* or *uncle*. Another class of examples are spatial terms like, e.g., *in*, *above*, *behind*. There are terms involving some order like, e.g., *larger* or *louder*. Although there are ternary relations and relations of even higher degree in natural language, our focus here will be on binary relations only.

The denotation of a binary relative term is a binary relation on D. The relative noun *brother* denotes the set of pairs $\langle a, b\rangle$ of objects of D such that a is a brother of b, the preposition *in* denotes the set of pairs $\langle a, b\rangle$ of objects of D such that a is in b, the adjective *larger* denotes the set of pairs $\langle a, b\rangle$ such that a is larger than b, and the transitive verb *hit* denotes the set of pairs $\langle a, b\rangle$ such that a hits b. It is therefore necessary to replace the set (15.9) by

$$\mathcal{P}(D) \cup \mathcal{P}(D \times D) \tag{15.10}$$

But this is not sufficient. Although we will now be able to derive a denotation for, e.g., *brothers and sisters*, no denotation can be derived for *John is a brother of Mary*, or *London is near Paris*. What is needed are operations that compute sets or binary relations from sets and binary relations. For a list and definition of such operations see Sect. 1.2. In the following we shall define some operations and illustrate them by grammatical constructions from the English language.

Image

The operation *image set* is defined in Sect. 1.2. From this we can derive the term

$$\bigcup_{x \in A} R(x) \tag{15.11}$$

and the term

$$\bigcap_{x \in A} R(x). \tag{15.12}$$

[Riguet 1948] proposed to use the term (15.11) to define a binary operation " taking a binary relation R and a set A as its arguments:

$$R``A = \bigcup_{x \in A} R(x). \tag{15.13}$$

This operation was called "coupe de première espèce" in [Riguet 1948]. It may be better known by the name *upper image* of set A under relation R. Using the element relation the result of this operation amounts to the set

$$\{y | (\exists x)(x \in A \wedge xRy)\}. \tag{15.14}$$

This operation can be illustrated by a certain type of relative clause in which the relative pronoun has the role of the object. An example would be *who Mary likes*.

The denotation of this clause is the set of all those human beings of the domain D such that Mary likes them, i.e. the set

$$\{y|mLy\} \, , \tag{15.15}$$

where L is the binary relation denoted by the transitive verb *like* and m is the individual denoted by *Mary*. Since

$$L``\{m\} = \{y|(\exists x)(x = m \wedge xLy)\}, \tag{15.16}$$

we are able to derive the denotation for the string *who Mary likes* from the denotations for *likes* and *Mary* by the operation defined in (15.13). The semantic tree for the string *who Mary likes* would then look as follows

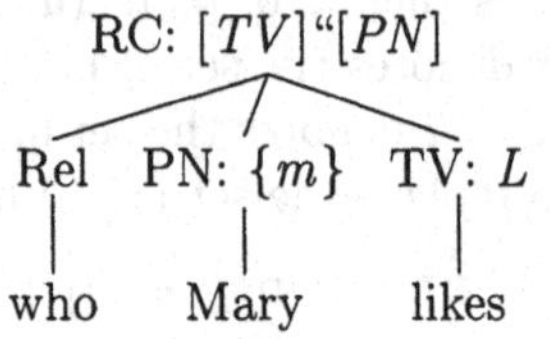

$$
\begin{array}{c}
\text{RC: } [TV]``[PN] \\
\diagup\quad\diagdown \\
\text{Rel}\quad \text{PN: }\{m\}\quad \text{TV: }L \\
|\qquad\quad |\qquad\qquad | \\
\text{who}\qquad \text{Mary}\qquad \text{likes}
\end{array}
\tag{15.17}
$$

where RC = relative clause, Rel = relative pronoun, PN = proper noun, and TV = transitive verb.

In Sect. 15.2 we have seen that nouns can be modified by property expressions. The semantic operation corresponding to modification is intersection. Intersection is therefore the operation that we have used in (15.8i). In (15.8i) properties were exhibited by adjectives. But adjectives are not the only syntactic category that may express a property. A property can also be expressed by a relative clause like, e.g., *who Mary likes*. An example would be *boys who Mary likes* where the noun *boys* is modified by the relative clause *who Mary likes*. The corresponding semantic tree would be:

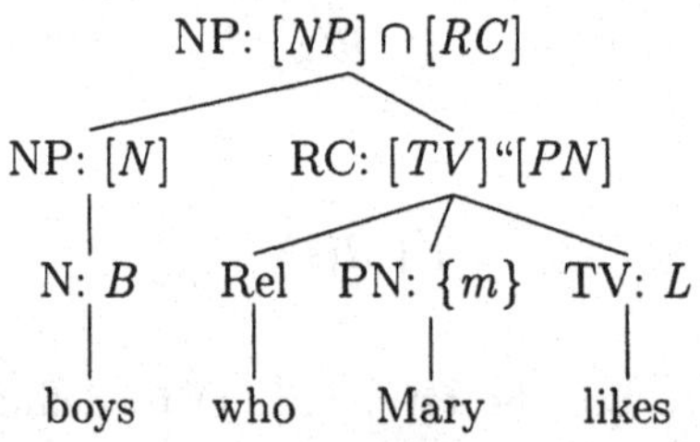

$$
\begin{array}{c}
\text{NP: } [NP]\cap[RC] \\
\diagup\qquad\qquad\diagdown \\
\text{NP: }[N]\qquad\qquad \text{RC: }[TV]``[PN] \\
|\qquad\qquad\qquad \diagup\quad\diagdown \\
\text{N: }B\qquad \text{Rel}\quad \text{PN: }\{m\}\quad \text{TV: }L \\
|\qquad\quad |\qquad\quad |\qquad\qquad | \\
\text{boys}\quad \text{who}\quad \text{Mary}\qquad \text{likes}
\end{array}
\tag{15.18}
$$

In a fashion completely parallel to (15.13) [Riguet 1948] also proposed to use the term (15.12) to define the operation

$$R[A] = \bigcap_{x\in A} R(x) \tag{15.19}$$

called "coupe de deuxième espèce". In terms of the element relation the result of this operation can be expressed by

$$\{y|(\forall x)(x \in A \to xRy)\} \, . \tag{15.20}$$

A nice illustration of the operation defined in (15.19) is the possessive case in English like, e.g., *Mary's* in *Mary's toys*. The possessive case marker *'s* can be thought of as denoting a binary relation P relating human beings with objects like, e.g., Mary with a set of toys. The denotation of *Mary's* can then be derived from the singleton $\{m\}$ and the binary relation P by

$$P[\{m\}] = \{y|(\forall x)(x \in \{m\} \to xPy)\} \ . \tag{15.21}$$

Common nouns can also be modified by possessive nouns like, e.g., *John's* in *John's toys*. Let *John's* denote the set of objects that belong to John. This set can be used to restrict the denotation of the noun phrase *toys*, i.e. the set T of toys. The denotation of the noun phrase *John's toys* can then be derived along the following semantic tree:

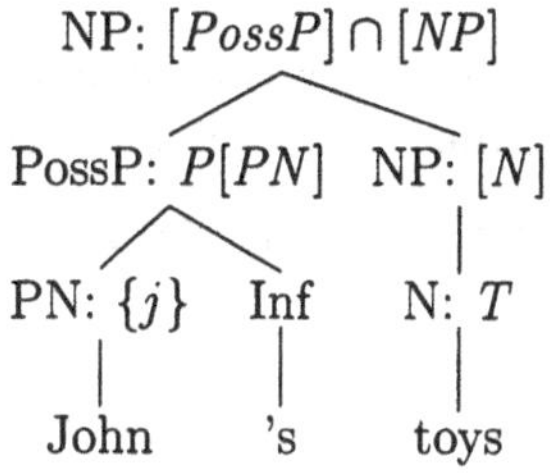

$$
\begin{array}{c}
\text{NP: } [PossP] \cap [NP] \\
\diagup \frown \\
\text{PossP: } P[PN] \quad \text{NP: } [N] \\
\diagup \diagdown \qquad | \\
\text{PN: } \{j\} \quad \text{Inf} \qquad \text{N: } T \\
| \qquad\quad | \qquad\quad | \\
\text{John} \quad\quad \text{'s} \quad\quad \text{toys}
\end{array}
\tag{15.22}
$$

We are now in a position to prove certain semantic facts about English. First, the phrases *John's toys and Mary's toys* and *John's and Mary's toys* are equivalent, since

$$[John's \ toys \ and \ Mary's \ toys] = (P[\{j\}] \cap T) \cup (P[\{m\}] \cap T). \tag{15.23}$$

The equivalence of these expressions is an immediate instance of the distributive law of Boolean algebra. Second, the phrases *John and Mary's toys* and *John's and Mary's toys* are not equivalent. In order to see this recall from Rule ii) of (15.8) that the denotation of a phrase of two nouns combined by *and* is the union of the constituent denotations. We therefore have

$$[John \ and \ Mary] = \{j\} \cup \{m\} \tag{15.24}$$

and, hence,

$$[John \ and \ Mary's \ toys] = P[\{j\} \cup \{m\}] \cap T. \tag{15.25}$$

By the same token, we have

$$[John's \ and \ Mary's \ toys] = (P[\{j\}] \cup P[\{m\}]) \cap T. \tag{15.26}$$

To show that the left hand sides of (15.25) and (15.26) are not equivalent it is sufficient to show that the respective right hand sides are not equivalent. Let us therefore assume that the set of toys T has a teddy bear t and a toy car c as its elements:

$$T = \{t, c\} \tag{15.27}$$

Let the teddy bear belong to both John and Mary and let the toy car belong only to John, i.e. our relation P of possession is defined

$$P = \{\langle j, t\rangle, \langle m, t\rangle, \langle j, c\rangle\}. \tag{15.28}$$

We then have

$$P[\{j\} \cup \{m\}] \cap T = \{t\} \tag{15.29}$$

and

$$(P[\{j\}] \cup P[\{m\}]) \cap T = \{t, c\} \cup \{t\} = \{t, c\} \tag{15.30}$$

Since $\{t\} \neq \{t, c\}$ it follows that *John and Mary's toys* and *John's and Mary's toys* have different meanings.

Peirce product

Recall from Sect. 1.2 the definition of *Peirce product*

$$R\!:\!A = \{x \in D | (\exists y)(y \in A \wedge xRy)\}. \tag{15.31}$$

Let O be the binary relation denoted by the transitive verb *own* and let H be the set of houses denoted by the noun *house*. Instantiating (15.31), we get

$$O\!:\!H = \{x \in D | (\exists y)(y \in H \wedge xOy)\}. \tag{15.32}$$

Therefore, we have

$$x \in O\!:\!H \qquad \text{iff} \qquad (\exists y)(y \in H \wedge xOy). \tag{15.33}$$

The right hand side is true just in case there is a house that x owns. The set $O\!:\!H$ can therefore serve as the denotation for the phrase *own some houses.*

Since the verb phrases *own some houses* and *own no houses* are contradictory, the respective denotations should be complements of each other:

$$[\textit{own no houses}] = \overline{O\!:\!H} \tag{15.34}$$

We thus have defined a semantic operation corresponding to the negative quantifier in object position.

Since any owner of no houses is identical to a not-owner of all houses we have

$$[\textit{own no houses}] = [\textit{not-own all houses}] \tag{15.35}$$

From the previous two equations we derive the equation

$$[\textit{not-own all houses}] = \overline{O\!:\!H}\ . \tag{15.36}$$

Replacing simultaneously the binary relation O by its complement relation $\overline{O}$ and *not-own* by its complement *own* yields

$$[\textit{own all houses}] = \overline{\overline{O}\!:\!H}\ . \tag{15.37}$$

We thus have defined a semantic operation corresponding to the universal quantifier in object position by

$$[\textit{own all houses}] = \overline{\overline{O}\!:\!H}. \tag{15.38}$$

Converse

The *converse* operation has the effect of turning around the order of the elements of a relation:

$$R^{\smile} = \{\langle y, x\rangle | \langle x, y\rangle \in R\} \tag{15.39}$$

As had already been observed by Peirce the standard use of this operation is to derive the meaning of a passive verb phrase from its active counterpart. Thus, if I is the binary relation denoted by the transitive verb *invite* then $I^{\smile}$ is the binary relation denoted by the passivized verb *is invited by.*

The denotation for the verb phrase *invited by some philosophers* is

$$I^{\smile} : P \tag{15.40}$$

where I is the binary relation denoted by *invite* and P is the set denoted by *philosophers.*

Theorem 15.2.1 For any subsets A, B of some set and any binary relation R on that set the following holds: If $B \cap \overline{(\overline{R^{\smile}} : A)} \neq \emptyset$ then $A \subseteq R : B$.

The validity of the argument

$$\frac{\textit{Some novels are liked by all people}}{\textit{All people like some novels}} \tag{15.41}$$

follows from Theorem 15.2.1. The theorem cannot be strengthened to a biconditional. Therefore the reverse argument

$$\frac{\textit{All people like some novels}}{\textit{Some novels are liked by all people}} \tag{15.42}$$

is not valid. So premise and conclusion do not have equivalent denotations. This is an interesting result, since it falsifies the widespread belief that passive sentences are synonymous with their active counterparts.

Lower image

The operation

$$R_{``}A = \overline{R : \overline{A}} \tag{15.43}$$

was introduced in [Suppes, Zanotti 1977] by the name of *lower image* of set A under relation R. This operation can be illustrated by a certain type of relative clause with the relative pronoun playing the role of an attribute like, e.g., *composers all of whose sons are composers*. The semantic tree for this expression is in Fig. 15.3 with RC = relative clause, UQ = universal quantifier, *RelPoss* = relative possessive pronoun, RN = relative noun, Cop = copula, VP = verb phrase, and P = preposition. The denotations involved are the set C of composers and the binary relation S of being a son of.

An element of the root denotation of the semantic tree in Fig. 15.3 is Leopold Mozart (at least in case we restrict our domain to adult people thus excluding Wolfgang Amadeus' five siblings who died at infant age). Although Johann Sebastian Bach is known to have at least four sons who were also composers, he does not belong in this class for the reason that he had seven more sons who are not known to have become composers themselves. On the other hand, George Frederick Handel and Franz Schubert certainly belong in this class although they

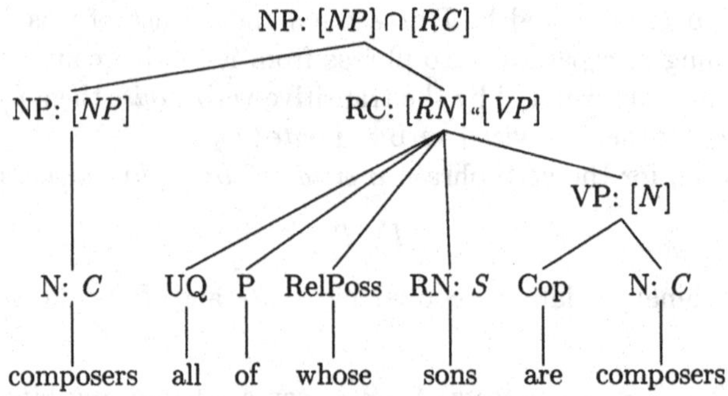

Fig. 15.3 Semantic tree with relative possessive pronoun

are not known to have had any children at all. This is odd but in line with the conventional interpretation of universal quantification in modern logic.

Notice that the relative clause denotation makes essential use of the lower image operation. Since by definition $S{}_{``}C = \overline{S``\overline{C}}$ and, since

$$\overline{S``\overline{C}} = \{x|(\forall y)[ySx \rightarrow y \in C]\}$$

we can see that this set reflects exactly our intuitive understanding of the phrase.

Restriction

So far we have only considered the modification of absolute terms. But there is also modification of relative terms. For instance, in kinship terminology the noun *brother* is defined by the phrase *male sibling.* The head of this phrase, which is *sibling,* is a relative term but the modifier of this phrase, which is *male,* is absolute. Therefore *sibling* denotes a binary relation and *male* denotes a set. But sets and binary relations cannot be intersected. The operation that is suitable for this case is *domain restriction* of a binary relation R by a set A defined as follows:

$$R \rceil A = R \cap (A \times D). \tag{15.44}$$

The relation B denoted by *brother* can then be defined

$$B = S \rceil M, \tag{15.45}$$

where S is the binary relation denoted by *sibling* and M is the set denoted by *male.* This operation provides the correct definition for brother: x is a brother of y iff x is a male sibling of y, and x is a male sibling of y iff x is a sibling of y and x is male.

An analogous operation of *range restriction* of R by A can be defined as follows:

$$R \upharpoonright A = R \cap (D \times A) \tag{15.46}$$

Initial segment

In (15.8*i*) a semantic function was given for combinations of a classifying adjective with a noun phrase. Not all adjectives are classifying though. Consider examples like, e.g., *old*, *high*, or *ugly*. These adjectives differ from classifying adjectives by deriving comparatives like *older than*, *higher than*, or *uglier than*. We therefore call them *comparison adjectives* or *intensive adjectives* (*IA*). Since comparison implies order, a comparison adjective denotes an ordering relation on D. But if a comparison adjective like, e.g., *high* denotes an ordering relation, then an operation is called for to derive the meaning of *high mountain* from the denotations for *high* and *mountain*. The notion that turns out to be useful here is the notion of an initial segment of a binary relation. Let A be some set, R some ordering (strict partial) relation R on A and c an element of A. Then the *initial segment* of A with respect to R and criterion object c is defined

$$IS(R, A, \{c\}) = (R:\{c\}) \cap A. \tag{15.47}$$

So the denotation of the phrase *high mountain* will be

$$(H : \{m\}) \cap M \tag{15.48}$$

where H is the relation denoted by *high*, M is the set denoted by *mountain* and m is some element of M. If a mountain is higher than this criterion mountain m (or at least as high as it) we call it a high mountain and if it is lower than it we do not call it a high mountain.

This solution nicely accounts for the fact that the order of adjectives modifying a noun sometimes matters. The expressions *Dutch high mountains* and *high Dutch mountains* may have different denotations.

15.3 Further refinements

In addition to the operations mentioned in the preceding section there are some operations of a very fundamental kind that have not been mentioned yet. These operations are *composition*, *identity*, and *domain*. The reason why they have not been mentioned is that they cannot be given a direct natural language interpretation. This is all the more surprising in the case of the operations of composition and identity. They will, however, be used to define more complicated operations for which very natural interpretations exist. We would like to mention here only the operations *Ref*, *Rec*, *Poss*, *RecPoss*, and *Id*. They are tailored specifically to the needs of certain natural language constructions. *Poss* is introduced as a counterpart for the possessive pronoun construction, *Ref* is intended to construe the semantics of verb phrases with a reflexive pronoun, *Rec* is introduced as a counterpart for the reciprocal pronoun construction, *RecPoss* is introduced as a counterpart for the reciprocal in the possessive case, *Id* is introduced as a counterpart for the identity pronoun construction.

Poss

From Sect. 1.2 we know that the operation *domR* computes the set of all elements of D related by R to some element of D. From the preceding section recall the definition of range restriction of a binary relation by a set. Let R be an arbitrary relation over D and let A be an arbitrary subset of D. Let P be the relation of possession in all models. We are then in a position to derive the term

$$Poss(R, A) = \overline{dom((P \restriction A) \cap \overline{R})}. \tag{15.49}$$

The operation *Poss* returns a subset of the universe D as its value.

We have already considered possessive phrases in Sect. 15.2. Of more interest is the question of how possessive anaphoric pronouns are interpreted like in, e.g.,

John likes his toys (15.50)

where the interpretation of the possessive pronoun *his* depends on the interpretation of the subject term of the sentence. For the verb phrase *likes his toys* we propose the tree

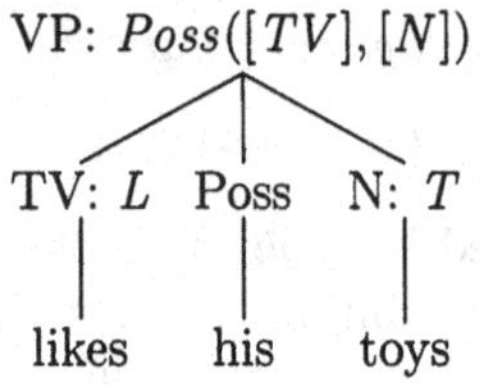

(15.51)

According to this tree the denotation of the verb phrase is

$$\overline{dom((P \restriction T) \cap \overline{L})} \tag{15.52}$$

where L is the denotation for the transitive verb *likes* and T is the denotation for the common noun *toys*. That (15.52) corresponds to the intuitive meaning of the verb phrase follows from the fact that

$$x \in \overline{dom((P \restriction T) \cap \overline{L})} \tag{15.53}$$

is true just in case the following condition holds:

$$(\forall y)[[T(y) \textit{ and } P(x, y)] \Rightarrow L(x, y)] \tag{15.54}$$

The trees for the phrases *like her toys* or *like their toys* would look very similar to (15.51). A proper treatment would have to introduce gender into the grammar.

One might be tempted to derive the interpretation of (15.50) by replacing *his toys* by *John's toys*. But that would be a mistake. For the sentence *John likes John's toys* could not be generalized to account for the contextual dependence of the interpretation of the possessive pronoun on the referent of the subject term. This can be seen if the subject term is replaced by a complex term like in, e.g., *John and Mary like their toys*, which is not equivalent to the sentence *John and Mary like John and Mary's toys*.

Ref

In Sect. 1.2 the notion of an identity relation I_X is defined. Let $X = D$, i.e. let us consider the "full" identity I_D over D. With the help of intersection, the full identity relation, and the domain operation we can derive the term

$$dom(R \cap I_D). \tag{15.55}$$

Let us refer to this operation by the name *Ref* for the reason that it corresponds to expressions with a reflexive pronoun like, e.g.

John and Mary like themselves. (15.56)

The semantic tree for the verb phrase of this sentence looks like this:

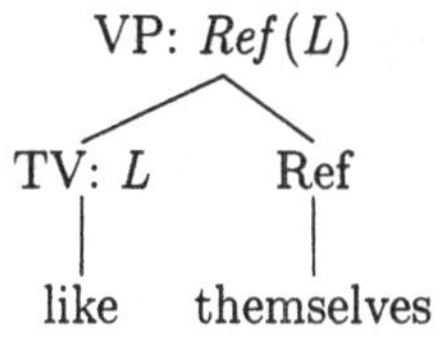

(15.57)

That (15.57) corresponds to the intuitive meaning of the verb phrase follows from the fact that $x \in dom(R \cap I_D)$ iff xRx.

Rec

Let R be an arbitrary relation on domain D. The operation $R \cap R^{\smile}$ is then the largest symmetric subrelation of R. For reasons that we shall not go into here we prefer to define an operation *Rec* by

$$Rec(R) = (R \cap R^{\smile}) - I \qquad (15.58)$$

rather than simply by $R \cap R^{\smile}$. This operation can be illustrated by the reciprocal pronoun construction. An example would be *John and Mary like each other.* The semantic tree for the verb phrase of this sentence is

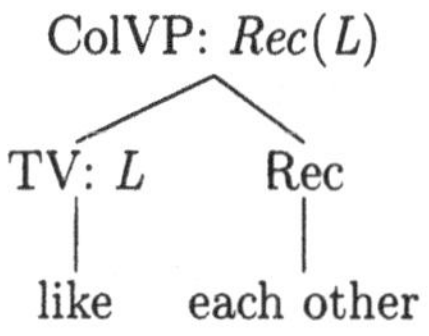

(15.59)

The structure of the sentence *John and Mary like each other* is quite similar to the structure of the sentence *John and Mary like themselves*: a compound subject term is followed by a verb phrase that consists of a transitive verb and some pronoun as object term. However, the semantic tree (15.59) is quite different: The predicate term in (15.57) has the label *VP* but the predicate term in (15.59) has the label *ColVP*. The purpose of different categories is to announce a difference in semantic type: Whereas an expression of category *VP* denotes a set, an expression of category *ColVP* denotes a binary relation.

Theorem 15.3.1 *Rec*(R) is a symmetrical relation.

This theorem guarantees the validity of the following argument:

$$\frac{\textit{John and Mary like each other}}{\textit{Mary and John like each other}} \qquad (15.60)$$

The validity of the argument

$$\frac{\textit{John and Mary like each other}}{\textit{John likes Mary}} \tag{15.61}$$

follows from the fact

$$\text{If } (A \times B) \cap R \neq \emptyset \text{ then } A \cap (R : B) \neq \emptyset,$$

since $((A \times B) \cap R) : B \subseteq ((A \times B) : B) \cap (R : B) = A \cap (R : B)$.

RecPoss

With the composition of two relations R and S introduced in Sect. 1.2 we define

$$RecPoss(R, A) = \overline{\overline{R};I_A;P^{\smile}} \cap \overline{P;I_A;\overline{R^{\smile}}} \cap \overline{I} \tag{15.62}$$

The operation *RecPoss* takes a binary relation and a set as its arguments and returns a binary relation.

This operation is designed for reciprocal pronoun occurring in the possessive case like in, e.g., *John and Mary like each other's books.* The semantic tree for the verb phrase *like each other's books* would look as follows:

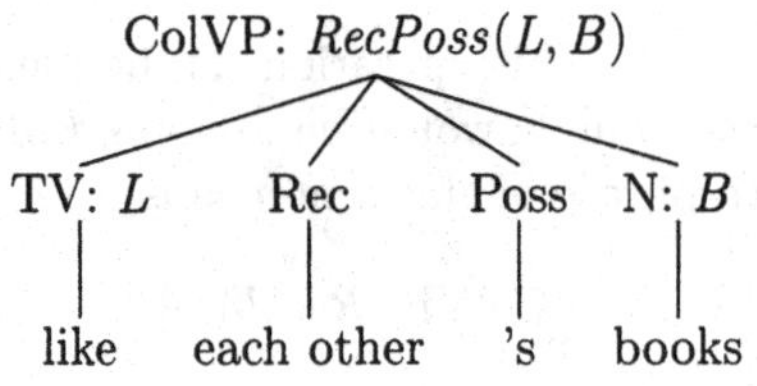

(15.63)

It has the denotation

$$\overline{\overline{L};I_B;P^{\smile}} \cap \overline{P;I_B;\overline{L^{\smile}}} \cap \overline{I}. \tag{15.64}$$

This is what it is supposed to denote, since

$$\langle x, y\rangle \in \overline{\overline{L};I_B;P^{\smile}} \cap \overline{P;I_B;\overline{L^{\smile}}}$$

is equivalent to

$$(\forall z)(Bx \rightarrow [yPz \rightarrow xLz] \wedge [xPz \rightarrow yLx]])$$

which is what the verb phrase *like each other's books* intuitively amounts to.

Theorem 15.3.2 *RecPoss*(R, A) is a symmetric relation for arbitrary sets A and relations R.

An immediate consequence of this theorem is that the argument

$$\frac{\textit{John and Mary like each other's books}}{\textit{Mary and John like each other's books}} \tag{15.65}$$

is valid.

Id

Id is an operation that turns a relation and a set into a relation. It is defined

$$Id(R,A) = \overline{(R;I_A;\overline{R^{\smile}})} \cap \overline{(\overline{R};I_A;R^{\smile})} \tag{15.66}$$

It can be illustrated by the expression *same* like occurring in

$$\textit{John and Mary read the same books} \tag{15.67}$$

Notice that the verb phrase *read the same books* of this semantic tree denotes a binary relation. The sentence (15.67) can be paraphrased by *John reads the same books as Mary*, which brings out better the feature that the sentence is relational. According to our analysis the denotation of the verb phrase *read the same books* is $Id(R, B)$. That $Id(R, B)$ indeed exhibits the meaning denoted by (15.67) can be seen from the fact that

$$\langle x, y\rangle \in \overline{R;I_B;\overline{R^{\smile}}} \cap \overline{\overline{R};I_B;R^{\smile}}$$

holds just in case $(\forall z)(Bz \to (xRz \leftrightarrow yRz))$ is true.

Theorem 15.3.3 $Id(R, A)$ is symmetric.

Theorem 15.3.4 $Id(R, A)$ is transitive.

For the definitions of the notions of symmetry and transitivity of a binary relation see Sect. 1.2. From 15.3.3 and 15.3.4 it follows that $Id(R, B)$ is an equivalence relation on the set $R : B$ of book readers. But if $Id(R, B)$ is an equivalence relation then the argument

John and Mary read the same books
Bill and Mary read the same books
―――――――――――――――――――――
John and Bill read the same books

should be valid, which is indeed the case.

15.4 Procedural semantics

So far our denotations for English words have been given in the fashion of standard model theory. This view is completely static: a world is conceived of as a set of objects with certain properties and relations. Applied to actions this view leads to certain absurdities. Consider the action described by the sentence *John adds 34711 and 9263*. Under this approach the verb *add* denotes some binary relation between a human individual and a sequence of numbers:

$$\{\langle j, \langle 34711, 9263\rangle\rangle\} \tag{15.68}$$

This analysis does not account for the fact that adding numbers is an operation that returns some result. But an account of successful addition should return the result of the addition. It has therefore been proposed to define natural language meanings in terms of procedures.

We now interpret the elements of the Boolean algebra as sets of states (of some agent) and the elements of the relation algebra as sets of state-transitions.

Following a suggestion of [Suppes 1973b], we view the agent as a machine that is able

- to move into four directions (left, right, up, down) in a gridlike environment
- to retain representations of perceived objects
- to recognize objects as objects of a certain kind, e.g. numbers.

Let us also assume that this machine has two registers. In line with [Suppes 1973a], we let the machine have a focus register F to keep track of the current location of the agent in the environment and a memory register M to keep track of objects that should be kept in memory for some time. Any state of the agent can then be represented by an ordered pair: the first component representing an object in visual focus and the second component representing an object in memory.

For purposes of illustration we use the case of arithmetical instructions like, e.g., column addition that require an agent to be able to identify objects like digits arranged in rows and columns like this

$$\begin{array}{ccc} & 3 & 4 \\ 1 & 7 & 8 \\ & & 7 \\ & 5 & 5 \\ - & - & - \end{array} \qquad (15.69)$$

It should be noted that the set of symbols is not just the set of digits 0,1,2,... together with a bar and a symbol b for blanks, but rather a set of occurrences or tokens of those symbols: the symbol 7 of the second row has to be distinguished from the symbol 7 in the third row. In the same way, empty spaces have to be kept distinct. The perceptual environment for arithmetic instruction (15.69) can then be represented as follows:

$$\begin{array}{ccc} b_1 & 3 & 4 \\ 1 & 7_1 & 8 \\ b_2 & b_3 & 7_2 \\ b_4 & 5_1 & 5_2 \\ -_1 & -_2 & -_3 \\ b_5 & b_6 & b_7 \end{array} \qquad (15.70)$$

Any arrangement of this kind can be identified as a finite geometry of two relations V ("vertically below") and H ("horizontally left of"), in the sense of [Crangle, Suppes 1987]. So we have $\langle 7,4\rangle \in V$ but $\langle 1,4\rangle \notin V$. Both V and H are strict ordering relations, cf. Sect. 1.2 We refer to 3 as the *left neighbour* of 4, to 8 as the *lower neighbour* of 4, to elements b,3,4 as the V*-maximal*, and to elements $-_1$, $-_2$, $-_3$ as the *minimal* elements of V.

Assume that the agent is located on the top square of the middle column. Assume further that the agent is in an initial state. Then the state of the agent would be $\langle 3,\varepsilon\rangle$ where ε indicates that the register is empty.

Assume now that our agent is given the command *Look!* The procedure expected of the agent would be that the content of the F register is changed to any other field, i.e. our agent would now be in state $\langle x, \varepsilon\rangle$. The exact destination of looking is given by additional wording. So the meaning of *look* has brought about the following state transition: $\langle\langle 3, \varepsilon\rangle, \langle x, \varepsilon\rangle\rangle$ It is then plausible to see that the meaning of the verb *look* can be identified with the set

$$\{\langle\langle f, m\rangle, \langle f', m\rangle\rangle\}$$

of state transitions.

Register M is the register that stores the content of the agent's memory. Let us assume that the agent has perceived the symbol in the first row and then gets the command to remember that number, then the agent's state would get changed to $\langle 3, 3\rangle$. So the procedure associated with the verb *remember* has effected the following state transition: $\langle\langle 3, \varepsilon\rangle, \langle 3, 3\rangle\rangle$ Again it is easy to see that *remember* can be identified with the following transition set:

$$\{\langle\langle f, m\rangle, \langle f, f\rangle\rangle\}$$

A minimal movement down can be defined by the following transition set:

$$\{\langle\langle f, m\rangle, \langle f', m\rangle\rangle | f' V f \text{ and } f' \text{ is a neighbour of } f\}$$

We let this set be the denotation for the adverb *down*.

A maximal movement up can be defined by the following transition set:

$$\{\langle\langle f, m\rangle, \langle f', m\rangle\rangle | f V f' \text{ and } f' \text{ is maximal}\}$$

We use this as the denotation for *top*.

An arbitrary movement in both up and down directions can be defined by the following transition set:

$$\{\langle\langle f, m\rangle, \langle f', m\rangle\rangle | f V f' \vee f' V f\}$$

We adopt this set as our denotation for the noun *column*.

Since the denotations of *top* and *column* are relational, we keep track of this by assigning these words relational categories: *top* the category RA (= relational adjective) or RN (= relational noun) and *column* the category RN. The relational character distinguishes both *top* and *column* from a noun like *number* that denotes the set of all states with a number symbol, i.e. a digit, in its first component:

$$\{\langle f, m\rangle : f \text{ is a number}\}$$

An overview of our toy lexicon is given in Table 15.1: This lexicon could easily be extended by adding corresponding entries for *bottom*, *up*, *left*, *right*, *leftmost*, *rightmost*, *bar* etc., for details, see [Böttner 1992a].

Composition

To derive non-primitive procedures from our lexical procedures we use the familiar operations from relation algebra. Composition of two relations R and S is no doubt the most important operation of relational algebra. It is defined as follows:

$$R;S = \{\langle x, y\rangle | (\exists z)(xRz \wedge zSy)\} \tag{15.71}$$

	Category	Denotation
look	V	$\{\langle\langle f,m\rangle,\langle f',m\rangle\rangle\}$
remember	V	$\{\langle\langle f,m\rangle,\langle f,f\rangle\rangle\}$
top	RN, RA	$\{\langle\langle f,m\rangle,\langle f',m\rangle\rangle \mid fVf'$ and f V-maximal$\}$
down	Adv	$\{\langle\langle f,m\rangle,\langle f',m\rangle\rangle \mid f'Vf$ and f' neighbour of $f\}$
column	RN	$\{\langle\langle f,m\rangle,\langle\langle f',m\rangle\rangle \mid fVf' \vee f'Vf\}$
number	N	$\{\langle f,m\rangle \mid f \textit{ is a number}\}$
spot	N	$\{\langle f,m\rangle \mid f \neq \varepsilon\}$

Table 15.1 Procedural Lexicon

If R and S are procedures the intuitive meaning of this operation is the sequencing of two procedures: first doing R and then S. If the two relations R and S are identical, i.e. if $R = S$ we may abbreviate $R;R$ by R^2. We define $R^0 = I$.

An example of this composition is the denotation of the phrase *column to the left*:

$$[\textit{left}];[\textit{column}]$$

Intuitively, this procedure first shifts the agent's focus to the field left of the field of the current focus and then to every state with fields of the same column in focus.

Transitive closure

This operation can be generalized to any finite number of operands and will then be called the nth *power* of R. The *transitive reflexive closure* R^* of a binary relation R is defined as follows:

$$R^* = \bigcup_{i \in I} R^i \tag{15.72}$$

The use of the closure operation is to define iteration operations. In the context of computation two types of iteration are conventionally distinguished: *while*-iteration and *until*-iteration. In *while*-iteration, the computation is triggered by a certain condition. In *until*-iteration the computation is started and continued until a certain condition is met to stop it. In [Böttner 1992a] we referred to these operations by symbols $\vdash$ and $\dashv$ with the understanding that the horizontal bar represents a process and the vertical bar represents a state. One should bear in mind though that in general there is no symmetry involved, i.e. $A \vdash R \neq R \dashv A$.

While-iteration

While-iteration is defined as a binary operation

$$A \vdash R = (R \upharpoonright A)^*; (I \upharpoonright \overline{A}) \tag{15.73}$$

This procedure can be described intuitively by the following steps:

1. Go to the top!
2. While you are on a field with no number in it, go one field down!

The operation can be illustrated by the phrase *top number*. It denotes the procedure

$$[top];(\overline{[number]} \vdash [down])$$

Assume the focus of the agent in environment (15.70) is on b_4. The procedure $[top]$ then takes the focus to b_1. Since no number in this field, the triggering condition of procedure $(\overline{[number]} \vdash [down])$ is met and the focus is shifted to the next field below the current field. Since there is a number in this field, the triggering condition of the procedure is no longer met and the procedure stops on the field with 1 in focus.

Until-iteration

The until iteration is defined as follows:

$$R \dashv A = R; (R \upharpoonright \overline{A})^*; (I \upharpoonright A) \tag{15.74}$$

The procedure denoted by this expression can be phrased by the instruction *Move down to the first square with a digit in it!*

An example of this operation is the expression *next number down*. It denotes the procedure

$$[down] \dashv [number] \tag{15.75}$$

If the agent's current focus is on 7_1 in (15.70), this procedure will shift the focus to the field immediately below this field, which is b_3. Since this field has no number in it, the condition is met for iterating the procedure. After applying $[down]$ once more, the focus is shifted to 5_1. Since this field has a number in it, the stopping condition is met and $[down]$ does not apply any more. The shift from 7_1 to 5_1 is exactly what one would expect our procedure to do.

As before we have not given explicit rules of grammar. A more elaborated grammar with a procedural semantics is given in [Böttner 1992a]. It derives semantic trees for instructions of considerable linguistic complexity like, e.g., *Look two numbers down! Look three spots to the left! Look down until you see a bar! Look at the top rightmost number! Write the ones digits of the number in the next space down! Continue looking down!* An example is the semantic tree for *Look at the top number!* represented in Figure 15.4.

15.5 Conclusion

In this chapter we have shown how relation algebra can be fruitfully applied to the semantics of natural language. We have described two approaches, a static one in the standard sense of model theory and a dynamic one in terms of states of an agent understanding natural language.

The construction of a relational grammar for some fragment of English appears to be straightforward and it appears that the approach can be applied to other natural languages as well. But things may not always be as easy as they may seem to be from the examples considered so far. This may become apparent if one tries to express the sentences

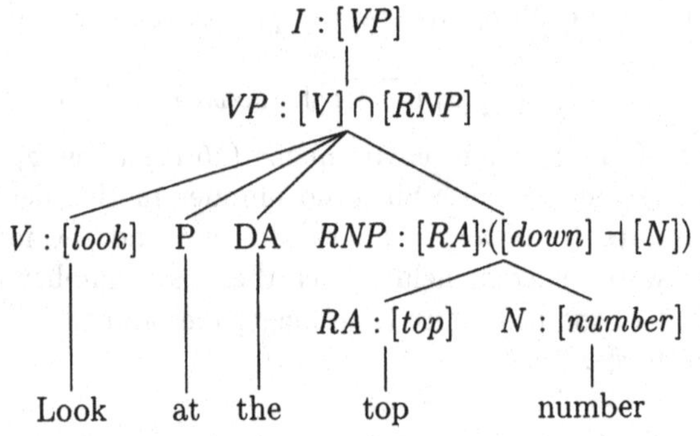

Fig. 15.4 Semantic Tree for *Look at the top number!*

i. *John bought five apples*
ii. *John is running under a tree*

in relation algebraic terms. Although i appears to be similar in structure to, e.g., *John bought some apples*, or *John bought red apples*, which both can be construed in terms of relational algebra, no simple solution along these lines can be given for cardinal expressions. The sentence ii has at least two different readings. According to one reading, the tree is the location of John's running. According to the other reading the tree is the destination of John's running. Whereas the locational meaning can easily be achieved in our model theory, it is not clear how the destinational meaning could be construed.

One major advantage of construing a natural language in terms of relational algebra is that natural languages can be viewed as equational languages (see Sect. 1.2). Equational languages are highly convenient for the purpose of computing inferences. Two caveats may be appropriate, though. First, the fragments of natural languages described by relational grammars are very small. Many important areas of natural language have not been dealt with. What is still missing, for instance, is the extension to relations of higher than binary degree, to mass nouns, or to adverbs. Also there has not yet been proposed a satisfactory treatment of tense. Second, relation algebra will certainly not be sufficient to account for all valid inferences in a natural language. A case in point is the semantics for place and time adverbials: they will require an analysis of space and time. For a first step into the analysis of space see [Crangle, Suppes 1989]. Intuitively, our world is made up of objects, persons, matter, colours, sounds, events, states, processes, actions. One might therefore have sincere doubts whether this abundant variety can be captured by an ontology only of sets and binary relations.

Bibliography

Wolfram Kahl, Thomas Ströhlein

The bibliography below is part of the "RelMiCS bibliography" which can be accessed through the RelMiCS home page
URL: `http://inf2-www.unibw-muenchen.de/relmics/`.

[Abrusci 1991] V. ABRUSCI. *Phase Semantics and Sequent Calculus for Pure Noncommutative Classical Linear Propositional Logic.* J. Symbolic Logic **56** 1403–1451, 1991.

[Aho, Beeri+ 1979] A. H. AHO, C. BEERI, J. D. ULLMAN. *The Theory of Joins in Relational Databases.* ACM Trans. Database Systems **4** 297–314, 1979.

[Alexiev 1994] V. ALEXIEV. *Applications of Linear Logic to Computation: An Overview.* Bull. of the IGPL **2** 77–107, 1994.

[Andréka, Monk+ 1991] H. ANDRÉKA, J. MONK, I. NÉMETI, eds. *Proc. of a Conf. on Algebraic Logic, Budapest, Aug. 8–12, 1988*, Colloq. Math. Soc. János Bolyai **54**. North-Holland, 1991.

[Andréka, Mikulás 1994] H. ANDRÉKA, S. MIKULÁS. *Lambek Calculus and its Relational Semantics: Completeness and Incompleteness.* J. Logic Lang. Inform. **3** 1–38, 1994.

[Armstrong 1974] W. W. ARMSTRONG. *Dependency Structures of Database Relationships.* In: 1974 IFIP Congress, pp. 580–583. North-Holland, 1974.

[Back 1981] R. J. R. BACK. *On Correct Refinement of Programs.* J. Comput. System Sci. **23** 49–68, 1981.

[Back, von Wright 1992] R. J. R. BACK, J. VON WRIGHT. *Combining Angels, Demons and Miracles in Program Specifications.* Theoret. Comput. Sci. **100** 365–383, 1992.

[Backhouse, de Bruin+ 1992] R. BACKHOUSE, P. J. DE BRUIN, P. HOOGENDIJK, G. MALCOLM, T. VOERMANS, J. V. D. WOUDE. *Polynomial Relators.* In M. NIVAT, C. RATTRAY, T. RUS, G. SCOLLO, eds., Proc.of the 2nd Conf.on Algebraic Methodology and Software Technology, AMAST'91, pp. 303–362. Springer, 1992.

[Backhouse, de Bruin+ 1991a] R. C. BACKHOUSE, P. J. DE BRUIN, P. HOOGENDIJK, G. MALCOLM, E. VOERMANS, J. VAN DER WOUDE. *Polynomial Relators.* Computing Science Notes 91/10, Eindhoven Univ. of Technology, Dept. of Mathematics and Computing Science, 1991.

[Backhouse, de Bruin+ 1991b] R. C. BACKHOUSE, P. J. DE BRUIN, G. MALCOLM, E. VOERMANS, J. VAN DER WOUNDE. *Relational Catamorphisms.* In [Möller 1991a], pp. 319–371.

[Backhouse, van der Woude 1993] R. C. BACKHOUSE, J. VAN DER WOUDE. *Demonic Operators and Monotype Factors.* Math. Structures Comput. Sci. **3** 417–433, 1993.

[Backhouse, Doornbos 1994] R. C. BACKHOUSE, H. DOORNBOS. *Mathematical Induction Made Calculational.* Computing Science Notes 94/16, Eindhoven Univ. of Technology, Dept. of Mathematics and Computing Science, 1994.

[Baeten, Weijland 1990] J. BAETEN, W. WEIJLAND. *Process Algebra*, Tracts in Theoretical Computer Science **18**. Cambridge Univ. Press, 1990.

[Barr 1979] M. BARR. *∗-Autonomous Categories*, Lect. Notes in Math. **752**. Springer, 1979.

[Baum, Haeberer+ 1992] G. A. BAUM, A. M. HAEBERER, P. A. VELOSO. *On the Representability of the Abstract Relational Algebra.* IGPL Newsletter **1**, 1992. European Foundation for Logic, Language and Information Interest Group on Programming Logic.

[Baum, Frias+ 1996] G. A. BAUM, M. F. FRIAS, A. M. HAEBERER, P. MARTÍNEZ LÓPEZ. *From Specifications to Programs: A Fork–algebraic Approach to Bridge the Gap.* In: Proceedings of Mathematical Foundations of Computer Science 1996 (MFCS '96),Cracow, Poland, Lect. Notes in Comput. Sci. **1113**, pp. 180–191. Springer, 1996.

[Beaver 1995] D. BEAVER. *Presupposition and Assertion in Dynamic Semantics.* PhD thesis, Univ.Edinburgh, 1995.

[Beeri, Bernstein 1979] C. BEERI, P. BERNSTEIN. *Computational Problems Related to the Design of Normal Form Relational Schemes.* ACM Trans. Database Systems **4** 30–59, 1979.

[Belkhiter, Desharnais+ 1993] N. BELKHITER, J. DESHARNAIS, A. JAOUA, T. MOUKAM. *Providing Relevant Additional Information to Users Asking Queries Using a Galois Lattice Structure.* In: 8th IEEE Internat. Sympos. on Computer and Information Sciences (ISCIS-8), pp. 594–604, Istanbul, 1993.

[Belkhiter, Bourhfir+ 1994] N. BELKHITER, C. BOURHFIR, M. M. GAMMOUDI, A. JAOUA, N. LE THANH, M. REGUIG. *Décomposition Rectangulaire Optimale d'une Relation Binaire: Application aux Bases de Données Documentaires.* Information Science and Operational Research J. **32** 34–54, 1994.

[van Benthem 1976] J. F. VAN BENTHEM. *Modal Correspondence Theory.* PhD thesis, Mathematisch Inst. & Inst. voor Grondslagenonderzoek, Univ. Amsterdam, 1976.

[van Benthem 1989] J. F. VAN BENTHEM. *Semantic Parallels in Natural Language and Computation.* In M. GARRIDO, ed., Logic Colloquium 1988. North-Holland, Amsterdam, 1989.

[van Benthem 1993a] J. F. VAN BENTHEM. *Logic and the flow of information.* In D. PRAWITZ, B. SKYRMS, D. WESTERSTÅHL, eds., Proc. 9th Internat. Congress of Logic, Methodology and Philosophy of Science, Uppsala 1991, pp. 693–724, Amsterdam, 1993. Elsevier.

[van Benthem 1993b] J. F. VAN BENTHEM. *Programming operations that are safe for bisimulations.* CSLI Research Report 93-197, Center for the Study of Language and Information, Stanford Univ., 1993. to appear in Logic Colloquium, 1994, North-Holland.

[van Benthem 1994] J. F. VAN BENTHEM. *Dynamic arrow logic.* In J. V. EIJCK, A. VISSER, eds., Logic and Information Flow. MIT Press, Cambridge, MA, 1994.

[van Benthem, Muskens+ to appear] J. F. VAN BENTHEM, R. MUSKENS, A. VISSER. *Dynamics.* In J. F. VAN BENTHEM, A. TER MEULEN, eds., Handbook of Logic and Language. Elsevier, Amsterdam, to appear.

[Bergeron, Hatcher 1995] M. BERGERON, W. S. HATCHER. *Models of Linear Logic.* In: Zapiski Nauchnykh Seminarov Peterburg. Otdel. Mat. Inst. Steklov (POMI) (Proc. of the Steklov Inst. of Mathematics, St. Petersburg Branch), Vol. 220, pp. 23–35, 1995.

[Berghammer, Zierer 1986] R. BERGHAMMER, H. ZIERER. *Relational Algebraic Semantics of Deterministic and Nondeterministic Programs.* Theoret. Comput. Sci. **43** 123–147, 1986.

[Berghammer, Schmidt 1991] R. BERGHAMMER, G. SCHMIDT. *The RELVIEW-System.* In [Choffrut, Jantzen 1991], pp. 535–536.

[Berghammer, Gritzner+ 1993] R. BERGHAMMER, T. F. GRITZNER, G. SCHMIDT. *Prototyping Relational Specifications Using Higher-Order Objects.* In [Heering, Meinke+ 1993], pp. 56–75.

[Berghammer, Haeberer+ 1993] R. BERGHAMMER, A. M. HAEBERER, G. SCHMIDT, P. A. VELOSO. *Comparing Two Different Approaches to Products in Abstract Relation Algebra.* In [Scollo 1993], pp. 167–176.

[Berghammer, Schmidt 1993] R. BERGHAMMER, G. SCHMIDT. *Relational Specifications.* In C. RAUSZER, ed., Proc. XXXVIII Banach Center Semester on Algebraic Methods in Logic and their Computer Science Applications, Banach Center Publ. **28**, pp. 167–190, Warszawa, 1993. Polish Academy of Sciences, Inst. of Computer Science.

[Berghammer, von Karger 1996] R. BERGHAMMER, B. VON KARGER. *Towards a Design Calculus for CSP.* Sci. Comput. Programming **26** 99–115, 1996.

[Bird 1989] R. S. BIRD. *Lectures on Constructive Functional Programming.* In [Broy 1989], pp. 151–216.

[Bird 1990] R. S. BIRD. *A Calculus of Functions for Program Derivation.* In D. A. TURNER, ed., Research Topics in Functional Programming, The UT Year of Programming Ser., Chapt. 11, pp. 287–308. Addison-Wesley, 1990.

[Bird, de Moor 1992] R. S. BIRD, O. DE MOOR. *From Dynamic Programming to Greedy Algorithms.* In B. MÖLLER, H. PARTSCH, S. SCHUMAN, eds., Formal Program Development: Proc. of an IFIP TG2/WG 2.1 State of the Art Seminar, Rio de Janeiro, Jan. 1992, Lect. Notes in Comput. Sci. **755**, pp. 43–61. Springer, 1992.

[Biskup 1978] J. BISKUP. *On the Complementation Rule for Multivalued Dependencies in Database Relations.* Acta Inform. **10** 297–305, 1978.

[Biskup 1980] J. BISKUP. *Inference of Multivalued Dependencies in Fixed and Undetermined Universes.* Theoret. Comput. Sci. **10** 93–106, 1980.

[Blackburn, Venema 1993] P. BLACKBURN, Y. VENEMA. *Dynamic squares.* Logic Group Preprint Series 92, Dept. of Philosophy, Utrecht Univ., 1993. To appear in J. Philos. Logic.

[Blackburn, de Rijke+ 1994] P. BLACKBURN, M. DE RIJKE, Y. VENEMA. *The algebra of modal logic.* CWI Report CS-R9463, CWI Amsterdam, 1994.

[Bojanowski, Iglewski+ 1994] J. BOJANOWSKI, M. IGLEWSKI, J. MADEY, A. OBAID. *Functional approach to Protocols Specification.* In: Proc. of the 14th Internat. IFIP Sympos. on Protocol Specification, Testing and Verification, PSTV'94, Vancouver, B.C., 7-10 June 1994, pp. 371–378, 1994.

[Boole 1847] G. BOOLE. *The Mathematical Analysis of Logic, Being an Essay Toward a Calculus of Deductive Reasoning.* Macmillan, Cambridge, 1847.

[Böttner 1992a] M. BÖTTNER. *State transition semantics.* Theoret. Linguist. **18** 239–286, 1992.

[Böttner 1992b] M. BÖTTNER. *Variable-free semantics for anaphora.* J. Philos. Logic **21** 375–390, 1992.

[Böttner 1994] M. BÖTTNER. *Open Problems in Relational Grammar.* In P. HUMPHREYS, ed., Patrick Suppes: Scientific Philosopher, Vol. 3, pp. 19–39. Kluwer, Dordrecht, 1994.

[Böttner 1996] M. BÖTTNER. *A collective extension of relational grammar.* J. of the Interest Group in Pure and Applied Logics , 1996. to appear.

[Boudriga, Elloumi+ 1992] N. BOUDRIGA, F. ELLOUMI, A. MILI. *On the Lattice of Specifications: Applications to a Specification Methodology.* Formal Aspects of Computing **4** 544–571, 1992.

[Brink 1977] C. BRINK. *On Birkhoff's Postulates for a Relation Algebra.* J. London Math. Soc. **15** 391–394, 1977.

[Brink 1979] C. BRINK. *Two Axiom Systems for Relation Algebras.* Notre Dame J. Formal Logic **20** 909–914, 1979.

[Brink 1988] C. BRINK. *On the Application of Relations.* South African J. of Philosophy **7(2)** 105–112, 1988.

[Brink 1993] C. BRINK. *Power Structures.* Algebra Universalis **30** 177–216, 1993.

[Brink, Britz+ 1993] C. BRINK, K. BRITZ, A. MELTON. *A note on fuzzy power relations.* Fuzzy Sets and Systems **54** 115–117, 1993.

[Brink, Gabbay+ 1995] C. BRINK, D. GABBAY, H. J. OHLBACH. *Towards Automating Duality.* Comput. Math. Appl. **29** 73–90, 1995.

[Brink, Rewitzky 1995] C. BRINK, I. REWITZKY. *Predicate Transformers as Power Operations.* Formal Aspects of Computing **7** 169–182, 1995.

[Brookes, Roscoe 1984] S. BROOKES, A. ROSCOE. *An improved failure model for communicating sequential processes.* In: Proc. of the NFS-SERC Seminar on Concurrency, Lect. Notes in Comput. Sci. **197**, pp. 281–305. Springer, 1984.

[Brown, Gurr 1993] C. BROWN, D. GURR. *A Representation Theorem for Quantales.* J. Pure and Applied Algebra **85** 27–42, 1993.

[Broy 1989] M. BROY, ed. *Constructive Methods in Computing Science*, NATO ASI Ser. F **F55**. Springer, 1989.

[Bunge 1967] M. BUNGE. *Scientific Research I, The Search for System.* Springer, Berlin, 1967.

[Burkhardt 1980] H. BURKHARDT. *Logik und Semiotik in der Philosophie von Leibniz.* Philosophia Verlag, München, 1980.

[Burstall, Darlington 1977] R. BURSTALL, J. DARLINGTON. *A transformation system for developing recursive programs.* J. Assoc. Comput. Mach. **24** 44–67, 1977.

[Chin, Tarski 1951] L. H. CHIN, A. TARSKI. *Distributive and Modular Laws in the Arithmetic of Relation Algebras.* Univ. California Publ. Math. **1** 341–384, 1951.

[Choffrut, Jantzen 1991] C. CHOFFRUT, M. JANTZEN, eds. *STACS 91, 8th Annual Sympos. on Theoretical Aspects of Computer Science*, Lect. Notes in Comput. Sci. **480**, Hamburg, 1991. Springer.

[Codd 1970] E. F. CODD. *A Relational Model of Data for Large Shared Data Banks.* Comm. ACM **13** 377–387, 1970.

[Crangle, Suppes 1987] C. CRANGLE, P. SUPPES. *Context-fixing semantics for instructable robots.* International Journal of Man-Machine Studies **27** 371–400, 1987.

[Crangle, Suppes 1989] C. CRANGLE, P. SUPPES. *Geometrical semantics for spatial prepositions.* Midwest Studies in Philosophy **XIV** 399–422, 1989.

[Davey, Priestley 1990] B. A. DAVEY, H. A. PRIESTLEY. *Introduction to Lattices and Order.* Cambridge Univ. Press, 1990.

[De Morgan 1856] A. DE MORGAN. *On the Symbols of Logic, the Theory of the Syllogism, and in Particular of the Copula, and the Application of the Theory of Probabilities to some Questions in the Theory of Evidence.* Trans. of the Cambridge Philosophical Society **9** 79–127, 1856. (read February 25, 1850) Reprinted in [De Morgan 1966].

[De Morgan 1864] A. DE MORGAN. *On the Syllogism: III, and on Logic in General.* Trans. of the Cambridge Philosophical Society **10** 173–230, 1864. (read February 8, 1858) Reprinted in [De Morgan 1966].

[De Morgan 1966] A. DE MORGAN. *On the Syllogism, and Other Logical Writings.* Yale Univ. Press, New Haven, 1966.

[de Roever, Jr. 1972] W. P. DE ROEVER, JR. *A Formalization of Various Parameter Mechanisms as Products of Relations within a Calculus of Recursive Program Schemes.* In: Théorie des Algorithmes, des Languages et de la Programmation, Séminaires IRIA, pp. 55–88. IRIA, Rocquencourt, 1972.

[Delobel 1978] C. DELOBEL. *Normalization and Hierarchical Dependencies in the Relational Data Model.* ACM Trans. Database Systems **2** 201–222, 1978.

[Demri, Orlowska+ 1994] S. DEMRI, E. ORLOWSKA, I. REWITZKY. *Towards Reasoning about Hoare Relations.* Ann. Math. Artificial Intelligence **12** 265–289, 1994.

[Demri, Orlowska 1996] S. DEMRI, E. ORLOWSKA. *Logical Analysis of Demonic Nondeterministic Programs.* Theoret. Comput. Sci. **166**, 1996. To appear.

[Desharnais, Belkhiter+ 1995] J. DESHARNAIS, N. BELKHITER, S. BEN MOHAMED SGHAIER, F. TCHIER, A. JAOUA, A. MILI, N. ZAGUIA. *Embedding a Demonic Semilattice in a Relation Algebra.* Theoret. Comput. Sci. **149** 333–360, 1995.

[Desrosiers, Iglewski+ 1995] B. DESROSIERS, M. IGLEWSKI, A. OBAID. *Utilisation de la méthode de traces pour la définition formelle d'un protocole de communication.* Electronic J. on Networks and Distributed Processing **2** 57–73, 1995.

[Dijkstra 1974] E. W. DIJKSTRA. *A simple axiomatic basis for programming language constructs.* Indag. Math. **36** 1–15, 1974.

[Dijkstra 1975] E. W. DIJKSTRA. *Guarded commands, nondeterminacy and formal derivation of programs.* Comm. ACM **18** 453–457, 1975.

[Dijkstra 1976] E. W. DIJKSTRA. *A Discipline of Programming.* Prentice-Hall, 1976.

[Dijkstra, Scholten 1990] E. W. DIJKSTRA, C. S. SCHOLTEN. *Predicate Calculus and Program Semantics.* Texts Monogr. Comput. Sci. Springer, 1990.

[Dijkstra, Feijen 1984] E. DIJKSTRA, W. FEIJEN. *Een Methode van Programmeren.* Academic Service, Den Haag, 1984. Also available as *A Method of Programming*, Addison-Wesley, Reading, Mass., 1988.

[Doornbos 1994] H. DOORNBOS. *A Relational Model of Programs Without the Restriction to Egli-Milner Monotone Constructs.* In E.-R. OLDEROG, ed., Programming Concepts, Methods and Calculi (ProCoMet '94), IFIP Transactions **A-56**, pp. 363–382. North-Holland, 1994.

[Doornbos, Backhouse 1995] H. DOORNBOS, R. BACKHOUSE. *Induction and recursion on datatypes.* In B. MÖLLER, ed., Mathematics of Program Construction, 3rd Internat. Conf., Lect. Notes in Comput. Sci. **947**, pp. 242–256. Springer, 1995.

[Doornbos 1996] H. DOORNBOS. *Reductivity arguments and program construction.* PhD thesis, Eindhoven Univ. of Technology, Dept. of Mathematics and Computing Science, 1996.

[Eijck 1994] J. v. EIJCK. *Presupposition failure — a comedy of errors.* Formal Aspects of Computing **6A** 766–787, 1994.

[Everett 1944] C. J. EVERETT. *Closure Operators and Galois Theory in Lattices.* Trans. Amer. Math. Soc. **55** 514–525, 1944.

[Everett, Ulam 1946] C. J. EVERETT, S. M. ULAM. *Projective Algebra I.* Amer. J. Math. **68** 77–88, 1946.

[Fischer, Ladner 1979] M. FISCHER, R. LADNER. *Propositional dynamic logic of regular programs.* J. Comput. System Sci. **18** 194–211, 1979.

[Floyd 1967] R. W. FLOYD. *Assigning meaning to programs.* In J. T. SCHWARTZ, ed., Mathematical Aspects of Computer Science, Proc. Sympos. in Appl. Math., pp. 19–32. Amer. Math. Soc., 1967.

[Fokkinga 1992] M. FOKKINGA. *Law and Order in Algorithmics.* PhD thesis, Twente Univ., 1992.

[Frappier 1995] M. FRAPPIER. *A Relational Basis for Program Construction by Parts.* PhD thesis, Univ. Ottawa, Computer Science Dept., 150 Louis Pasteur, Ottawa, ON, K1N 6N5, Canada, 1995.

[Frappier, Mili+ 1995] M. FRAPPIER, A. MILI, J. DESHARNAIS. *Program Construction by Parts.* In B. MÖLLER, ed., Mathematics of Program Construction (MPC'95), Lect. Notes in Comput. Sci. **947**, pp. 257–281. Springer, 1995.

[Frias, Aguayo+ 1993] M. F. FRIAS, N. AGUAYO, B. NOVAK. *Development of Graph Algorithms with Fork Algebras.* In: Proc. of the XIX Latinamerican Conf. on Informatics, pp. 529–554, 1993.

[Frias, Aguayo 1994] M. F. FRIAS, N. AGUAYO. *Natural Specifications vs. Abstract Specifications. A Relational Approach.* In: Proc. of SOFSEM '94, Milovy, Czech Republic, pp. 17–22, 1994.

[Frias, Baum 1995] M. F. FRIAS, G. A. BAUM. *On the Exact Expressiveness and Probability of Fork Algebras.* In: Abstracts of the 10th Latinamerican Sympos. on Mathematical Logic, Colombia, 1995.

[Frias, Baum+ 1995] M. F. FRIAS, G. A. BAUM, A. M. HAEBERER, P. A. VELOSO. *Fork Algebras are Representable.* Bull. of the Sect. of Logic, Univ. of Łódź **24** 64–75, 1995.

[Frias, Gordillo 1995] M. F. FRIAS, S. GORDILLO. *Semantical Optimization of Queries in Deductive Object-Oriented Databases.* In: Proc. of ADBIS'95, Moscow, pp. 55–72. Springer, 1995.

[Frias, Haeberer+ 1995a] M. F. FRIAS, A. M. HAEBERER, P. A. VELOSO. *On the Metalogical Properties of Fork Algebras.* Bull. Symbolic Logic **1** 364–365, 1995.

[Frias, Haeberer+ 1995b] M. F. FRIAS, A. M. HAEBERER, P. A. VELOSO, G. A. BAUM. *Representability of Fork Algebras.* Bull. Symbolic Logic **1** 234–235, 1995.

[Gärdenfors 1988] P. GÄRDENFORS. *Knowledge in Flux.* MIT Press, Cambridge, MA, 1988.

[Gargov, Passy+ 1987] G. GARGOV, S. PASSY, T. TINCHEV. *Modal Environment for Boolean Speculations.* In D. SKORDEV, ed., Mathematical Logic and Applications, pp. 253–263, New York, 1987. Plenum Press.

[Girard 1987] J.-Y. GIRARD. *Linear Logic.* Theoret. Comput. Sci. **50** 1–102, 1987.

[Girard 1989] J.-Y. GIRARD. *Towards a Geometry of Interaction.* In J. W. GRAY, A. SCEDROV, eds., Categories in Computer Science and Logic, Contemporary Mathematics **92**, pp. 69–108. Amer. Math. Soc., 1989.

[Girard 1995] J.-Y. GIRARD. *Linear Logic: Its Syntax and Semantics.* In J.-Y. GIRARD, Y. LAFONT, L. REGNIER, eds., Advances in Linear Logic, Workshop on Linear Logic, 1993, pp. 1–42. Cambridge Univ. Press, 1995.

[Givant 1994] S. R. GIVANT. *The Structure of Relation Algebras Generated by Relativizations*, Contemporary Mathematics **156**. Amer. Math. Soc., Providence, 1994.

[Goldblatt 1987] R. GOLDBLATT. *Logics of Time and Computation.* CSLI Publications, Stanford, 1987.

[Gries 1981] D. GRIES. *The Science of Programming.* Springer, New York, 1981.

[Gritzner, Berghammer 1996] T. F. GRITZNER, R. BERGHAMMER. *A Relation Algebraic Model of Robust Correctness.* Theoret. Comput. Sci. **159** 245–270, 1996.

[Groenendijk, Stokhof 1991] J. GROENENDIJK, M. STOKHOF. *Dynamic Predicate Logic.* Linguistics and Philosophy **14** 39–100, 1991.

[Groeneveld, Veltman 1994] W. GROENEVELD, F. VELTMAN. *Inference Systems for Update Semantics.* Manuscript, ILLC, Amsterdam, 1994.

[Gurevich 1991] Y. GUREVICH. *Evolving algebras: A tutorial introduction.* Bull. of the European Association for Theoretical Computer Science (EATCS) **43** 264–286, 1991.

[Haeberer, Veloso 1991] A. M. HAEBERER, P. A. VELOSO. *Partial Relations for Program Derivation: Adequacy, Inevitability and Expressiveness.* In [Möller 1991a], pp. 319–371.

[Haeberer, Baum+ 1994] A. M. HAEBERER, G. A. BAUM, G. SCHMIDT. *On the smooth calculation of relational recursive expressions out of first-order non-constructive specifications involving quantifiers.* In D. BJØRNER, M. BROY, I. POTTOSIN, eds., Formal Methods in Programming and Their Applications, Proc. Internat. Conf. Novosibirsk, Jun 28–Jul 3, 1993, Lect. Notes in Comput. Sci. **735**, pp. 403–420. Springer, 1994.

[Halmos 1962] P. R. HALMOS. *Algebraic Logic.* Chelsea, New York, 1962.

[Harel 1984] D. HAREL. *Dynamic Logic.* In D. GABBAY, F. GUENTHNER, eds., Handbook of Philosophical Logic, Vol. II, pp. 497–604. Reidel, Dordrecht, 1984.

[Hattensperger, Berghammer+ 1993] C. HATTENSPERGER, R. BERGHAMMER, G. SCHMIDT. *RALF — A Relation-Algebraic Formula Manipulation System and Proof Checker. Notes to a System Demonstration.* In [Scollo 1993], pp. 405–406.

[Heering, Meinke+ 1993] J. HEERING, K. MEINKE, B. MÖLLER, T. NIPKOW, eds. *Higher-Order Algebra, Logic and Term Rewriting, 1st Internat. Workshop, HOA '93, Amsterdam, Sept. 1993, Selected Papers*, Lect. Notes in Comput. Sci. **816**. Springer, 1993.

[Hehner 1984] E. HEHNER. *Predicative Programming, Parts I and II.* Comm. ACM **27** 134–151, 1984.

[van Heijenoort 1967] J. VAN HEIJENOORT. *From Frege to Gödel: A Source Book in Mathematical Logic, 1879–1931.* Harvard Univ. Press, Cambridge, MA, 1967.

[Heim 1983] I. HEIM. *File change semantics and the familiarity theory of definites.* In R. BÄUERLE, C. SCHWARZE, A. V. STECHOW, eds., Meaning, Use and Interpretation of Language. de Gruyter, Berlin, 1983.

[Heinle 1995] W. HEINLE. *Expressivity and Definability in Extended Modal Languages.* Shaker, Aachen, 1995.

[Heninger, Kallander+ 1978] K. HENINGER, J. KALLANDER, D. L. PARNAS, J. SHORE. *Software Requirements for the A-7E Aircraft.* NRL Memorandum Report 3876, United States Naval Research Laboratory, Washington DC, 1978.

[Heninger 1980] K. HENINGER. *Specifying Software Requirements for Complex Systems: New Techniques and their Application.* IEEE Trans. on Software Engineering **6** 2–13, 1980.

[Henkin, Monk+ 1971] L. HENKIN, J. D. MONK, A. TARSKI. *Cylindric Algebras, Part I.* North-Holland, Amsterdam, 1971.

[Henkin 1973] L. HENKIN. *Internal Semantics and Algebraic Logic.* In H. LEBLANC, ed., Truth, Syntax, and Modality, Studies in Logic **68**, pp. 111–127. North-Holland, Amsterdam, 1973.

[Henkin, Monk+ 1985] L. HENKIN, J. D. MONK, A. TARSKI. *Cylindric Algebras, Part II.* North-Holland, Amsterdam, 1985.

[Hennessy, Ashcroft 1976] M. HENNESSY, E. ASHCROFT. *The Semantics of Nondeterminism.* In: Third ICALP, Edinburgh, pp. 478–493, 1976.

[Hester, Parnas+ 1981] S. HESTER, D. PARNAS, D. UTTER. *Using Documentation as a Software Design Medium.* Bell System Tech. J. **60** 1941–1977, 1981.

[Hitchcock, Park 1973] P. HITCHCOCK, D. PARK. *Induction rules and termination proofs.* In M. NIVAT, ed., Proc. Automata, Languages and Programming (ICALP '72), Rocquencourt, France, July 1972, pp. 225–251. North-Holland, 1973.

[Hoare 1969] C. HOARE. *An axiomatic basis for computer programming.* Comm. ACM **12** 578–580, 1969.

[Hoare, He 1986a] C. HOARE, J. HE. *The Weakest Prespecification, Part I.* Fund. Inform. **4** 51–54, 1986.

[Hoare, He 1986b] C. HOARE, J HE. *The Weakest Prespecification, Part II.* Fund. Inform. **4** 217–252, 1986.

[Hoare+ 1987] C. HOARE et al. *Laws of Programming.* Comm. ACM **30** 672–686, 1987. Corrigenda in **30**, 9, p. 770.

[Hoare, He 1987] C. HOARE, J. HE. *The Weakest Prespecification.* Inform. Process. Lett. **24** 127–132, 1987.

[Hoare, von Karger 1995] C. HOARE, B. VON KARGER. *Sequential Calculus.* Inform. Process. Lett. **53** 123–130, 1995.

[Hollenberg, Vermeulen 1994] M. HOLLENBERG, K. VERMEULEN. *Counting variables in a dynamic setting.* Technical report, Dept. of Philosophy, Utrecht Univ., 1994.

[Huntington 1933a] E. V. HUNTINGTON. *Boolean Algebra. A Correction.* Trans. Amer. Math. Soc. **35** 557–558, 1933.

[Huntington 1933b] E. V. HUNTINGTON. *New Sets of Independent Postulates for the Algebra of Logic, with Special Reference to Whitehead and Russell's Principia Mathematica.* Trans. Amer. Math. Soc. **35** 274–304, 1933.

[Iglewski, Madey 1996] M. IGLEWSKI, J. MADEY. *Software Engineering Issues Emerged from Critical Control Applications.* In: 2nd IFAC Workshop on Safety and Reliability in Emerging Control Technologies, Daytona Beach, FL, 1-3 November 1995. Elsevier, 1996.

[Ishiguro 1990] H. ISHIGURO. *Leibniz' Philosophy of Logic and Language.* Duckworth, London, 2 edition, 1990.

[Janicki 1995] R. JANICKI. *Towards a Formal Semantics of Parnas Tables.* In: Proc. of the 17th Internat. Conf. on Software Engineering, Seattle, WA, pp. 231–240, 1995.

[Jaoua 1987] A. JAOUA. *Recouvrement avant de Programmes sous les Hypothèses de Spécifications Déterministes et non-déterministes.* Diss. de Doctorat d'Etat dès sciences, Univ. de Toulouse, 1987.

[Jaoua, Beaudry 1989] A. JAOUA, M. BEAUDRY. *Difunctional Relations: A Formal Tool for Program Design.* Rapport de recherche no 55, Département de Mathématique et d'Informatique, Univ. de Sherbrooke, Québec, Canada, 1989.

[Jaoua, Boudriga+ 1991] A. JAOUA, N. BOUDRIGA, J. L. DURIEUX, A. MILI. *Regularity of Relations: A Measure of Uniformity.* Theoret. Comput. Sci. **79** 323–339, 1991.

[Jaoua, Ounalli+ 1994] A. JAOUA, H. OUNALLI, N. BELKHITER. *Automatic Entity Extraction From an n-ary Relation: Towards a General Law for Information Decomposition.* In: Joint Conf. on Information Sciences (JCIS), pp. 92–95, Pinehurst, Duke Univ., NC, 1994.

[Jaspars, Krahmer 1995] J. JASPARS, E. KRAHMER. *Unified dynamics.* Technical Report CS-R95, CWI, Amsterdam, 1995.

[Jónsson, Tarski 1948] B. JÓNSSON, A. TARSKI. *Boolean Algebras with Operators.* Bull. Amer. Math. Soc. **54** 79–80, 1948. Abstract 88.

[Jónsson, Tarski 1951] B. JÓNSSON, A. TARSKI. *Boolean Algebras with Operators, Part I.* Amer. J. Math. **73** 891–939, 1951.

[Jónsson, Tarski 1952] B. JÓNSSON, A. TARSKI. *Boolean Algebras with Operators, Part II.* Amer. J. Math. **74** 127–167, 1952.

[Jónsson 1959] B. JÓNSSON. *Representation of Modular Lattices and of Relation Algebras.* Trans. Amer. Math. Soc. **92** 449–464, 1959.

[Jónsson 1982] B. JÓNSSON. *Varieties of Relation Algebras.* Algebra Universalis **15** 273–298, 1982.

[Jónsson 1988] B. JÓNSSON. *Relation Algebras and Schröder Categories.* Discrete Math. **70** 27–45, 1988.

[Kahl 1996a] W. KAHL. *Algebraic Graph Derivations for Graphical Calculi.* In G. AUSIELLO, A. MARCHETTI-SPACCAMELA, eds., Graph Theoretical Concepts in Computer Science, Proc. of WG '96, June 12-14, Caddenabbia, Como, Italy, Lect. Notes in Comput. Sci. Springer, 1996.

[Kahl 1996b] W. KAHL. *Algebraische Termgraphersetzung mit gebundenen Variablen.* Reihe Informatik. Herbert Utz Verlag Wissenschaft, München, 1996. also Doctoral Diss. at Univ. der Bundeswehr München, Fakultät für Informatik.

[Kamp, Reyle 1993] H. KAMP, U. REYLE. *From Discourse to Logic.* Kluwer, Dordrecht, 1993.

[Kanazawa 1994] M. KANAZAWA. *Completeness and decidability of the mixed style of inference with composition.* In P. DEKKER, M. STOKHOF, eds., Proc. of the 9th Amsterdam Colloq., pp. 377–391, Amsterdam, 1994. ILLC.

[Kawahara 1973] Y. KAWAHARA. *Relations in Categories with Pullbacks.* Mem. Fac. Sci. Kyushu Univ. Ser. A **27** 149–173, 1973.

[Kawahara 1978] Y. KAWAHARA. *A Relation Theoretic Proof of a Tripleability Theorem over Exact Categories.* Bull. Kyushu Inst. Tech. Math. Natur. Sci. **25** 31–40, 1978.

[Kawahara 1988] Y. KAWAHARA. *Applications of Relational Calculus to Computer Mathematics.* Bull. Inform. Cybernet. **23** 67–78, 1988.

[Kawahara 1990] Y. KAWAHARA. *Pushout-Complements and Basic Concepts of Grammars in Toposes.* Theoret. Comput. Sci. **77** 267–289, 1990.

[Kawahara, Mizoguchi 1992] Y. KAWAHARA, Y. MIZOGUCHI. *Categorical Assertion Semantics in Topoi.* Adv. Software Sci. Tech. **4** 137–150, 1992.

[Keenan, Faltz 1978] E. KEENAN, L. FALTZ. *Logical Types for Natural Language.* UCLA Occasional Papers in Linguistics , 1978.

[Keenan, Faltz 1985] E. KEENAN, L. FALTZ. *Boolean Semantics for Natural Language.* Reidel, Dordrecht, 1985.

[Konikowska 1987] B. KONIKOWSKA. *A formal language for reasoning about indiscernibility.* Bull. Polish Acad. Sci. Math. **35** 239–249, 1987.

[Konikowska 1994] B. KONIKOWSKA. *A logic for reasoning about similarity.* In E. ORLOWSKA, ed., Reasoning with incomplete information. 1994. In preparation for publication.

[Kozen 1981] D. KOZEN. *On the duality of dynamic algebras and Kripke models.* In: Logic of Programs 1981, Lect. Notes in Comput. Sci. **651**, pp. 1–11, Berlin, 1981. Springer.

[Kripke 1963] S. KRIPKE. *Semantical analysis of modal logic I.* Z. Math. Logik Grundlag. Math. **9** 67–96, 1963.

[Kripke 1965] S. KRIPKE. *Semantical analysis of intuitionistic logic.* In J. CROSSLEY, M. DUMMETT, eds., Formal Systems and Recursive Functions, Amsterdam, 1965. North-Holland.

[Ladkin, Maddux 1994] P. B. LADKIN, R. D. MADDUX. *On Binary Constraint Problems.* J. Assoc. Comput. Mach. **41** 435–469, 1994.

[Lambek 1993] J. LAMBEK. *From Categorial Grammar to Bilinear Logic.* In K. DOŠEN, P. SCHROEDER-HEISTER, eds., Substructural Logics, pp. 207–238. Oxford Univ. Press, 1993.

[Lincoln 1992] P. LINCOLN. *Linear Logic.* ACM SIGACT News **23** 29–37, 1992.

[Lincoln, Mitchell+ 1992] P. LINCOLN, J. MITCHELL, A. SCEDROV, N. SHANKAR. *Decision Problems for Propositional Linear Logic.* Ann. Pure Appl. Logic **56** 239–311, 1992.

[LinLogList] Electronic forum on linear logic. To ask for registration, send e-mail to `linear-request@cs.stanford.edu`.

[Löwenheim 1915] L. LÖWENHEIM. *Über Möglichkeiten im Relativkalkül.* Math. Ann. **76** 447–470, 1915. English translation in [van Heijenoort 1967].

[Lyndon 1950] R. C. LYNDON. *The Representation of Relational Algebras.* Ann. of Math. (2) **51** 707–729, 1950.

[Lyndon 1956] R. C. LYNDON. *The Representation of Relation Algebras, II.* Ann. of Math. (2) **63** 294–307, 1956.

[Lyndon 1961] R. C. LYNDON. *Relation Algebras and Projective Geometries.* Michigan Math. J. **8** 21–28, 1961.

[Maddux, Tarski 1976] R. D. MADDUX, A. TARSKI. *A Sufficient Condition for the Representability of Relation Algebras.* Notices Amer. Math. Soc. **23** A–447, 1976. Reprinted in Alfred Tarski: Collected Papers (4 vols.), Birkhäuser.

[Maddux 1996] R. D. MADDUX. *Relation-Algebraic Semantics.* Theoret. Comput. Sci. **160** 1–85, 1996.

[Maddux 1976] R. D. MADDUX. *Some Nonrepresentable Relation Algebras.* Notices Amer. Math. Soc. **23** A–431, A–557, 1976.

[Maddux 1978a] R. D. MADDUX. *Some Sufficient Conditions for the Representability of Relation Algebras.* Algebra Universalis **8** 162–172, 1978.

[Maddux 1978b] R. D. MADDUX. *Topics in Relation Algebras.* Univ. of California Press, Berkeley, 1978. Doctoral Diss.

[Maddux 1982] R. D. MADDUX. *Some Varieties Containing Relation Algebras.* Trans. Amer. Math. Soc. **272** 501–526, 1982.

[Maddux 1983] R. D. MADDUX. *A Sequent Calculus for Relation Algebras.* Ann. Pure Appl. Logic **25** 73–101, 1983.

[Maddux 1985] R. D. MADDUX. *Finite Integral Relation Algebras.* In: Universal Algebra and Lattice Theory, Proc. of the Southeastern Conf. in Universal Algebra and Lattice Theory, July 11–14, 1984, Lect. Notes in Math. **1149**, pp. 175–197. Springer, 1985.

[Maddux 1989] R. D. MADDUX. *Nonfinite Axiomatizability Results for Cylindric and Relation Algebras.* J. Symbolic Logic **54** 951–974, 1989.

[Maddux 1991a] R. D. MADDUX. *Introductory Course on Relation Algebras, Finite-dimensional Cylindric Algebras, and their Interconnections.* In [Andréka, Monk+ 1991], pp. 361–392.

[Maddux 1991b] R. D. MADDUX. *The Origin of Relation Algebras in the Development and Axiomatization of the Calculus of Relations.* Studia Logica **50** 421–455, 1991.

[Maddux 1991c] R. D. MADDUX. *Pair-dense Relation Algebras.* Trans. Amer. Math. Soc. **328** 83–131, 1991.

[Maddux 1992] R. D. MADDUX. *A working relational model: The derivation of the Dijkstra-Scholten predicate transformer semantics from Tarski's axioms for the Peirce-Schröder calculus of relations.* Technical report, Dept. of Mathematics, Iowa State Univ., Ames, Iowa 50011, USA, 1992. Superseded by [Maddux 1993].

[Maddux 1993] R. D. MADDUX. *A working relational model: The derivation of the Dijkstra-Scholten predicate transformer semantics from Tarski's axioms for the Peirce-Schröder calculus of relations.* South African Computer J. **9** 92–130, 1993.

[Maddux 1994a] R. D. MADDUX. *A Perspective on the Theory of Relation Algebras.* Algebra Universalis **31** 456–465, 1994.

[Maddux 1994b] R. D. MADDUX. *Undecidable semiassociative relation algebras.* J. Symbolic Logic **59** 398–418, 1994.

[Maier 1983] D. MAIER. *The Theory of Relational Databases.* Computer Science Press, Rockville, MD, 1983.

[Marx] M. MARX. *Dynamic arrow logic with pairs.* In M. MARX, L. POLOS, eds., Arrow Logic and Multi-Modal Logic, Studies in Logic, Language and Information. CSLI Publications, Stanford. to appear.

[Marx, Venema] M. MARX, Y. VENEMA. *Multi-Dimensional Modal Logic.* Kluwer. to appear.

[McKenzie 1970] R. N. W. MCKENZIE. *The Representation of Integral Relation Algebras.* Michigan Math. J. **17** 279–287, 1970.

[Meertens 1986] L. MEERTENS. *Algorithmics: Towards Programming as a Mathematical Activity.* In J. W. DE BAKKER, M. HAZEWINKEL, J. K. LENSTRA, eds., Proc. CWI Sympos. on Mathematics and Computer Science, pp. 289–334. North-Holland, 1986.

[Mendelzon 1979] A. O. MENDELZON. *On Axiomatizing Multivalued Dependencies in Relational Databases.* J. Assoc. Comput. Mach. **26** 37–44, 1979.

[Mikulás 1992] S. MIKULÁS. *The completeness of the Lambek calculus with respect to relational semantics.* Itli prepublications, Inst. for Language, Logic and Information, Amsterdam, 1992.

[Mikulás, Sain+ 1992] S. MIKULÁS, I. SAIN, A. SIMON. *Complexity of the Equational Theory of Relational Algebras with Projection Elements.* Bull. of the Sect. of Logic, Univ. of Łódź **21** 103–111, 1992.

[Mikulás 1995] S. MIKULÁS. *Taming Logics.* PhD thesis, ILLC Diss. Series 1995–12, 1995.

[Mili 1983] A. MILI. *A Relational Approach to the Design of Deterministic Programs.* Acta Inform. **20** 315–328, 1983.

[Mili 1985] A. MILI. *Towards a Theory of Forward Error Recovery.* IEEE Trans. on Software Engineering **11** 735–748, 1985.

[Mili, Desharnais+ 1987] A. MILI, J. DESHARNAIS, F. MILI. *Relational Heuristics for the Design of Deterministic Programs.* Acta Inform. **24** 239–276, 1987.

[Mili, Mili 1992] F. MILI, A. MILI. *Heuristics for Constructing While Loops.* Sci. Comput. Programming **18** 67–106, 1992.

[Mills 1975] H. D. MILLS. *The New Math of Computer Programming.* Comm. ACM **18** 43–48, 1975.

[Mills, Basili+ 1987] H. D. MILLS, V. R. BASILI, J. D. GANNON, R. G. HAMLET. *Principles of Computer Programming. A Mathematical Approach.* Allyn and Bacon, 1987.

[Mirkowska 1977] G. MIRKOWSKA. *Algorithmic logic and its application in the theory of programs.* Fund. Inform. **1** 1–17, 147–165, 1977.

[Möller 1991a] B. MÖLLER, ed. *Constructing Programs From Specifications — Proc. of the IFIP TC2 Working Conf. on Constructing Programs From Specifications.* IFIP WG 2.1, North-Holland, 1991.

[Möller 1991b] B. MÖLLER. *Relations as Program Development Language.* In [Möller 1991a], pp. 319–371.

[Monk 1964] J. D. MONK. *On Representable Relation Algebras.* Michigan Math. J. **11** 207–210, 1964.

[Monk 1969] J. D. MONK. *Nonfinitizability of Classes of Representable Cylindric Algebras.* J. Symbolic Logic **34** 331–343, 1969.

[Monk 1971] J. D. MONK. *Provability with Finitely Many Variables.* Proc. Amer. Math. Soc. **27** 353–358, 1971.

[Montague 1974] R. MONTAGUE. *Formal Philosophy.* Yale Univ. Press, New Haven, 1974.

[de Moor, Swierstra 1992] O. DE MOOR, D. SWIERSTRA. *Virtual Data Structures.* Presented at IFIP WG 2.1 state of the art summer school, Itacuruçá Island, Brazil, Jan. 10-23, 1992. to appear., 1992.

[Morgan, Robinson 1987] C. MORGAN, K. ROBINSON. *Specification Statements and Refinement.* IBM J. Res. Dev. **31** 49–68, 1987.

[Morris 1987] J. M. MORRIS. *A Theoretical Basis for Stepwise Refinement and the Programming Calculus.* Sci. Comput. Programming **9** 287–306, 1987.

[Moshier 1995] M. MOSHIER. *Featureless HPSG*, 1995. Unpublished manuscript.

[Moss, Johnson 1995] L. MOSS, D. JOHNSON. *Dynamic interpretations of constraint-based grammar formalisms.* J. Logic Lang. Inform. **4** 61–79, 1995.

[Nguyen 1991] T. T. NGUYEN. *A Relational Model of Demonic Nondeterministic Programs.* Internat. J. Found. Comput. Sci. **2** 101–131, 1991.

[Nicolas 1978] J. M. NICOLAS. *Mutual Dependencies and some Results on Undecomposable Relations.* In: 4th Internat. Conf. on Very Large Data Bases, pp. 360–367, Berlin, 1978.

[Ohlbach, Schmidt 1995] H. J. OHLBACH, R. SCHMIDT. *Functional translation and second-order frame properties of modal logics.* Technical Report MPI-I-95-2-002, Max-Planck-Inst., Stuttgart, 1995.

[Ore 1942] O. ORE. *Theory of Equivalence Relations.* Duke Math. J. **9** 573–627, 1942.

[Orlowska 1983] E. ORLOWSKA. *Semantics of vague concepts.* In G. DORN, P. WEINGARTNER, eds., Foundations of Logic and Linguistics. Problems and Solutions. Selected contributions to the 7th Internat. Congress of Logic, Methodology, and Philosophy of Science, Salzburg 1983, pp. 465–482, London, New York, 1983. Plenum Press.

[Orlowska 1984] E. ORLOWSKA. *Reasoning About Database Constraints*, PAS Reports **543**. Polish Academy of Sciences, Inst. of Computer Science, Warsaw, 1984.

[Orlowska, Pawlak 1984] E. ORLOWSKA, Z. PAWLAK. *Representation of nondeterministic information.* Theoret. Comput. Sci. **29** 27–39, 1984.

[Orlowska 1985] E. ORLOWSKA. *Logic of nondeterministic information.* Studia Logica **44** 93–102, 1985.

[Orlowska 1988] E. ORLOWSKA. *Kripke models with relative accessibility and their application to inferences from incomplete information.* In G. MIRKOWSKA, H. RASIOWA, eds., Mathematical Problems in Computation Theory, Banach Center Publications **21**, pp. 329–339, 1988.

[Orlowska 1989] E. ORLOWSKA. *Interpretation of Dynamic Logic and its Extensions in the Relational Calculus.* Bull. Polish Acad. Sci. Math., Sect. on Logic **18** 132–137, 1989.

[Orlowska 1991] E. ORLOWSKA. *Relational interpretation of modal logics.* In [Andréka, Monk+ 1991], pp. 443–471.

[Orlowska 1992] E. ORLOWSKA. *Relational Proof Systems for Relevant Logics.* J. Symbolic Logic **57** 1425–1440, 1992.

[Orlowska 1993] E. ORLOWSKA. *Dynamic Logic with Program Specifications and its Relational Proof System.* J. Appl. Non-Classical Logics **3** 147–171, 1993.

[Orlowska 1994] E. ORLOWSKA. *Relational Semantics for Non-classical Logics: Formulas are Relations.* In J. WOLENSKI, ed., Philosophical Logic in Poland., pp. 167–186. Kluwer, 1994.

[Orlowska 1995] E. ORLOWSKA. *Temporal Logics — in a Relational Framework.* In L. BOLC, A. SZALAS, eds., Time and Logic — A Computational Approach., pp. 249–277. Univ. College London Press, 1995.

[Ounalli, Jaoua+ 1994] H. OUNALLI, A. JAOUA, N. BELKHITER. *Rectangular Decomposition of n-ary Relations.* In: 7th SIAM Conf. on Discrete Mathematics, Albuquerque, NM, 1994.

[Paredaens 1980] J. PAREDAENS. *Transitive Dependencies in a Database Scheme.* RAIRO Informatique/Computer Science **14** 149–163, 1980.

[Park 1981] D. PARK. *Concurrency and automata on infinite sequences.* In: Proc. 5th GI Conf., pp. 167–183, New York, 1981. Springer.

[Parkinson 1966] G. H. R. PARKINSON. *Leibniz: Logical Papers.* Clarendon Press, Oxford, 1966.

[Parnas 1983] D. L. PARNAS. *A Generalized Control Structure and its Formal Definition.* Comm. ACM **26** 572–581, 1983.

[Parnas, Asmis+ 1991] D. L. PARNAS, G. ASMIS, J. MADEY. *Assessment of Safety-Critical Software in Nuclear Power Plants.* Nuclear Safety **32** 189–198, 1991.

[Parnas 1992] D. L. PARNAS. *Tabular Representation of Relations.* Technical Report CRL Report 260, McMaster Univ., Communications Research Laboratory, TRIO (Telecommunications Research Inst. of Ontario), 1992.

[Parnas 1994a] D. L. PARNAS. *Inspection of Safety Critical Software using Function Tables.* In [Pehrson, Simon 1994], pp. 270–277.

[Parnas 1994b] D. L. PARNAS. *Mathematical Descriptions and Specification of Software.* In [Pehrson, Simon 1994], pp. 354–359.

[Parnas, Madey+ 1994] D. L. PARNAS, J. MADEY, M. IGLEWSKI. *Precise Documentation of Well-Structured Programs.* IEEE Trans. on Software Engineering **20** 948–976, 1994.

[Parnas, Madey 1995] D. L. PARNAS, J. MADEY. *Functional Documentation for Computer Systems Engineering (Version 2).* Sci. Comput. Programming **25** 41–61, 1995. also CRL Report 237, McMaster Univ., Communications Research Laboratory and TRIO (Telecommunications Research Inst. of Ontario), Sept. 1991, pp.14.

[Pawlak 1991] Z. PAWLAK. *Rough sets.* Kluwer, Dordrecht, 1991.

[Pehrson, Simon 1994] B. PEHRSON, I. SIMON, eds. 13th *World Computer Congress 94*, Vol. 1. Elsevier, 1994.

[Peirce 1870] C. S. PEIRCE. *Description of a Notation for the Logic of Relatives, Resulting from an Amplification of the Conceptions of Boole's Calculus of Logic.* Memoirs of the American Academy of Sciences **9** 317–378, 1870. Reprint by Welch, Bigelow and Co., Cambridge, MA, 1870, pp. 1–62. Also reprinted in [Peirce 1933] and [Peirce 1984].

[Peirce 1880] C. S. PEIRCE. *On the Algebra of Logic.* Amer. J. Math. **3** 15–57, 1880. reprinted in [Peirce 1933].

[Peirce 1883] C. S. PEIRCE, ed. *Studies in Logic by Members of the Johns Hopkins University.* Little, Brown, and Co., Boston, 1883.

[Peirce 1885] C. S. PEIRCE. *On the Algebra of Logic: A Contribution to the Philosophy of Notation.* Amer. J. Math. **7** 180–202, 1885. reprinted in [Peirce 1933].

[Peirce 1933] C. S. PEIRCE. *C. S. Peirce Collected Papers.* Harvard Univ. Press, Cambridge, 1933. ed. by C. Hartshorne and P. Weiss.

[Peirce 1984] C. S. PEIRCE. *Writings of Charles S. Peirce, A Chronological Edition.* Indiana Univ. Press, Bloomington, 1984. edited by Edward C. Moore, Max H. Fisch, Christian J. W. Kloesel, Don D. Roberts, and Lynn A. Ziegler.

[Peters, Parnas 1994] D. PETERS, D. L. PARNAS. *Generating a Test Oracle from Program Documentation.* In: Proc. of the 1994 Internat. Sympos. on Software Testing and Analysis (ISSTA), August 17-19, 1994, pp. 58–65, 1994.

[Plotkin 1976] G. PLOTKIN. *A Powerdomain Construction.* SIAM J. Comput. **5** 452–487, 1976.

[Pomykala 1988] J. POMYKALA. *On definability in the nondeterministic information system.* Bull. Polish Acad. Sci. Math. **36** 193–210, 1988.

[Ponse, de Rijke+ 1995] A. PONSE, M. DE RIJKE, Y. VENEMA, eds. *Modal Logic and Process Algebra,* CSLI Lecture Notes **53**, Stanford, 1995. CSLI Publications.

[Pratt 1979] V. PRATT. *Models of program logics.* In: Proc. of the 20th IEEE Sympos. on Foundations of Computer Science, pp. 115–122, 1979.

[Pratt 1992] V. PRATT. *Origins of the Calculus of Binary Relations.* In: 7th Annual Sympos. on Logic in Computer Science, pp. 248–254, Santa Cruz, CA, 1992. IEEE Computer Society Press.

[Pratt 1993] V. PRATT. *The Second Calculus of Binary Relations.* In A. M. BORZYSZKOWSKI, S. SOKOLOWSKI, eds., Mathematical Foundations of Computer Science (MFCS), Lect. Notes in Comput. Sci. **81**, pp. 142–155, Gdańsk, Poland, 1993. Springer.

[Pratt 1976] V. PRATT. *Semantical considerations on Floyd-Hoare logic.* In: Proc. 17th Annual IEEE Sympos. on Foundations of Computer Science, pp. 109–121, 1976.

[Raju, Majumdar 1988] K. RAJU, A. MAJUMDAR. *Fuzzy Functional Dependencies and Lossless Join Decomposition of Fuzzy Relational Database Systems.* ACM Trans. Database Systems **13** 129–166, 1988.

[Rasiowa, Sikorski 1963] H. RASIOWA, R. SIKORSKI. *The Mathematics of Metamathematics.* Polish Science Publishers, Warsaw, 1963.

[Rasiowa, Skowron 1985] H. RASIOWA, A. SKOWRON. *Approximation logics.* In W. BIBEL, K. JANTKE, eds., Mathematical Methods of Specification and Synthesis of Software Systems, pp. 123–139, Berlin, 1985. Akademie Verlag.

[Rasiowa, Marek 1989] H. RASIOWA, W. MAREK. *On reaching consensus by groups of intelligent agents.* In Z. RAS, ed., Methodologies for Intelligent Systems 4th Proc. of ISMIS'89, pp. 134–243. North-Holland, 1989.

[Rauszer, Skowron 1992] C. RAUSZER, A. SKOWRON. *The discernibility matrices and functions in information systems.* In R. SLOWINSKI, ed., Intelligent Decision Support. Handbook of Applications and Advances in the Rough Set Theory, pp. 331–362. Kluwer, Dordrecht, 1992.

[Richardson 1953] M. RICHARDSON. *Solutions of Irreflexive Relations.* Ann. of Math. (2) **58** 573–590, 1953.

[Rietman 1995] F. J. RIETMAN. *A Relational Calculus for the Design of Distributed Algorithms.* PhD thesis, Dept. of Computing Science, Utrecht Univ., 1995.

[Riguet 1948] J. RIGUET. *Relations Binaires, Fermetures, Correspondances de Galois.* Bull. Soc. Math. France **76** 114–155, 1948.

[Riguet 1950] J. RIGUET. *Quelques propriétés des relations difonctionnelles.* C. R. Acad. Sci. Paris Ser. A-B **230** 1999-2000, 1950.

[de Rijke 1992] M. DE RIJKE. *A System of Dynamic Modal Logic.* CSLI Research Report 92-170, Stanford Univ., 1992. To appear in J. Philos. Logic.

[de Rijke 1993] M. DE RIJKE. *Extending Modal Logic.* PhD thesis, ILLC Dissertation series 1993-4, 1993.

[de Rijke 1994a] M. DE RIJKE. *The logic of Peirce algebras.* Technical Report CS-R9467, CWI, Amsterdam, 1994. To appear in J. Logic Lang. Inform.

[de Rijke 1994b] M. DE RIJKE. *Meeting some neighbours.* In J. V. EIJCK, A. VISSER, eds., Logic and Information Flow, pp. 170–195. MIT Press, Cambridge, MA, 1994.

[Rissanen 1978] J. RISSANEN. *Theory of Relations for Databases – A Tutorial Survey.* In: Mathematical Foundations of Computer Science 1978, Proc. of 7th Sympos. on Mathematical Foundations of Computer Science, Lect. Notes in Comput. Sci. **64**, pp. 537–551. Springer, 1978.

[Sain, Németi 1994] I. SAIN, I. NÉMETI. *Fork Algebras in Usual as well as in Non-well-founded Set Theories.* Preprint, Mathematical Inst. of the Hungarian Academy of Sciences, 1994.

[Schein 1970] B. M. SCHEIN. *Relation Algebras and Function Semigroups.* Semigroup Forum **1** 1–61, 1970.

[Schmidt 1977] G. SCHMIDT. *Programme als partielle Graphen.* Habil. Thesis, FB Mathematik der Technischen Univ. München, Bericht 7813, 1977. English as [Schmidt 1981a; Schmidt 1981b].

[Schmidt 1981a] G. SCHMIDT. *Programs as Partial Graphs I: Flow Equivalence and Correctness.* Theoret. Comput. Sci. **15** 1–25, 1981.

[Schmidt 1981b] G. SCHMIDT. *Programs as Partial Graphs II: Recursion.* Theoret. Comput. Sci. **15** 159–179, 1981.

[Schmidt, Ströhlein 1985] G. SCHMIDT, T. STRÖHLEIN. *Relation Algebras — Concept of Points and Representability.* Discrete Math. **54** 83–92, 1985.

[Schmidt, Ströhlein 1989] G. SCHMIDT, T. STRÖHLEIN. *Relationen und Graphen.* Mathematik für Informatiker. Springer, Berlin, 1989.

[Schmidt, Ströhlein 1993] G. SCHMIDT, T. STRÖHLEIN. *Relations and Graphs, Discrete Mathematics for Computer Scientists.* EATCS-Monographs on Theoretical Computer Science. Springer, 1993.

[Schouwen, Parnas$^+$ 1993] A. V. SCHOUWEN, D. L. PARNAS, J. MADEY. *Documentation of Requirements for Computer Systems.* In: Proc. of '93 IEEE Internat. Sympos. on Requirements Engineering, San Diego, CA, 4 - 6 January, 1993, pp. 198–207, 1993.

[Schröder 1895] E. SCHRÖDER. *Vorlesungen über die Algebra der Logik (exacte Logik).* Teubner, Leipzig, 1895. Vol. 3, Algebra und Logik der Relative, part I, 2nd edition published by Chelsea, 1966.

[Scollo 1993] G. SCOLLO, ed. *Proc. 3rd Internat. Conf. Algebraic Methodology and Software Technology, June 21 - 25,* Enschede, 1993. Springer.

[Sekerinski 1993] E. SEKERINSKI. *A Calculus for Predicative Programming.* In R. S. BIRD, C. C. MORGAN, J. C. P. WOODCOCK, eds., 2nd Internat. Conf. on the Mathematics of Program Construction, Lect. Notes in Comput. Sci. **669**. Springer, 1993.

[Shen 1995] H. SHEN. *Implementation of Table Inversion Algorithms.* M. Eng. thesis, McMaster Univ., Communications Research Laboratory, 1995.

[Ströhlein 1970] T. STRÖHLEIN. *Untersuchungen über kombinatorische Spiele.* Doctoral diss., Technische Univ. München, 1970.

[Suppes 1973a] P. SUPPES. *Facts and Fantasies of Education.* In M. C. WITTROCK, ed., Changing Education: Alternatives from Educational Research, pp. 6–45. Prentice-Hall, Englewood Cliffs, N.J., 1973.

[Suppes 1973b] P. SUPPES. *Semantics of context-free fragments of natural languages.* In J. HINTIKKA, J. M. E. MORAVCSIK, P. SUPPES, eds., Approaches to Natural Languages, pp. 370–394. Reidel, Dordrecht, 1973.

[Suppes 1976] P. SUPPES. *Elimination of Quantifiers in the Semantics of Natural Languages by the use of Extended Relation Algebras.* Rev. Internat. Philos. **30** 243–259, 1976.

[Suppes, Zanotti 1977] P. SUPPES, M. ZANOTTI. *On using random relations to generate upper and lower probabilities.* Synthese **36** 427–440, 1977.

[Suppes, Macken 1978] P. SUPPES, E. MACKEN. *Steps toward a variable-free semantics of attributive adjectives, possessives, and intensifying adverbs.* In K. E. NELSON, ed., Children's Language, Vol. 1, pp. 81–115. Gardner Press, New York, 1978.

[Suppes 1979a] P. SUPPES. *Logical inference in English: A preliminary analysis.* Studia Logica **38** 375–391, 1979.

[Suppes 1979b] P. SUPPES. *Variable-free semantics for negations with prosodic variation.* In R. HILPINEN, I. NIINILUOTO, M. P. HINTIKKA, eds., Essays in Honor of Jaakko Hintikka, pp. 49–59. Reidel, Dordrecht, 1979.

[Suppes 1991] P. SUPPES. *Language for Humans and Robots.* Blackwell, Oxford, 1991.

[Tarski 1941] A. TARSKI. *On the Calculus of Relations.* J. Symbolic Logic **6** 73–89, 1941.

[Tarski 1953] A. TARSKI. *Some Metalogical Results Concerning the Calculus of Relations.* J. Symbolic Logic **18** 188–189, 1953.

[Tarski 1955] A. TARSKI. *Contributions to the Theory of Models, III.* Indag. Math. **17** 56–64, 1955.

[Tarski, Givant 1987] A. TARSKI, S. GIVANT. *A Formalization of Set Theory without Variables,* Amer. Math. Soc. Colloq. Publ. **41**. Amer. Math. Soc., Providence, 1987.

[le Thanh 1986] N. LE THANH. *Contribution à l'étude de la Généralisation et de l'Association dans une Base de Données Relationnelle: les Isodépendances et le Modèle b-relationnel.* Diss. de Doctorat d'Etat des sciences, Univ. de Nice, 1986.

[Troelstra 1992] A. S. TROELSTRA. *Lecture on Linear Logic,* CSLI Lecture Notes **29**. CSLI Publications, Stanford, CA, 1992.

[Troelstra 1993] A. S. TROELSTRA. *Tutorial on Linear Logic.* In K. DOŠEN, P. SCHROEDER-HEISTER, eds., Substructural Logics, pp. 327–356. Oxford Univ. Press, 1993.

[Ullman 1982] J. D. ULLMAN. *Principles of Database Systems.* Computer Science Press, 1982. 2nd Edition.

[Ullman 1988] J. D. ULLMAN. *Principles of Database and Knowledge-Base Systems.* Computer Science Press, 1988.

[Vakarelov 1987] D. VAKARELOV. *Abstract characterization of some knowledge representation systems and the logic NIL of nondeterministic information.* In P. JORRAND, V. SGUREV, eds., Artificial Intelligence II, Methodology, Systems, Applications, pp. 255–260. North-Holland, Amsterdam, 1987.

[Vakarelov 1991a] D. VAKARELOV. *Logical Analysis of Positive and Negative Similarity Relations in Property Systems.* In M. DEGLAS, D. GABBAY, eds., Proc. of the 1st World Conf. on the Fundamentals of Artificial Intelligence, pp. 491–500, Paris, France, 1991. Angkor.

[Vakarelov 1991b] D. VAKARELOV. *A modal logic for similarity relations in Pawlak knowledge representation systems.* Fund. Inform. **15** 61–79, 1991.

[Vakarelov 1989] D. VAKARELOV. *Modal Logics for Knowledge Representation Systems.* In A. R. MEYER, M. A. TAITSLIN, eds., Proc. of the Sympos. on Logical Foundations of Computer Science, Lect. Notes in Comput. Sci. **363**, pp. 257–277, Berlin, 1989. Springer.

[Veloso, Haeberer 1991] P. A. VELOSO, A. M. HAEBERER. *A Finitary Relational Algebra for Classical First-Order Logic.* Bull. Polish Acad. Sci. Math., Sect. on Logic **20** 52–62, 1991.

[Veloso, Haeberer$^+$ 1995] P. A. VELOSO, A. M. HAEBERER, M. F. FRIAS. *Fork Algebras as Algebras of Logic.* Bull. Symbolic Logic pp. 265–266, 1995.

[Veltman] F. VELTMAN. *Defaults in Update Semantics.* J. Philos. Logic . to appear.

[Venema 1994] Y. VENEMA. *Completeness through flatness.* In D. GABBAY, H. J. OHLBACH, eds., Temporal Logic, 1st Internat. Conf., ICTL'94, Lect. Notes in Comput. Sci. **827**, pp. 149–164, Berlin, 1994. Springer.

[Venema 1995] Y. VENEMA. *A crash course in arrow logic.* In M. MARX, L. POLOS, eds., Arrow Logic and Multi-Modal Logic, Studies in Logic, Language and Information. CSLI Publications, Stanford, 1995.

[Visser, Vermeulen 1995] A. VISSER, K. VERMEULEN. *Dynamic Bracketing and Discourse Representation.* Technical report, Dept. of Philosophy, Utrecht Univ., 1995.

[Walukiewicz 1995] I. WALUKIEWICZ. *Completeness of Kozens axiomatization of the propositional μ-calculus.* In: Annual Sympos. on Logic in Computer Science. IEEE Computer Society Press, 1995.

[Whitehead, Russell 1910] A. N. WHITEHEAD, B. RUSSELL. *Principia Mathematica, Volume I.* Cambridge Univ. Press, Cambridge, England, 1910.

[Wilder, Tucker 1995] A. WILDER, J. TUCKER. *System Documentation Using Tables — A short course.* CRL Report 306, McMaster Univ., Communications Research Laboratory, TRIO (Telecommunications Research Inst. of Ontario), 1995. Also published as Report CSR 11-95, Computer Science Dept., Univ. of Wales, Swansea, 1995.

[Yetter 1990] D. YETTER. *Quantales and (Noncommutative) Linear Logic.* J. Symbolic Logic **55** 41–64, 1990.

[Zierer 1988] H. ZIERER. *Programmierung mit Funktionsobjekten: Konstruktive Erzeugung semantischer Bereiche und Anwendung auf die partielle Auswertung.* PhD thesis, Technische Univ. München, Fakultät für Informatik, 1988.

[Zierer 1991] H. ZIERER. *Relation-Algebraic Domain Constructions.* Theoret. Comput. Sci. **87** 163–188, 1991.

[Zucker 1996] J. ZUCKER. *Transformations of Normal and Inverted Function Tables.* Formal Aspects of Computing , 1996. to appear (Also as CRL Report No. 291, August 1994, McMaster University, Communications Research Laboratory and Telecommunications Research Inst. of Ontario.).

Symbol Table

The symbol table below does not attempt to list all symbols used in this book. In particular, symbols which appear in only one chapter are not listed.

Sets

$\triangleq$	equality by definition	1
$\in$	set membership	1
$\emptyset$	empty set	1
U	universal set	1
$\{x \in X : P(x)\}$	set comprehension	1
$\{x : x \in X \text{ and } P(x)\}$	set comprehension	1
$\times$	Cartesian product	2
π_i	projection function	5
$\subseteq$	set inclusion	1
$\cup$	set union	1
$\cap$	set intersection	1
$-$	set difference	1
$\overline{}$	set complement	1
$\mathcal{P}(\_)$	powerset	1
$\bigcup$	union of a collection of sets	2
$\bigcap$	intersection of a collection of sets	2
$(\_,\_)$	(ordered) pair	2
$\mathbb{B}$	set of booleans	2
$\mathbb{N}$	set of natural numbers	2
$\mathbb{Q}$	set of rationals	2
$\mathbb{R}$	set of reals	2
$\mathbb{Z}$	set of integers	2
$\_^{+}$	transitive closure	3
$\_^{*}$	reflexive transitive closure	3

Relations

$dom R$	domain of relation R	3
$ran R$	range of relation R	3
I_X	identity on X	3
$R^{\smile}$	converse of relation R	3
$R \backslash S$	right residual	3
R/S	left residual	3
$R : Y$	Peirce product	3
$R(x)$	image set	3

R^n	exponentiation	3
$R;S$	composition	5
$Rel(U)$	full relation algebra over U	9

Partially ordered structures

$\sqsubseteq$	less-or-equal	6
$\perp\!\!\!\perp$	bottom	6
$\top\!\!\!\top$	top	6
$\bigsqcap$	arbitrary meet	7
$\bigsqcup$	arbitrary join	7
$\sqcap$	binary meet	7
$\sqcup$	binary join	7
μ	least fixed point	11
ν	greatest fixed point	11

Abstract relation algebra

$\mathbb{I}$	identity	8,40
$\overline{}$	complementation	8,40
$\sim$	monotype complementation	159
$\_^{\smile}$	converse	8,40
$\_;\_$	composition	8,40
$\_\backslash\_$	right residual	3
$\_/\_$	left residual	3
$\_:\_$	Peirce product	3
R^i	exponentiation	3
$\top\!\!\!\top$	top	6
$\perp\!\!\!\perp$	bottom	6
$\dagger$	relative addition	25
∇	fork	31
$\vartriangle$	split	155
$\underline{\nabla}$	fork of relations	55
$[\_,\_]$	fork in Chapt. 8	118
$\star$	"pair naming function"	31,56
$\_ \lhd \_ \rhd \_$	conditional	121
$\Vert$	parallel composition	123
π, ρ	product projections	50,55
$\ll, \gg$	product projections in Chapt. 10	154
ι, κ	injection functions in Chapt. 9	146
$tag0, tag1$	injection functions in Chapt. 10	154
$\mathsf{syq}(\_,\_)$	symmetric quotient	52
ε	direct power	52
$\blacktriangledown$	co-junc	155
$\_^{c}$	cylcindrification	9

Categories

source f	source of f	14
target f	target of f	14
$\mathsf{Mor}_{\mathbf{C}}$	class of morphisms of $\mathbf{C}$	14
$\mathsf{Obj}_{\mathbf{C}}$	class of objects of $\mathbf{C}$	14
id_A	identity morphism for object A	14

Logic

$\vdash$	logically derives	15
$\models$	models	15
$\Diamond$	possibility operator	20
$\Box$	necessity operator	20
$\neg$	negation	16
$\vee$	disjunction	16
$\wedge$	conjunction	16
$\rightarrow$	implication	16
$\leftrightarrow$	equivalence	16
$\exists$	existential quantifier	17
$\forall$	universal quantifier	17

Programs

wp	weakest precondition	50
wlp	weakest liberal precondition	161
$[\!]$	nondeterministic choice	125
.	image of elements under concrete relations	200
$t[\_]$	projection on components	201
$\rightleftharpoons$	difunctional dependency	201

Index

Addresses of Contributors

Roland Backhouse, Department of Mathematics and Computing Science, Eindhoven University of Technology, P.O. Box 513, 5600 MB Eindhoven, The Netherlands, `rolandb@win.tue.nl`

Gabriel Baum, Departamento de Informática, Universidad Nacional de La Plata, C.C.11, Correo Central, 1900, La Plata, Buenos Aires, República Argentina, `gbaum@info.unlp.edu.ar`

Nadir Belkhiter, Département d'Informatique, Université Laval, Québec QC, GIK 7P4, Canada, `Nadir.Belkhiter@ift.ulaval.ca`

Rudolf Berghammer, Institut für Informatik und Praktische Mathematik, Universität Kiel, Preusserstraße 1–9, 24105 Kiel, Germany, `rub@informatik.uni-kiel.de`

Patrick Blackburn, Computerlinguistik, Universität des Saarlandes, Postfach 1150, 66041 Saarbrücken, Germany, `patrick@coli.uni-sb.de`

Michael Böttner, Max Planck-Institut für Psycholinguistik, 6500 AH Nijmegen, The Netherlands, `boettner@mpi.nl`

Chris Brink, Department of Mathematics and Applied Mathematics, University of Cape Town, Rondebosch 7700, South Africa, `cbrink@maths.uct.ac.za`

Jules Desharnais, Département d'Informatique, Université Laval, Québec QC, GIK 7P4, Canada, `Jules.Desharnais@ift.ulaval.ca`

Henk Doornbos, Department of Mathematics and Computing Science, Eindhoven University of Technology, P.O. Box 513, 5600 MB Eindhoven, The Netherlands, `henkd@win.tue.nl`

Marcelo Frias, Departamento de Informática, Pontifícia Universidade Católica do Rio de Janeiro, Rua Marquês de São Vicente 225, 22453-900, Rio de Janeiro, RJ, Brazil, `mfrias@inf.puc-rio.br`

Antonetta van Gasteren, Department of Mathematics and Computing Science, Eindhoven Univ. of Technology, P.O. Box 513, 5600 MB Eindhoven, The Netherlands, `netty@win.tue.nl`

Armando Haeberer, Departamento de Informática, Pontifícia Universidade Católica do Rio de Janeiro, Rua Marquês de São Vicente 225, 22453-900, Rio de Janeiro, RJ, Brazil, `armando@inf.puc-rio.br`

Claudia Hattensperger, Fakultät für Informatik, Universität der Bundeswehr München, 85577 Neubiberg, Germany, `claudia@informatik.unibw-muenchen.de`

Wolfgang Heinle, Institut für Informatik und Angewandte Mathematik, Universität Bern, Neubrückstraße 10, 3012 Bern, Switzerland, `heinle@IAM.unibe.ch`

Bernard Hodgson, Département de mathématiques et de statistique, Université Laval, Québec QC, GIK 7P4, Canada, `bhodgson@mat.ulaval.ca`

Ryszard Janicki, Department of Computer Science and Systems, McMaster University, Hamilton Ontario, L8S 4K1, Canada, `janicki@maccs.dcss.McMaster.CA`

Ali Jaoua, Département des Sciences de l'Informatique, Université de Tunis, Campus Universitaire, 1060 Tunis, Tunisia, c/o `Kamel.Benrhouma@cck.rnrt.tn`

Peter Jipsen, Department of Mathematics and Applied Mathematics, University of Cape Town, Rondebosch 7700, South Africa, `pjipsen@maths.uct.ac.za`

Wolfram Kahl, Fakultät für Informatik, Universität der Bundeswehr München, 85577 Neubiberg, Germany, kahl@informatik.unibw-muenchen.de

Burghard von Karger, Institut für Informatik und Praktische Mathematik, Universität Kiel, Preusserstraße 1–9, 24105 Kiel, Germany, bvk@informatik.uni-kiel.de

Roger Maddux, Department of Mathematics, Iowa State University, 400 Carver Hall, Ames, Iowa 50011-2066, USA, maddux@iastate.edu

Ali Mili, Department of Computer Science, University of Ottawa, Ottawa ON, K1N 6N5, Canada, amili@csi.uottawa.ca

Théodore Moukam, Département d'Informatique, Université Laval, Québec QC, GIK 7P4, Canada

John Mullins, Department of Computer Science, University of Ottawa, Ottawa ON, K1N 6N5, Canada, mullins@csi.uottawa.ca

Thanh Tung Nguyen, Avenue des Glycines 62, 1950 Kraainem, Belgium, tn@info.ucl.ac.be

Ewa Orlowska, Institute of Telecommunications, Szachowa 1, 04-894 Warsaw, Poland, orlowska@plearn.edu.pl

Habib Ounalli, Département des Sciences de l'Informatique, Université de Tunis, Campus Universitaire, 1060 Tunis, Tunisia

David Parnas, Department of Electrical and Computer Engineering, McMaster University, Hamilton Ontario, L8S 4K1, Canada, parnas@triose.eng.mcmaster.ca

Maarten de Rijke, Department of Computer Science, University of Warwick, Coventry CV4 7AL, England, Maarten.de.Rijke@dcs.warwick.ac.uk

Holger Schlingloff, Fakultät für Informatik, Technische Universität München, 80290 München, Germany, schlingl@informatik.tu-muenchen.de

Gunther Schmidt, Fakultät für Informatik, Universität der Bundeswehr München, 85577 Neubiberg, Germany, schmidt@informatik.unibw-muenchen.de

Thomas Ströhlein, Fakultät für Informatik, Technische Universität München, 80290 München, Germany, stroehle@informatik.tu-muenchen.de

Paulo Veloso, Departamento de Informática, Pontifícia Universidade Católica do Rio de Janeiro, Rua Marquês de São Vicente 225, 22453-900, Rio de Janeiro, RJ, Brazil, veloso@inf.puc-rio.br

Yde Venema, Department of Computer Science, Free University Amsterdam, de Boelelaan 1081, 1081 HV Amsterdam, The Netherlands, yde@cs.vu.nl

Michael Winter, Fakultät für Informatik, Universität der Bundeswehr München, 85577 Neubiberg, Germany, thrash@informatik.unibw-muenchen.de

Jeffery Zucker, Department of Computer Science and Systems, McMaster University, Hamilton Ontario, L8S 4K1, Canada, zucker@maccs.dcss.McMaster.CA

Springer-Verlag
and the Environment

WE AT SPRINGER-VERLAG FIRMLY BELIEVE THAT AN international science publisher has a special obligation to the environment, and our corporate policies consistently reflect this conviction.

WE ALSO EXPECT OUR BUSINESS PARTNERS – PRINTERS, paper mills, packaging manufacturers, etc. – to commit themselves to using environmentally friendly materials and production processes.

THE PAPER IN THIS BOOK IS MADE FROM NO-CHLORINE pulp and is acid free, in conformance with international standards for paper permanency.